ALGEBRAIC ANALYSIS OF SOCIAL NETWORKS

Wiley Series in Computational and Quantitative Social Science

Computational Social Science is an interdisciplinary field undergoing rapid growth due to the availability of ever increasing computational power leading to new areas of research.

Embracing a spectrum from theoretical foundations to real world applications, the Wiley Series in Computational and Quantitative Social Science is a series of titles ranging from high level student texts, explanation and dissemination of technology and good practice, through to interesting and important research that is immediately relevant to social / scientific development or practice. Books within the series will be of interest to senior undergraduate and graduate students, researchers and practitioners within statistics and social science.

Behavioral Computational Social Science
Riccardo Boero

Tipping Points: Modelling Social Problems and Health
John Bissell (Editor), Camila Caiado (Editor), Sarah Curtis (Editor), Michael Goldstein (Editor), Brian Straughan (Editor)

Understanding Large Temporal Networks and Spatial Networks: Exploration, Pattern Searching, Visualization and Network Evolution
Vladimir Batagelj, Patrick Doreian, Anuška Ferligoj, Nataša Kejžar

Analytical Sociology: Actions and Networks
Gianluca Manzo (Editor)

Computational Approaches to Studying the Co-evolution of Networks and Behavior in Social Dilemmas
Rense Corten

The Visualisation of Spatial Social Structure
Danny Dorling

ALGEBRAIC ANALYSIS OF SOCIAL NETWORKS

Models, Methods & Applications using R

J. Antonio Rivero Ostoic

WILEY

Registered Offices
John Wiley & Sons, Inc., 111 River Street, Hoboken, NJ 07030, USA
John Wiley & Sons Ltd, The Atrium, Southern Gate, Chichester, West Sussex, PO19 8SQ, UK

Editorial Office
9600 Garsington Road, Oxford, OX4 2DQ, UK

For details of our global editorial offices, customer services, and more information about Wiley products visit us at www.wiley.com.

Wiley also publishes its books in a variety of electronic formats and by print-on-demand. Some content that appears in standard print versions of this book may not be available in other formats.

Library of Congress Cataloging-in-Publication Data

Names: Ostoic, J. Antonio Rivero, author.
Title: Algebraic analysis of social networks : models, methods &
    applications using R / J Antonio Rivero Ostoic.
Description: Hoboken, NJ : Wiley, [2020] | Series: Wiley series in
    computational and quantitative social science | Includes index.
Identifiers: LCCN 2020025443 (print) | LCCN 2020025444 (ebook) | ISBN
    9781119250388 (cloth) | ISBN 9781119250326 (adobe pdf) | ISBN
    9781119250395 (epub)
Subjects: LCSH: Social networks–Mathematical models. | R (Computer program
    language)
Classification: LCC HM741 .O77 2020 (print) | LCC HM741 (ebook) | DDC
    302.301/5118–dc23
LC record available at https://lccn.loc.gov/2020025443
LC ebook record available at https://lccn.loc.gov/2020025444

Cover Design: Wiley
Cover Image: Courtesy of the author

Set in 10/12pt TimesLTStd by SPi Global, Chennai, India
Printed and bound by CPI Group (UK) Ltd, Croydon, CR0 4YY

10 9 8 7 6 5 4 3 2 1

to gitte, santiago, maria-ines

Contents

List of Figures

List of Tables

Preface

This book is about network analysis with algebra with a mixture of theory, methods and practice. Although there is an emphasis on the study of complex systems like multiplex, multimodal, and multilevel networks, it also covers elementary structures with the genesis of algebraic approaches for the analysis of kinship networks from the 1940s. In kinship systems, for example, the ties of descent between parents and children carry one dimension, and the rules of marriage relations distinguish the type of structure the network has.

In the case of multiplex networks, linking actors or linking relations is a matter of order structure. For a given social arrangement, actor connections correspond to a system of first order, while relationships among relations represent a second-order structure with a higher level of abstraction. The semigroup of relations captures this second-order structure where the ties are the object set, and the composition of the ties is the associated operation.

For signed networks, in which valences are special cases of multiplex relations, there is the classical application in terms of structural balance theory by semiring objects that rather have two associated operations. Different kinds of semigroups and semirings objects serve to define empirical data with R function implementations having a systematic description of the code at the end of each chapter.

An algebraic approach for dealing with affiliation networks is found in Formal concept analysis framework, which allows producing cross-domain connections into an ordered structure, and for studying affiliation systems having valued ties or many-valued contexts. Semigroups are also useful for constructing valued paths in one-mode networks, and particular types of semirings are for the reduction of this kind of systems. A set of algebraic tools like this allows making analyses of multilevel structures that are also multiplex, and it may pave the way for an integrated algebraic approach to these kinds of complex structures.

As with many endeavours, this manuscript started as a PhD project at the University of Southern Denmark seven years ago. Then took form at the University of Melbourne with my special gratitude to Philippa Pattison for introducing me to relational algebra both from her work and at the Social Networks Laboratory. The writing of the book ended at Aarhus University with additional chapters and the development of the related software; first at the School of Business and Social Sciences and then at the School of Culture and Society. I acknowledge the support of all these institutions. My special recognition to the people who were directly or indirectly involved in this project, particularly to the people and organizations that took part in the survey for the Incubator networks dataset.

Thanks to my family, who supported me from the beginning to the end.

Antonio Rivero Ostoic
Aarhus, November 2020

Abbreviations

iff	if and only if
poset	partially ordered set
SE	Structural equivalence
AE	Automorphic equivalence
RE	Regular equivalence
GE	Generalized equivalence
LRE	Local Role equivalence
CE	Compositional equivalence
JNTHOM	Joint Homomorphic reduction
CSS	Common Structure Semigroup

The acronyms of $\mathscr{X}_{G20}$ organizations and their country abbreviations are given in Appendix A.

Symbols

Sets

A, B, X, Y, Z	sets
a, b, x, y, z	elements of a set
$\varnothing$	empty set
$X \times Y$	Cartesian product of X and Y
$\langle x, y \rangle$	unordered pair
(x, y)	ordered pair
$\{x, y\}$	list with two elements
f, g, h	functions

Networks

X	(empirical) network
$\mathscr{X}$	relational system of X
$\mathscr{X}_t^+$	collection of t relational systems
$\mathscr{G}$	graph
$\mathscr{G}^d$	directed graph
$\mathscr{G}^+$	multigraph (or multiplex network)
$\mathscr{G}^\sigma$	signed graph (or network)
$\mathscr{G}^B$	bipartite graph (or affiliation network)
$\mathscr{G}^V$	valued graph (or network)
$\mathscr{G}^M$	multilevel graph (or network)
$\mathcal{H}$	graph image
$\mathcal{N}, \mathcal{M}$	set (or subset) of actors or nodes
$\mathscr{E}$	set of ties or edges
$\mathscr{R}, \mathscr{T}$	collection of relations

R	relation type
r	number of relation types
n	number of actors or nodes
i, j, k, l, n_i, m_i	actors or nodes
e_i	tie, edge, or path, chain
e_t	number of ties or edges in $\mathscr{E}$
k	length of a path or a chain
$\mathbf{A}$	adjacency matrix
a_{ij}	value of a cell in $\mathbf{A}$ (the ith row and jth column)
$\mathbf{A}(R)$	adjacency matrix for R
$\mathbf{A}(\mathscr{R})$	array for $\mathscr{R}$
$\mathbf{A}^\alpha$	diagonal matrix
$\mathbf{A}^\sigma$	valency matrix
$\mathbf{C}$	incidence matrix

Signed Networks

$\mathscr{X}^\sigma$	signed structure of $\mathscr{X}$
$\mathscr{V}$	set of valences
υ	valence
υ_t	number of valences in $\mathscr{V}$
p, n, a, o	positive, negative, ambivalent, absent relation

Bundle classes

B^N	null
B^A	asymmetric
B^R	reciprocal
B^E	tie entrainment
B^X	tie exchange
B^M	mixed
B^F	full

Relational Algebra

$\mathbf{W}$	free word algebra (algebraic system)
$\mathbf{W}_k$	word algebra of order k
Σ	alphabet (object set made of each type of $\mathscr{R}$ in $\mathscr{G}^+$)
Σ^*	set of all words over Σ
F	set of operations defined on Σ
$*, \cdot$	binary operations
$R_1 \circ R_2$	composition of relations R_1 and R_2

R^C	complement of R
R^{-1}, R^T	converse, transpose of R
R^t	transitive relation
E, F, R	concrete (primitive) relations
I	identity relation
U	universal relation
$\varnothing$	empty relation
$\equiv$	equivalence or a congruence relation
$\leq$	partial order relation
$\cong$	isomorphism
π	partition of X or $\mathbf{X}$
$[x]$	subset of equivalent elements in X
R^+	the set of primitive and compound relations
w	the number of elements in R^+
$\mathbf{R(W)}$	Relation-Box for $\mathbf{W}$
$\mathbf{R_X(W_k)}$	Relation-Box for X with $\mathbf{W}$ of order k
R_l^+	Relation Plane of actor l
w	amount of relation types in R^+
$R_{l_{x,j}}^*$	Role Relation of actor l with actor j and relation x
R_l^*	Role Set of actor l
$\mathbf{H}_l$	Person Hierarchy of actor l
$\mathscr{H}_X$	Cumulated Person Hierarchy of network X
$\mathbf{A}_X^\alpha$	diagonal matrix with attributes in network X

Algebraic Structures

S	semigroup
T	subsemigroup or a quotient semigroup
G	group
H	subgroup
I	ideal
x	element in G
s	element in S
a, b, c	*of an algebraic structure*: regular elements
e	identity element
1	neutral element (and identity)
0	zero or absorbing element
S^1	S with identity adjoined
S^0	S with zero adjoined

$F(S)$	free semigroup
$S(R)$	semigroup of relations
$FS(R)$	free semigroup of relations generated by Σ^*
$\mathscr{S}_X$	positional system of network X
$\mathscr{P}_X$	partial order structure of X
$\mathscr{Q}_X$	Role structure of $\mathscr{S}_X$
Q	semiring

Decomposition

h	homomorphism
$h(\mathscr{X})$	homomorphic reduction of network $\mathscr{X}$
$h(\mathscr{S})$	homomorphism on the reduced structure of $\mathscr{X}$
$h(\mathscr{Q})$	homomorphic reduction of relational system $\mathscr{Q}$
L	object set of a lattice or a semilattice
$\vee, \wedge$	union and intersection of two elements in L
$L(S)$	Lattice of Homomorphisms of the Semigroup
$L_\pi(\mathscr{Q})$	Congruence lattice in $\mathscr{Q}$
π_i	π-relation i
π_a	atom
π_a^*	meet-complement of the atom
F	factor
S/π_i	Factor on the semigroup S by partition i
$S(R)/\pi_i$	the homomorphic image of $S(R)$ mod π_i

Formal Concept Analysis

G	set of objects
M	set of attributes
A	extent
B	intent
I	incidence relation
$\mathbb{C}$	formal concept
$\mathbb{K}$	formal context
$\mathbf{B}$	set of all concepts

About the Companion Website

This book is accompanied by a companion website:

www.wiley.com/go/ostoic/algebraicanalysis

The website includes datasets and code.

Scan this QR code to visit the companion website.

About the Companion Website

1

Structural Analysis with Algebra

To find bonds, patterns, and stories that tell us that we are not just product of random.

1.1 Preliminaries

We start with the main rationale behind the structural study of networks and social phenomena, and this initial Chapter provides a basic language and vocabulary required for diverse types of algebraic analyses of social networks. Much of the mathematical formulae are based on classic and standard books about algebraic combinatorics like Kim and Roush (1983), Dunn and Hardegree (2001), or Maddux (2006); within the social networks the literature includes Boyd (1991), Freeman (1992), Pattison (1993), Wasserman and Faust (1994), and Degenne and Forsé (1999).

The arrangement of actors in a social network can be expressed in mathematical terms with the fundamentals from set theory. A *set*, which is the basic structure upon which all other structures are built, is a collection of objects called the *elements* or *members* of the set. Sets can furthermore be described by using what is called the "set build notation." For example, $\{x \mid x > 0 \text{ and } x < 4\}$ reads "the set of all x such that x is greater than zero and less than four." This would be $x = \{1, 2, 3\}$ if x is a member of the set of natural numbers, written as $x \in \mathbb{N}$, and where $\mathbb{N}$ in this case describes completely the set.

In order to avoid the so-called Russell's paradox, which states that a "set of all sets which do not contain themselves does not exist," it is necessary to define the universe of the element in question, and $x \in \mathbb{N}$ does in this example, otherwise if $x \notin \mathbb{N}$ (i.e. in the opposite situation) it should have been added to the expression $x \in X$ for instance where X represents the universe.

Curls and angles are used in the notation for unordered sets and parentheses for the ordered ones. A finite set is called a *list* and we refer to the set as a *family* of objects (rather than merely a collection) when the set sequence is important. Basic operations on sets are the union [resp. intersection], which for sets A and B are denoted by $A \cup B$ [resp. $A \cap B$], and where $A \cup B = \{x \mid x \in A \text{ or } x \in B\}$ [resp. $A \cap B = \{x \mid x \in A \text{ and } x \in B\}$]; and also complement $A^C = \{x \in X \mid x \notin A\}$.

Algebraic Analysis of Social Networks: Models, Methods and Applications using R,
First Edition. J. A. R. Ostoic. Companion website: www.wiley.com/go/ostoic/algebraicanalysis.
© 2021 John Wiley & Sons Ltd. Published 2021 by John Wiley & Sons Ltd.

To define a relation in terms of sets, an ordered pair such as (x, y) refers to a directed linkage from an element x to an element y, where $x \in X$ and $y \in Y$. The overall relation set results from the Cartesian product of the sets X and Y, which is the set $X \times Y$ of all ordered pairs (x, y) from it, and this product set is said to be the context of the relation.

A binary relation R between X and Y is then defined as a triple (X, R, Y), where $R \subseteq X \times Y$; and '$\subseteq$' means "is a subset of." Since the relation is defined on an ordered pair, that is a pair of objects with an order associated with them; then actually the relation is a "binary" relation. In the text, however, the term "relation" refers to a relationship with a binary operation, and this to avoid confusion with a special type of tie having just the binary values 0 and 1.

In set theory, X is known as the *domain*, and Y as the *codomain* of a relation. When the domain and the codomain are equal, the ties occur among a single set of entities, which typically are the social actors, and a hence a one-mode (social) network is defined as a domain with a set of relations on such domain. However, sometimes the entire social system needs to be described by more than a single set of entities, and the domain and the codomain result not being equal as with two-mode network data.

After this short preamble, we are able to define in formal terms a social network as a *relational system*:

$$\mathscr{X} = \langle\, X, \mathscr{G}, \mathbf{A} \,\rangle.$$

Each element in this triple stands for a different type of representation of the social network ranging from a concrete account of the system in the first element, to more abstracts versions of the structure in the other two elements.

Loosely speaking, the term "relational system" constitutes the widest notion to refer to diverse representation forms of social networks. We will see later on that this concept applies as well to different types of social networks, and other structures that are derived from these social systems. However, first we take a closer look at each of the above components in $\mathscr{X}$.

In the relational system, X represents the structure of the simplest social network and comprises a triple of sets:

$$X = \langle\, \mathscr{N}, \mathscr{M}, \mathscr{E} \,\rangle,$$

which is a collection of social actors $\mathscr{N} = \{\, i \mid i \text{ is an actor} \,\}$, together with a set of ties $\mathscr{E} = \{\, \langle i,j \rangle \mid i \text{ 'has a tie to' } j \,\}$. For one-mode networks $\mathscr{M} = \mathscr{N}$, and for two-mode systems $\mathscr{M} = \{\, i \mid i \text{ is an event} \,\}$, $\mathscr{M} \neq \mathscr{N}$. This simple structure of the network is called an *algebraic structure* (Freeman, 1992) or simply a *network* (Shier, 1991), and thus X represents the empirical structure of the social network.

1.2 Graphs

1.2.1 Graphs and Digraphs

The empirical structure can be described in a more abstract way by using the elements of graph theory (Harary, 1994; König, 1936). Hence, a network can similarly be represented by a graph $\mathscr{G}$ where the actors are depicted as points in the graph, and lines connecting the points correspond to the ties between the actors:

$$\mathscr{G} = \langle\, \mathscr{N}, \mathscr{M}, \mathscr{E} \,\rangle.$$

In graph theoretic terms, the points are referred to as *vertices* or *nodes*, and the lines as *edges*. When the ties have direction, the graph is directed, $\mathscr{G}^d$, and it is called a *digraph*. $\langle i, j \rangle$ then becomes an ordered pair of nodes, i.e. (i, j), and the directed edges are depicted in the digraph as arcs pointing the direction of the tie. In consequence, a digraph is a labelled graph with a family of relations, as opposed to an unlabeled graph with just a list of relations.

A graph is considered as *complete* when each pair of nodes is joined by an edge; otherwise the graph is not complete. Although for convenience in this case both the sets of actors and ties, and the sets of points and lines are formally represented by $\mathscr{N}$ and $\mathscr{E}$, in principle different characters should be assign the level of abstraction. Nevertheless, for our purpose a social network comprises a domain that is a collection of social entities, $\mathscr{N} = \{ n_1, n_2, \ldots, n_g \}$, $\mathscr{M} = \{ m_1, m_2, \ldots, m_h \}$, and a set of relations, $\mathscr{E} = \{ e_1, e_2, \ldots, e_t \}$, being g as the number of actors or nodes in $\mathscr{N}$, h the number of events in $\mathscr{M}$, and t as the total edges or ties in $\mathscr{E}$ with $t = g + h$.

1.2.2   Multigraphs

Yet until now all these formal representations are for a *simple* social network, and this is because a single type of relation is linking the set of actors in the system. However, social networks are far more complex and typically there is more than one class of relations playing on the same set of actors. To capture the multiplicity of ties, this extra information requires additional notation.

A network having more than one type of relation is called a *multiple* or *multiplex network* and its corresponding graph is a multiple graph or *multigraph*:

$$\mathscr{G}^+ = \langle \mathscr{N}, \mathscr{E}, \mathscr{R} \rangle$$

where $\mathscr{R}$ represents a collection of relations whose elements are $R_1, R_2, \ldots, R_r$, where r is total number of relational types R in the social system. Each relational type then is represented by a distinct set of edges in $\mathscr{E}$.

1.2.3   Signed Graph

Special types of multiplex networks are systems having edges attached with a "sign" or a "valence" to capture the ties with an opposite sense. This type of ties represents affective ties such as "liking" or "disliking," or instrumental relations like "cooperation" and "competition" among economic organizations for instance.

A graph with lines having different valences or signs on the edges constitutes a *signed graph* that is defined as:

$$\mathscr{G}^\sigma = \langle \mathscr{N}, \mathscr{R}, \mathscr{V} \rangle.$$

With $\mathscr{G}^\sigma$ there is a set $\mathscr{V} = \{ v_1, v_2, \ldots, v_l \}$ of *valences* attached to $\mathscr{R}$, and where v_l stands for the total number of valences that a tie or a line can possible bear. For instance, when a relationship is either *positive* or *negative* then l equals 2. However, the meaning that a relation as labeled as "positive" or "negative" does not imply that one relation is "good" and the other is "bad" in a moral sense; it only indicates that there is a change of a *sign* in the valence of the tie.

Moreover, the absence of a tie is also considered as a different valence, and a relationship can be a mixture of different signs as well. In this latter case, a relation is neither positive nor negative, but rather an *ambivalent* type of tie, which means that the value of l is greater than 2. Note that a signed graph contemplates at least two kinds of ties and therefore a signed graph by definition is not a simple network, but it is rather a special case of a multiplex network structure.

1.2.4   Bipartite Graph

A graph $\mathcal{G}$ can be partitioned into s subsets of nodes $\mathcal{N}_1, \mathcal{N}_2, \ldots, \mathcal{N}_s$ such that each line in $\mathcal{G}$ are between a node in $\mathcal{N}_i$ and a node in $\mathcal{N}_j$, where $i \neq j$. The graph representing a network with this condition, which is termed as s-partite graph (Wasserman and Faust, 1994; Harary, 1994, (1st ed. 1969)), correspond to a one-mode network.

Hence, networks with a single class of actors or nodes have s = 1, and these are partitionable as well. In this case, however, the partition of the system is made according to rules of a chosen type of equivalence among a single class of network members, which are the actors.

When s = 2, then a two-partite graph or the *two-mode* network represents *affiliation networks* in which there is another class of network members that are called "events". Moreover, "multi-mode" networks are then those systems where s ⩾ 2.

The *bipartite graph* $\mathcal{G}^B$ that is formally defined as

$$\mathcal{G}^B = \langle\, \mathcal{N}, \mathcal{M}, \mathcal{E}\, \rangle$$

$\mathcal{G}^B$ is then made of 2 sets of entities, $\mathcal{N}_1 = \mathcal{N}$ and $\mathcal{N}_2 = \mathcal{M}$, which typically stand respectively for actors and events, or something else related to them that is not another actor from the actor set.

Bipartite graphs can be multiplex as well when there is a set of relations $\mathcal{R}$ attached to the structure

$$\mathcal{G}^B = \langle\, \mathcal{N}, \mathcal{M}, \mathcal{E}, \mathcal{R}\, \rangle$$

1.2.5   Valued Graph

When the set of ties in a system carries on a *value* or *weight* score that reflects the strength of the relationship between two connected actors, then the type of system corresponds to a *valued network*, also known as *weighted network*. Weights are typically measured in real numbers or $\mathcal{W} \in \mathbb{R}$, and the relational content of the tie corresponds to a qualitative piece of information.

Valued graphs represent a valued network, and a *valued graph* $\mathcal{G}^V$ has a set of values V attached to $\mathcal{G}$. Formally, this kind of structure is defined as

$$\mathcal{G}^V = \langle\, \mathcal{N}, \mathcal{E}, \mathcal{W}\, \rangle$$

with $\mathcal{W} = \{v_1, v_2, \ldots, v_g\}$, and for two-mode networks v_h as well.

As with bipartite graphs, valued graphs that are multiplex also add a set of relations $\mathcal{R}$ to the structure

$$\mathcal{G}^V = \langle\, \mathcal{N}, \mathcal{E}, \mathcal{W}, \mathcal{R}\, \rangle.$$

1.2.6 Multilevel Graph

Structures that are more complex are found in multilevel graphs that are special types of graphs having two or more related sets of interrelated entities. Formally, a *multilevel graph* $\mathscr{G}^M$ with sets $\mathscr{N}$ and $\mathscr{M}$ and a value set $\mathscr{W}$ is defined as

$$\mathscr{G}^M = \langle\, \mathscr{N}, \mathscr{M}, \mathscr{E}_{\mathscr{N}}, \mathscr{E}_{\mathscr{M}}, \mathscr{E}_{\mathscr{N} \times \mathscr{M}}, \mathscr{W}\, \rangle.$$

Hence, multilevel graphs are "a kind of" bipartite graphs where the two sets, actors and events, are or can be related within each other. Typically the ties between the sets are dichotomous and the relationships can have different intensities.

In addition, multilevel graphs that are multiplex supplement a set of relations $\mathscr{R}$ to the structure $\mathscr{G}^{M+}$

$$\mathscr{G}^M = \langle \mathscr{N}, \mathscr{M}, \mathscr{E}_{\mathscr{N}}, \mathscr{E}_{\mathscr{M}}, \mathscr{E}_{\mathscr{N} \times \mathscr{M}}, \mathscr{W}, \mathscr{R} \rangle$$

1.3 Matrices

The third element of the relational system $\mathscr{X}$, is another representation of the social network that is in a matrix form. In this case, $\mathbf{A}$ is a two-dimensional array size $n \times n$ called an *adjacency matrix*, which is associated with a relation R among the actor set in the network.

$$\mathbf{A} = \langle\, a_{ij}\, \rangle, \qquad i \neq j$$

where a_{ij} records the value of a tie from actor i to actor j on the relation. This is actually the same as saying that (i, j) is an element of $\mathscr{E}$. In this case, since there is a direct tie between two actors, they are considered to be *adjacent* to one another and become neighbor nodes.

In the adjacency matrix, the actors in the rows are the *senders*, whereas those actors in the columns are the *receivers* of the ties. When the ties of the network have no direction then $a_{ij} = a_{ji}$, resulting in a symmetric matrix and in a graph with undirected edges. Often a_{ii} is undefined by design for all actors i, which means that an actor cannot relate to himself. In this case, the entries in the *diagonal* of the matrix or $\mathbf{A}^{\alpha}$ are set to zero and ignored, and the nodes of the graph have no loops.

For a dichotomous relation—that is when a link either exists or not between two parts—the values of a_{ij} are simply represented by 0's and 1's, and the entries in the sociomatrix are defined as:

$$a_{ij} = \begin{cases} 1 & \text{if actor } i \text{ has a relation to actor } j, \\ 0 & \text{otherwise.} \end{cases}$$

1.3.1 Affiliation Matrix

For two-mode network data, the adjacency matrix $\mathbf{A}$ is called *affiliation matrix* with size $n \times m$. The entry the values for a dichotomous relation in this two-dimensional array are:

$$a_{ij} = \begin{cases} 1 & \text{if actor } i \text{ is affiliated to event } j, \\ 0 & \text{otherwise.} \end{cases}$$

1.3.2 Multiple Relations

For multiplex networks, an adjacency matrix $\mathbf{A}(R)$ is defined for each relation-type R:

$$\mathbf{A}(R) = \langle\, a_{ijR} \,\rangle, \qquad i \neq j$$

where a_{ijR} records the value of a tie from actor i to actor j on relation R. The entries for dichotomous relations in the case of a multiplex network are:

$$a_{ijr} = \begin{cases} 1 & \text{if } r \text{ relates actor } i \text{ to actor } j, \\ 0 & \text{otherwise.} \end{cases}$$

As a result of this, there are r sociomatrices with size $n \times n$, one for each relation-type for all the actors in X, and the multigraph $\mathscr{G}^+$ contains multiple edges sharing the corresponding endpoints. A three-dimensional array $\mathbf{A}(\mathscr{R})$ of size $n \times n \times r$ can serve to represent a network with multiple relations where every "slice" of the vector of matrices corresponds to a relationship class. Such matrix is arranged and labelled identically in order to have a coherent representation of the entire relational system under study.

1.3.3 Incidence Matrix

The connection between nodes and edges corresponds to an *incidence relation,* and this type of association is recorded in a rectangular matrix $n \times e$ of actors by ties that is called an *incidence matrix* $\mathbf{C}$ defined as

$$a_{ik}^{\mathbf{C}} = \begin{cases} 1 & \text{if node } i \text{ is related to edge } k, \\ 0 & \text{otherwise.} \end{cases}$$

Undirected and directed graphs are represented by "unoriented" and "oriented" incidence matrices respectively.

A relationship between the incident matrix $\mathbf{C}$ and the adjacent matrix $\mathbf{A}$ is

$$\mathbf{C} \circ \mathbf{C}^T = \mathbf{A}^\alpha - \mathbf{A}$$

where the left side of the expression represents the relational composition of the incidence matrix with its transpose (the two operations are defined later on in this chapter), whereas $\mathbf{A}^\alpha$ stands for the diagonal of the adjacent matrix.

1.3.4 Valency Matrix

For signed graphs, a special type of adjacency matrix $\mathbf{A}^\sigma$ known as the *valency matrix* is defined in the following manner:

$$\mathbf{A}^\sigma = \langle\, a_{ij}^\sigma \,\rangle, \qquad i \neq j$$

where a_{ij}^σ records the valency v of a tie from i to j, and here the entries are:

$$a_{ij}^\sigma = \begin{cases} \text{p} & \text{if actor } i \text{ has a positive relation to actor } j, \\ \text{n} & \text{if actor } i \text{ has a negative relation to actor } j, \\ \text{a} & \text{if actor } i \text{ has an ambivalent relation to actor } j, \\ \text{o} & \text{if there is no tie between } i \text{ and } j. \end{cases}$$

Hence, the network can have up to four valences and two actors or nodes in the system can be p-adjacent, n-adjacent, a-adjacent, or they have any tie in common. This last valence is applied as well when there is a relation of indifference between the actors that are aware of each other.

1.3.5 Different Systems

It might be possible that more than one social network is under study, and in this case there is a separate triple in $\mathcal{X}$ representing each relational system, $\mathcal{X}^+ = \mathcal{X}_1, \mathcal{X}_2, \ldots , \mathcal{X}_t$, being t the total number of networks under study. Then a separate matrix $\mathbf{A}(R\mathcal{X}^+)$ is required for each type of relation in every network:

$$\mathbf{A}(R\mathcal{X}) = \langle\, a_{ijR\mathcal{X}} \,\rangle, \qquad i \neq j$$

where $a_{ijR\mathcal{X}^+}$ records the value of a tie from actor i to actor j in relation R, and in the relational system $\mathcal{X}^+$.

Once more, the entries in $\mathbf{A}(R\mathcal{X})$ for a dichotomous relation can be:

$$a_{ijR\mathcal{X}^+} = \begin{cases} 1 & \text{if } R \text{ relates actor } i \text{ to actor } j \text{ on network } \mathcal{X}^+\,, \\ 0 & \text{otherwise.} \end{cases}$$

1.3.6 Graph and Matrix Representations

$\mathcal{G}$ and $\mathbf{A}$ usually stand for the social networks and these representation forms are referred as *graph theoretic* and *sociometric* notations by Wasserman and Faust (1994). There are advantages and disadvantages of each kind of representation but it is worthy to mention that sociometric notation permits to inclusion of measurements on actors in the analysis together with different social ties. However, the multiplicity of ties and the combinations of indirect relations can result quickly cumbersome with both the graph theoretic and sociometric notations. For the analysis of different kinds of relations and combinations of these, the algebraic notation results a better alternative and therefore the rest of the chapter has more on algebraic notation.

For instance, Figure 1.1 gives both graph and matrix representations for a configuration made of three connected nodes or a triad. The triad is then given as a simple graph and a symmetric matrix in Figure 1.1a, whereas Figure 1.1b shows a digraph where arcs pointing the direction of the ties from one node to another represent relations and the matrix is in this case asymmetric.

At this point, it is possible to distinguish some network effects like mutuality in dyads, i.e. configurations made of two elements, or in higher level structures a phenomenon like transitivity, which is the tendency that relates two actors that have a common neighbor. Later on, we will see how these particular properties of the configurations and others serve to describe the group structures under study.

The representation in Figure 1.1c characterizes a signed graph where there are two levels in the relations that are represented by directed edges with different shapes. In this case, solid arcs stand for positive relations and a dashed arc points a negative tie, whereas the absence of a tie is represented by o in the signed matrix. Note that if the graph of Figure 1.1c stood for a general directed multigraph, then both the positive and negative relations would have needed a separate adjacency matrix for a complete representation of the system.

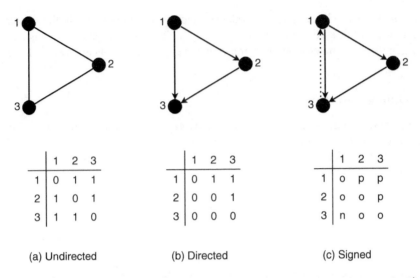

	1	2	3
1	0	1	1
2	1	0	1
3	1	1	0

	1	2	3
1	0	1	1
2	0	0	1
3	0	0	0

	1	2	3
1	o	p	p
2	o	o	p
3	n	o	o

(a) Undirected   (b) Directed   (c) Signed

Figure 1.1   Three types of networks with graph and matrix representations. *A shape in $\mathscr{G}$, $\mathscr{G}^d$, and $\mathscr{G}^\sigma$ corresponds to the content in their respective matrices.*

1.4   Chains, Paths, and Other Graph Properties

There are several concepts and properties of graphs and digraphs that are useful for the analysis and characterization of social networks, and the involve systems with either a single type or different types of relations.

We start by defining an alternating sequence of nodes and edges in a graph $\mathscr{G}$:

$$e_i = n_0, e_0, n_1, \ldots, e_k, n_k$$

that is called a *chain* when the sequence in $\mathscr{G}$ begins in a node n_0 and ends in node n_k. These are known as the initial and terminal nodes respectively of the sequence; that is $e_i = \langle n_0, n_k \rangle$. On the other hand, if the sequence of nodes and edges is placed in a directed graph $\mathscr{G}^d$, then the arrangement is called a *path* $e_i = (n_0, n_k)$ and the order of the elements is important. A *semipath* in the graph disregards directionality, which implies that $e_i = (n_0, n_k)$ or (n_k, n_0). Not all chains are paths but the converse is always true.

A chain, path, or semipath is called *open* or *simple* if and only if $n_0 \neq n_k$, whereas the sequence is called *closed* when $n_0 = n_k$. A simple closed chain, path, and semipath are called a *circuit*, a *cycle*, and a *semicycle* respectively. Now, k represents the *length* of the sequence, and it is the total number of edge occurrences in the chain, path, or semipath that can be either open or closed.

A graph or a digraph is said to be *connected* if and only if there is a chain, semipath, or path of relations between each pair of nodes in the configuration, which implies that the graph is *unilaterally*, *weakly*, or *strongly* connected respectively. In any of these cases, each node is *reachable* from the other nodes in the graph. As a result, a connected graph with either of the kinds of connectedness just mentioned has a single *component* and no *isolated* nodes, which are defined as nodes having no edges attached in the network. On the other hand, a graph is *disconnected* when it has either more than one component of connected nodes or isolated nodes, or both of them. A trivial graph containing just one node and edge is regarded as a strongly connected graph and not an isolated node.

1.5   Algebra of Relations

Several concepts and properties of graphs and digraphs are useful for the analysis and the characterization of social networks either with single or different types of relations. However, in the case of systems made of multiple relations, a fundamental matter of study is the *algebra* of network relations.

Algebra is a mathematical branch that deals with symbolic relations, and it can be effectively applied in the analysis of systems made of multiple relations. In algebraic notation the relations between actors are written in capital letters instead for subscripts. For instance, if $a_{ijR} = 1$ such that i and j are in $\mathbf{A}$ and $R \subseteq \mathbf{A} \times \mathbf{A}$, means that there exists a relation R between actor i and actor j. If $i \neq j$ in $\mathscr{G}_d$ then actor i "chooses" actor j in a relation, and this implies that $(i,j) \in R$. Such expression can be written just as iRj using the infix notation, which is sometimes used in arithmetic where the operator relation is written between the parts or operands on which they act.

1.5.1   Generators and Compounds

When $r > 1$; that is if the network has different kinds of relationships, each relation-type in $\mathscr{G}^+$ constitute a *primitive* or *generator* relation where different operations can be applied to it. This means that in a multiple network is possible to characterize the interaction among relations whenever the appropriated operation is used. As a way to elaborate this idea, we consider each relation type as a *letter* on an *alphabet* $\mathbf{\Sigma}$, which is a finite set of r ties; that is

$$\mathbf{\Sigma} = \{\ R_1, R_2,\ \dots\ , R_r\ \}.$$

Then one can apply operations on a given set with some defined rules and in this way create a system in the algebraic sense. Hence, an algebraic system or "algebra" is defined as a set of *objects*, together with a set of *operations* on it. For instance, the alphabet and all products over $\mathbf{\Sigma}$ produce an algebraic system known as "free word algebra," which is an unconstrained class of algebra that permits the construction of representations of multiplex social networks in algebraic form.

The *free word algebra* is represented as:

$$\mathbf{W} = \langle\ \mathbf{\Sigma}^*\ ,\ \boldsymbol{F}\ \rangle,$$

where $\mathbf{W}$ stands for the algebraic object, $\mathbf{\Sigma}^*$ for the unconstrained object set, and $\boldsymbol{F}$ for the operations set.

Hence, since in the algebra $\mathbf{W}$ each single relationship is a generator that constitutes a letter of the alphabet of multiple relations, this suggests that the combination of letters will make a "word" or "string" that is a product of the *composition* of relations. Either words can be made from different types of letters or by reusing the same letters where the number of letters determinates the *length* of the word, which in principle can be infinite even just with one generator. However, in a limited population some of these relations will connect precisely the same pair of individual actors, which means that the number of unique compound relations will also be finite (Lorrain and White, 1971; Wasserman and Faust, 1994, p. 495).

It is also possible to "constrain" the free word algebra and limit this object to words until certain length k. This means that we obtain the algebra $\mathbf{W}^k$, which is a truncated version of $\mathbf{W}$, and we use for its analysis a *partial algebra* rather than a full algebra. The difference now is that in $\mathbf{W}^k$ the object set in the system is made by the single relationships and by a collection of compound relations up until length k. In either of the two cases, we can define algebraic structures useful for the analysis of multiplex networks depending on the imposition of certain conditions on the strings.

1.6   Operations on Social Networks

Operations defined on pairs of elements from social networks are fundamental in making up the algebraic structures that describe collections of ties in the social systems, and the structure made by these relations can have either a single type of tie or multiple kinds of relationships. Different types of relationships can be differentiated by looking at their patterns of tie overlapping and also through the categorization of indirect relations (White, 1992, p. 95; also Pattison 1993), and where tie overlapping is described by the *intersection* of elements, i.e. $R_1 \cap R_2$, and the categorization of indirect relations by applying the tie composition, respectively.

1.6.1   Binary Operation on Relations

There is natural association between operations and relations where operations are regarded as special sort of relations. Operations are functions that assign elements from a domain X to a codomain Y, and for instance a function f on X and Y will be written as $f : X \rightarrow Y$ using the arrow notation. If we consider the domain of social relations X, the function will have a single input, namely the relationship itself, which is in this case a unary operation written as $f : X \rightarrow X$.

A *binary operation*, on the other hand, is a function that assigns the elements of the Cartesian product of the relations to a third set, that is $f : X \times X \rightarrow X$. This means that while the unary operation has one operand or argument, the binary operation takes two arguments. However, an operation can also have no arguments and it can be a "nullary" operation that produces a constant like the empty set of relations. Thus the set of operations F in $\mathbf{W}$ has a number of operands that determines its "arity," and where a n-ary operation is a function that assigns an element of a set to each n-tuple of elements of the set. We state this formally by defining an n-ary product R^n on a set $X \mid R^n \subseteq X \times \dots \times X$ (n times) as the generalization of the direct (aka "Cartesian") product of sets.

Some examples of binary operations are the union and intersection of sets, and there are other binary operations crucial for the analysis of multirelational systems. Before looking at these, we introduce first two important unary operations for the analysis of social networks, which are the complement and the converse of a relation.

In formal terms, the (Boolean) *complement* of a tie R in a system X is defined as:

$$R^\complement = \{\ \langle\ i,j\ \rangle \mid j, i \in X \text{ and } \langle\ j, i\ \rangle \notin R\ \}.$$

Hence, the complement of R consists of all ordered pairs of elements of the universe of discourse that do not belong to R (Maddux, 2006, p. 5). For example, if iRj means that "actor j is R related to actor i" then iR^Cj implies that "actors j is *not* R related to actor i."

For directed networks, R^{-1} represents the *converse* of a relation R that is produced by reversing the order of the members of a pair, and this is formally defined as:

$$R^{-1} = \{ (i,j) \mid (j,i) \in R \}.$$

The converse relation serves to define the asymmetry of ties in X, and it is useful to identify passive subjects of relational actions. For instance, if R means "supervising" then R^{-1} will mean "being supervised." Arabie et al. (1978) point out that the converse operation in social relations serves to generate *relational contrast* in the social system, and it is often very useful in the modeling process of its "relational structure", which is the configuration where relations are tied.

Furthermore, the converse relation of an entire network is represented by the *transpose* of the adjacency matrix, $\mathbf{A}^T$, which is obtained by exchanging the matrix rows for its columns. That is, for each a_{ij} in $\mathbf{A}$ we get a_{ji}.

The categorization of indirect relations plays an essential part in the analysis of networks of multiple relations, and this is because it combines relationships that can be primitives or not into indirect links are occurring in the system. For this reason, we are going to take next the composition of ties in a more detailed manner.

1.6.2   Relational Composition

The *relational composition* is a binary operation that is useful in the interpretation of the organizing principles of multiplex networks. The importance of this operation lays in the fact that by combining the different types of a tie occurring in the network, we create an abstract structure that represents the entire relational structure of the system.

A combination of ties then creates an indirect link called a *compound* relation between two nodes in the system through *right multiplication*. The right multiplication procedure adjoins a generator strings to the right of the path label where both equations and orderings—defined later on in this chapter—are preserved. This kind of indirect tie can result either in a relational path or in a chain of relations depending on whether the network is directed or undirected, and it can combine relations at different levels in the system.

The operation of relational composition is denoted by the symbol ' $\circ$ ', and in order to illustrate this operation consider, for example, the relations $R_1 = F$ and $R_2 = E$, representing e.g. friendship and enmity. A combination of these link-types makes a new compound relation where a tie $i(F \circ E)j$ is present if an actor k exists such that iFk and kEj. This implication can be written by dropping the symbol for the operation of composition as $i(FE)j$ or simply by $iFEj$, which means that there is a compound relationship made of F and E among actors i and j, and it is read as "actor j is 'the enemy of a friend' of actor i." In this sense, the relational composition involves at least three subjects in the relational path, namely actors i, j, and k in the example, and by convention it reads from right to left.

To have a more formal characterization of the composition operation, let $R_1 \subset X \times Z$ and $R_2 \subset Z \times Y$ (where $\subset$ means "is a proper subset of"). The composition of R_1 and R_2 is a binary relation between X and Y that for all $x \in X$, $y \in Y$ is given by:

$$R_1 \circ R_2 = \{ (x,y) \mid \text{exists } z \in Z \text{ such that } (x,z) \in R_1 \text{ and } (z,y) \in R_2 \}.$$

As a result, $R_1 \circ R_2$ denotes the product of R_1 and R_2 or the compound relation between x and y that in this case corresponds to the *matrix product* of the arrays representing R_1 and R_2.

It may be possible for the relational composition that no ordered pairs of actors exist on a particular compound relation. This happens when there are no elements x, y, and z such as $(x,z) \in R_1$ and $(z,y) \in R_2$; thus the relation $R_1 \circ R_2 = \varnothing$, meaning that such compound relation is empty or undefined for x and y. Since we represented collection of relation types in a multiplex network by $\mathscr{R}$, the set of primitive and compound relations are going to be represented by $\mathscr{R}^+$, whereas r^+ will denote the number of elements in $\mathscr{R}^+$.

However, the composition of two operands such as different kinds of relations can be based in other types of operations than the matrix product. For instance, the max − min product is used to create compound relations of weighted or valued ties, but we are going to take a look at these operations when we introduce these types of networks later on.

As illustration, Figure 1.2 presents some operations performed on the graphs depicted in Figure 1.1, and Figure 1.2a, for example, represents the complement of the undirected and loopless graph in Figure 1.1a. The converse of ties in the digraph in Figure 1.1b is shown in Figure 1.2b, and finally Figure 1.2c represents the relational composition of the positive ties and the negative tie in Figure 1.1c where actor 1 ends up having a pair of 2-step paths that combine the two types of valences.

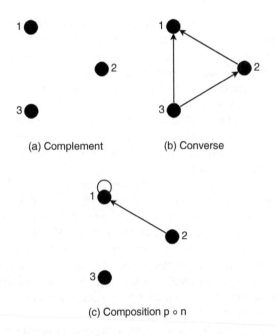

(a) Complement   (b) Converse

(c) Composition p ∘ n

Figure 1.2   Different operations on $\mathscr{G}$, $\mathscr{G}^d$, and $\mathscr{G}^o$ from Figure 1.1

1.7   Types and Properties of Relations

We characterized an algebraic structure as a relational system with a family of operations defined on a set where operations are sorts of relations with their own properties and rules that must conform in order to describe properly the structures they made. In this section, we are going to differentiate types of relations in general and their properties for two basic reasons. First properties on relations are fundamental in the characterization of the different types of functions operating in such structures, and second because some specific types of relations with defined properties are building blocks in the establishment of the algebraic structures that represent multiplex networks.

In this sense, we start with three standard relations that have been already introduced either directly or indirectly:

1.   the *empty* relation $\varnothing$
2.   the *universal* relation U
3.   the *identity* relation I

The empty relation is the smallest relation and it is simply the empty set. Since in this case relations are sets of pairs, the empty relation never holds for all the pairs in the set, i.e. $R = \varnothing$. The universal or full relation U is the widest possible relation and it is the Cartesian product of the set, $R = X \times X$, which always holds for all the pairs in set X. Lastly the identity relation I holds for pairs whose first and second elements are identical, that is $R = \{ (x_i, x_i) \mid \text{for all } x \in X \}$.

In a graph format, the identity relation is represented by loops in the nodes, whereas in a matrix form, the empty and the identity relations are the zero-matrix and the identity-matrix respectively; the former with all its entries being zero, and the latter has ones in its main diagonal and zeros elsewhere. The universal relation on the other hand has 1s in all the entries of the matrix including in the diagonal. However, these particular structures assume matrix multiplication in the relational composition, and this is because for matrix addition operation, for example, it is the zero-matrix that represents the identity element.

The three above mentioned standard relations can in turn be characterized in terms of relational composition, and when they are concatenated for instance with a "concrete" relation R, it results in the following identities:

$$\varnothing \circ \mathsf{R} = \varnothing$$
$$\mathsf{I} \circ \mathsf{R} = \mathsf{R}$$
$$\mathsf{U} \circ \mathsf{R} \circ \mathsf{U} = \mathsf{U}, \qquad \text{iff} \quad \mathsf{R} \neq \varnothing$$

where "iff" is a shortening of "if and only if," a logical operator that is also represented by the symbol '$\Leftrightarrow$'.

As a result, the product of a relational composition with an empty relation equals the empty set; the identity relation acts as a neutral element in a composition, and finally the universal relation includes all other relations except for the empty relation. Sometimes each one of these relations have the universe attached as $\varnothing_X, \mathsf{I}_X, \mathsf{U}_X, \mathsf{R}_X$, but for simplicity we are going to avoid this when the statements are clear by the context.

It is time now to see some important properties for the existing relation R that are key conditions for making up some algebraic structures useful for the analysis of multiplex networks:

$$
\begin{aligned}
\textit{reflexivity} &\Leftrightarrow R \cup I \subseteq R \\
\textit{symmetry} &\Leftrightarrow R^{-1} = R \\
\textit{antisymmetry} &\Leftrightarrow R \cap R^{-1} = I \\
\textit{transitivity} &\Leftrightarrow R \circ R \subseteq R \\
\textit{invertibility} &\Leftrightarrow R \circ R^{-1} = I
\end{aligned}
$$

Since the identity relation I is included in the reflexive property it means that the element is related to itself. Although the reflexive relation seems a trivial one, it is in fact a consequence of transitivity and reciprocity (Lorrain, 1975, p. 20). Reciprocity is reflected by the symmetric property where a relation equals its inverse, whereas the transitive property implies that the extension of the relation exactly corresponds to the relation itself, and in this sense the notion of transitivity implies that actions of the elements in a system can have "indirect" effects on one another (Berkowitz, 1982).

In the antisymmetric property there is no reciprocity but self-relations can occur. If loops were not allowed by definition, then the relation R is meant to be called *asymmetric* rather than antisymmetric whenever $R \cap R^{-1} = \varnothing$. The distinction between antisymmetry and asymmetry in the properties of binary relations serves for the definition of particular types of relations used in the algebraic analysis of relational systems. On the other hand, the asymmetry of social ties indicates that the relationship between a pair of actors has a single direction. Finally, an identity relation or a loop in the graph is produced by the composition of a relation with its converse R^{-1}, which is also called the *inverse* relation of R.

Relations can have one or more of such properties or none of these, and the combination of the properties permit us to identify special types of binary relations relevant for the analysis of different systems. For instance the statement "to be similar to" on a set of people is a sort of relation that is both symmetric and reflexive, and it is known as a *tolerance*. Such kind of relation is symmetric because the relation can go in both directions and it is reflexive because one is similar to him/her/itself.

1.8 Equivalence and Ordering

Two particularly important kinds of relations for the study of social systems are *equivalences* and *ordering* relations, and this is because they serve to classify the actors and relations of the network in different ways. In an equivalence relation two elements are the same in some respect such as "being alike" or "being in the same team" on a set of people.

Ordering relations, on the other hand, make a comparison of lesser to greater elements, i.e. "greater than" or "less than," which are represented (as seen before) by $>$ or $<$, respectively. Orderings are important to establish hierarchies in the system and also in performing asymmetric clustering (Boyd, 1980), which is a procedure that transforms network data.

1.8.1 Equivalence

The specific properties of the equivalence relation and the related concept of partition of the system are now expressed in mathematical terms. Both concepts are central in the modeling of social networks, since they are used to reduce the complexity of the systems.

Formally, a binary relation R defined on set X is an *equivalence relation* $\equiv$ whenever is:

$x \equiv x$,	for all $x \in X$	(reflexive)
$x \equiv y$	implies $y \equiv x$	(symmetric)
$x \equiv y$	and $y \equiv z$ imply $x \equiv z$	(transitive)

where the expression $x \equiv y \bmod R$ indicates that there is an equivalence relation between x and y under R, and R is a subset of $X \times X$; that is $(x, y) \in R$.

To define the classes of equivalent elements, let $[x]$ denote a subset of X of elements y such that $x \equiv y$. By the reflexivity property we have that $x \in [x]$, and by the symmetry and transitivity properties we have that if $y, z \in [x]$ then $y \equiv z$. This means that $[x]$ is a collection of equivalent elements, which makes it possible to produce a *partition* or a disjoint sets of equal elements in the network that characterizes the "relational structure" in a simplified form. Hence, each equivalence relation establishes an associated partition of the involved elements of the set, and every partition corresponds to a determinate type of equivalent relation.

Formally, a partition π of a set X is any collection $\{ X_i \mid i \in P \}$ of nonempty subsets of X, where π is an indexed set, and which satisfies the following two conditions:

$X_i \cap X_j = \emptyset$, for $i, j \in \pi$ with $i \neq j$	(*mutually exclusive*)
$X = \bigcup \{ X_i \mid i \in \pi \}$	(*collectively exhaustive*).

The mutually exclusive condition means that nothing can belong simultaneously to both subsets, and the collectively exhaustive condition means that everything must belong to one subset or the other subset. In other words, a partition of a set X is a decomposition of X into non-overlapping and non-empty subsets or "parts" whose union is all of X and whose intersection is empty.

1.8.2   Partial Order

In the case of ordering relations, special attention is given to the partial ordering since it helps to define the set of distinct relations in a multiplex network represented by an algebraic structure.

The conditions for a *partial order relation* $\leq$ imply that the relation is:

$x \leq x$,	for all $x \in X$	(reflexive)
$x \leq y$	and $y \leq x$ imply $x = y$	(antisymmetric)
$x \leq y$	and $y \leq z$ imply $x \leq z$	(transitive)

We note that the partial order, as any type of ordering, lacks symmetry, and that is the main difference from the equivalence relation, which is also reflexive and transitive.

A partial ordering on a set determines a *partially ordered set* or *poset* for short, and for a given collection of elements, it is possible to obtain a partial ordered structure, which can be represented by a diagram depicting ordering relation between certain pairs of elements of the set. Often the collection of poset elements makes certain types of algebraic structures with specific characteristics, which allow us to get an insight into the relational system involved. Algebraic structures are covered later on in this chapter and in more depth in the following chapters.

As seen, both the equivalence relations and the orderings have the transitive property, which for equivalence means that if two elements are equivalent to a third one this implies that they are also equivalent between themselves. For an ordering relation the order-preserving condition holds; that is if, for example, x is less than y and y is less than z, then x is less than z; or more formally $x < y$ and $y < z$ imply $x < z$.

Besides transitivity, equivalence relations are also reflexive and symmetric, whereas orderings relations are on the contrary in general irreflexive and asymmetric. There are, however, special types of ordering relations, and for instance a *quasi-order* is a transitive relation that is also reflexive, whereas a quasi-order that is antisymmetric is termed a *partial order*.

Strictly speaking a partial order relation that is reflexive is a "weak" partial order, and with the irreflexive property becomes a "strong" partial order. Partial order differs from a *total* or *lineal* order since a total order relation is said to be connected, complete, and satisfies the "law of trichotomy," which means that for all $x, y \in X$; either $x > y$, or $y < x$, or $y = x$ holds.

1.8.3   Hierarchy

With the definition of a poset and the equivalence it is possible to define the notion of hierarchy in the network, which is a form for gradation where superordinate and subordinate elements occupy different levels in the systems.

Thus, for a set of relations R, two subsets $X, Y \in R$ form a *hierarchy* whenever one of the following conditions hold:

$$1.\ X \cap Y = \emptyset$$
$$2.\ X \cap Y = X \quad \Leftrightarrow \quad Y \leq X$$
$$3.\ X \cap Y = Y \quad \Leftrightarrow \quad X \leq Y$$

That is, in case one of the subsets contains the other subset, or when these subsets form a partition of the system, which means that they can be pairwise disjoint as well.

1.9   Functions

A function is a particular type of binary relation used in set theory to compare structures in which a quantity in the domain uniquely determines a second quantity in the codomain. The domain is what can go into the function, the codomain is what may possibly come out of it, and what actually comes out of a function is the *range*. Thus each element of the domain is related to a point in the range, like a cartographical map corresponds to a geographical region, and that is why functions are also called "maps" or "mappings".

For instance, $f : X \rightarrow Y$, reads the function f with X as its domain and Y as its range. If for $x \in X, y \in Y, (x, y) \in f$, then $f(x) = y$, where y is called the *image* of x under f and x is the argument or *preimage* of y.

Important classes of functions are:

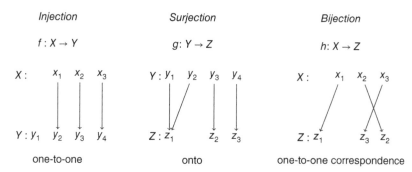

Injection	Surjection	Bijection
$f: X \to Y$	$g: Y \to Z$	$h: X \to Z$
one-to-one	onto	one-to-one correspondence

Thus for each element of the range there is "at most" one pair (x, y) in f, "at least" one pair (y, z) in g, and "exactly" one pair (x, z) in h. More precisely, while in the injective function f elements in the range Y are mapped into by a unique element of the domain X, with the surjective function g there are some elements in the domain Y matching the elements in the range Z, meaning that the image equals its codomain. On the other hand, the bijection is both one-to-one and onto, and in this function there is exactly one element in the domain for every element in the range, which means that the bijective function can be inverted; that is $h^{-1} = Z \to X$.

A bijection on a finite set as in h is called a *permutation*, and this function is actually the rearrangement of the elements in the set. Permutations are important in the study of cohesive subgroups in social systems and for the reduction of the structure into sets of equivalent actors.

Since functions are special types of binary relations, they can also be "concatenated." In this sense, if $f : X \to Y$ and $g : f(x) \to Z$ then $g \circ f : X \to Z$ by setting $(g \circ f)(x) = g(f(x))$, $x \in X$. The sequence of f and g in the expression $g \circ f$ is important, because the composition $f \circ g$ might not been even defined. If so, it is usually that $f \circ g \neq g \circ f$, which means that these relations are not *commutative*. However, the composition of functions can be *associative*, and this means that $h \circ (g \circ f) = (h \circ g) \circ f$. In such case the brackets can be omitted and the expression is written as $h \circ g \circ f$, or simply by juxtaposition as hgf.

Commutativity and associativity are axioms or laws that operations and relations conform, and in order to put in formal terms these and other important axioms, we define for all f, g, h of F the following equalities:

$$
\begin{aligned}
\text{Commutativity:} \quad & fg = gf \\
\text{Associativity:} \quad & f(gh) = (fg)h \\
\text{Idempotence:} \quad & ff = f \\
\text{Absorption:} \quad & fg = f \quad \text{or} \quad fg = g
\end{aligned}
$$

Besides commutativity and associativity, the *absorption* law is an identity between two functions or operations whose product equals only one of them, and it is usually present in algebraic structures having an ordering among its elements. In the case of *idempotence* we should note that this axiom is actually a transitive relation when the equality holds, and this means that the operation f can be applied several times to the relation without changing the result.

When idempotence is present in relational composition of a social network, it represents stability or invariance in a system (Boyd, 1991, p. 29). Idempotence is normally applied in relational systems to perform the transitive closure operation, which is a special kind of a transitive relation that is important to find out the connectivity or reachability in the network. Different types of closures are defined in Chapter 2 Algebraic Structures.

On the other hand, commutativity and associativity are properties that are useful in building scalable systems. Associativity enables for instance parallelism in distributed computing, while commutativity allows us to ignore order in the involved data. In the same way, we can apply the principles of commutativity and associativity in the analysis of social networks, where we employ as well as some other special types of functions.

Before introducing other special types of functions useful for the analysis of networks, it is worth mentioning that a commutative diagram can represent compound functions. Commutative diagrams map the compositions of functions and the objects involved in these, and below is the commutative diagram representing relations between the three classes of functions injection, surjection, and bijection introduced before:

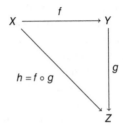

1.9.1   Identity and Empty Functions

As in the case of the standard relations, there are also special types of functions. For instance the *identity function* maps each element of a given set to itself, i.e. $f(x) = x, x \in X$; whereas the *empty function* is a unique function from the empty set to a give other set. In this sense, the empty function is a constant that satisfies the definition of "vacously" since the expression $x \in \emptyset$ is never true and then absolutely anything follows (Boyd, 1991, p. 22). Each empty function is thus a *constant function* that produces a constant value c that is defined as $f(x) = c$, for each $x \in X$.

The set of all functions from the domain X to the codomain Y is known as the *function set* and it is denoted by Y^X. This generalization can, however, be limited to the set of two logical or "truth" values, which is 1 for true and 0 for false, i.e. $\mathbf{V} = 0, 1$. Consequently for a subset of X, this set corresponds to the *characteristic* or *indicator function* for $A \subseteq X$ that indicates the membership of an element $\chi_A : X \rightarrow \mathbf{V}$ provided that:

$$\chi_A(x) = \begin{cases} 1, & \text{when } x \in A, \\ 0, & \text{when } x \notin A. \end{cases}$$

As a result of this, 2^X corresponds to the collection of all subsets of the set X, and the set of all binary relations or ordered pairs on X will be $2^{X \times X}$, this latter expression is also denoted by $\mathbf{R}_X$.

An important fact is that in a social system with multiple relations, it can be the case that two different ties whether they are compound or not coincide on the same ordered pairs. So, in order to keep them separate with different labels, it is necessary that the set of ordered pairs $\mathbf{R}_X$ is indexed by the alphabet set Σ, which means that the function set $(2^{X \times X})^\Sigma$ will represent a family of relations indexed by Σ of relational labels.

1.9.2 Transformations

Transformations are special types of function, which is an action on a set of elements to itself. If the arrangement of the object set remains invariant after the transformation, then the function represents "symmetry". Symmetry is closely related to group structure and group theory where are two types of transformations involved, namely *rotations* and *reflections*.

These two kinds of transformations, rotations and reflections, play an important role in certain types of permutation groups that are algebraic structure made of the set of all permutations on a given set. We look at some special types of permutation groups in Chapter 2, and in Chapter 3 a group structure representing a human social system with the use of permutation matrices.

Formally, a *transformation* on a set X is a function $f : X \to X$, and the *composition* of two transformations f and g, defined as $(f \circ g)(x) = f(g(x))$, means "apply f to the result of g." The associativity property applies in the composition of transformation functions, whereas commutativity generally does not apply in this case.

1.10 Homomorphism and Congruence

A **homomorphism** h is a structure-preserving mapping between algebraic objects that belongs to the *Universal Algebra*, which is the general theory of classes of algebraic structures; i.e., systems that maintain the operation in the involved algebraic objects.

For two algebraic objects S and T, a homomorphism $h : S \to T$ has h as a function from S to T whenever for each $a, b \in S$,

$$h(ab) = h(a)\, h(b).$$

Since homomorphisms are functions that preserve the structure, the products, which in this case result from juxtaposition, are preserved under the homomorphism and, for the two algebraic objects, the operation on the left-hand side corresponds to S while the one on the right-hand side is in T. If $h(a) = h(b)$, then a and b are said to be in the same *congruence class* and this is denoted as an equivalence relation by $a \equiv b$.

If S and T are semigroups (algebraic objects defined in Chapter 2), the function $h : S \to T$ is a *semigroup homomorphism*; and if h performs a simplification of the given system structure, then the mapping is a *homomorphic reduction*. Consequently, the semigroup obtained T, which is the *homomorphic image* of S, is called as a *quotient semigroup* under the equivalence relation on the semigroup that is stable under the operation, and hence the homomorphic reduction of an algebraic structure is a special kind of homomorphism that is surjective. On the other hand, a homomorphism that is a one-to-one mapping, which is also known as monomorphism, permits one to identify how closely the elements of systems are related to each other.

A homomorphism that is a one-to-one correspondence is called an *isomorphism* when is invertible, and two structures are said to be isomorphic when they are structurally identical, differing only in the labels of their elements. Thus, if h is an isomorphism defined on S, then S

and $h(S)$ are isomorphic and this is written as

$$S \cong h(S).$$

Besides, an isomorphism of a system onto itself is called an *automorphism*, and this means that the function is actually a permutation of the elements of the set; here the mapping acts as an identity function, i.e. $h : S \to S$, where h is an automorphism.

1.10.1   Congruence Relations

An equivalence relation $\equiv$ on a semigroup S is *right compatible* if there exists elements $a, b, c \in S^1$ such that $ac \equiv bc$ holds; similarly, $\equiv$ is *left compatible* if $ca \equiv cb$ holds. Any equivalence relation on S that is both left and right compatible is an equivalence relation on the semigroup S; however, one can relax this condition to be just considering a *right-* or *left congruence* whenever the equivalence relation is either simply right or left compatible (Howie, 1996, p. 22; Wu, 1984, p. 293). A right or left congruence will induces respectively a right- or left homomorphism and thus a *right-* or *left homomorphic images* of the algebraic structure.

In formal terms, a **congruence relation** $\equiv_h$ that corresponds to the homomorphism h on a semigroup S is an equivalent relation such that for each $a, b \in S$:

$$(a, b) \in \equiv_h \text{ and } (h(a), h(b)) \in \equiv_h \text{ implies } (a\, h(a), b\, h(b)) \in \equiv_h$$

that corresponds to the *substitution property* of the relation. Besides, any congruence relation is reflexive, symmetric, and transitive, which are the properties of the equivalence relation (Hartmanis and Stearns, 1966; Pattison, 1993).

Furthermore, the equivalence classes of $\equiv_h$ on S define a partition π on the elements of S where:

$$(a\equiv_h b) \in \pi \text{ and } (h(a)\equiv_h h(b)) \in \pi \text{ implies } (ab\equiv_h h(a)\, h(b)) \in \pi.$$

Although the congruence relation resembles the equivalence notion, any congruence also preserves the operation between the correspondent classes in the algebraic structure, which is not the case with the ordinary equivalence relation.

Each homomorphism determines a congruence relation, and conversely every congruence relation induces a homomorphism. Therefore, we define particular homomorphisms useful for the analysis of social networks in the following chapters, which will allow us performing the task of unfolding the crucial characteristics of multiplex social networks while keeping its essence in the process.

1.10.2   Kernel of a Homomorphism

Before introducing particular homomorphisms for the analysis of social networks, we define the "kernel of a homomorphism", which is intimately related to the image of the network relational system. For instance, the kernel of a semigroup S is the minimal ideal of S. Thus in a mapping h from S to T, the kernel of a h is a set K of all elements in S that are carried by h to the neutral element of T.

Formally, if 1 represents the neutral element of T, the *kernel K* of a homomorphism h in S is defined as:

$$K = \{ \ a \in S \mid h(a) = 1 \ \}.$$

That is, the homomorphism of element a produces the identity element of the algebraic structure.

1.11 Structural Analysis with Algebra: Summary

This initial Chapter focused on the introduction of fundamental concepts and network representations such as graphs and matrices. Different sorts of algebraic analyses of social networks benefit from these representation forms and from the algebra of relations introduced with formal definitions.

The equivalence and ordering of relations allowed us to establish a network structure as an ordered system made of classes of actors. As a result, we were able to look at the gradation of elements occupying different levels in the systems. We also looked at different types of properties, operations, and functions that are useful for the modeling of social networks. For instance, the relational composition operation plays a central role in the establishment of relational structures of different types of multiplex networks where relational structures link the network ties.

1.12   Learning Structural Analysis by Doing

1.12.1   Getting Started

```
# install the packages from CRAN
R> install.packages("multiplex", "multigraph")
# or their beta versions from GitHub
R> devtools::install_github("mplex/multiplex", ref = "beta")
R> devtools::install_github("mplex/multigraph", ref = "beta")
```

1.12.2   Matrices

```
# load the "multiplex" package
R> library("multiplex")

# create a network with three nodes
R> net <- transf(c("1, 2", "1, 3", "2, 3"))
# create a multiple network with two types of relations
R> net2 <- zbind(net, transf("3, 1"))
```

```
# adjacency matrix representing 'net'
R> net

  1 2 3
1 0 1 1
2 0 0 1
3 0 0 0
```

```
# symmetrize the adjacency matrix
R> mnplx(net, directed = FALSE)

  1 2 3
1 0 1 1
2 1 0 1
3 1 1 0
```

```
# coerce network 'net2' into a "Signed" class object
R> signed(net2)
```

```
$val
[1] 1  0  -1

$s
   1 2 3
1  0 1 1
2  0 0 1
3 -1 0 0

attr(,"class")
[1] "Signed"
```

1.12.3   Graphs

```
# load the "multigraph" package
R> library("multigraph")

# define two scopes with node / edge / graph characteristics
R> scp <- list(cex = 12, vcol = 1, lwd = 9, ecol = 1, rot = -30)
R> scps <- c(scp, signed = TRUE, bwd = .5, swp = TRUE)
```

```
# Fig. 1.1.a.  plot graph with customized format and (default) circular layout
R> multigraph(net, scope = scp, directed = FALSE)

# Fig. 1.1.b.  same as 1.1.a but as (default) directed graph
R> multigraph(net, scope = scp)

# Fig. 1.1.c.  same as 1.1.b but as signed graph
R> multigraph(signed(net2), scope = scps)
```

```
# Fig. 1.2.a.  the complement of Fig. 1.1.a
R> multigraph(1 - mnplx(net, directed = FALSE), scope = scp)

# Fig. 1.2.b.  the converse of Fig. 1.1.b
R> multigraph(t(net), scope = scp)

# Fig. 1.2.c.  the composition of relations in Fig. 1.1.c
R> multigraph(net2[,,1] %*% net2[,,2], scope = scp, loops = TRUE)
```

2

Algebraic Structures

2.1 Algebraic Structure Definition

Central in the analysis of multiplex social networks are different sorts of *algebraic structures*. This is not only because algebraic "structures" are capable of representing several types of network relations, but also because the set of rules governing the system allows the relations to be represented and combined into a meaningful ensemble.

The next sections serve to introduce different types of algebraic systems that are useful for the analysis of different sorts of social networks. We will also look at significant properties of the concepts introduced previously, which characterize the algebraic system.

A nonempty set of elements X together with a binary operation $*$ on the set is an *abstract algebraic structure*

$$\langle\, X,\, *\, \rangle.$$

Algebraic systems with more operations also exist, and the involved operations can follow one or more axioms. Specific algebraic objects, usually with one and two operations, serve to identify and describe properties of particular network structures, and it is possible to have a hierarchy among the elements of an algebraic system that is partially ordered.

2.1.1 Closure

An important feature in any algebraic structure is that it is a *closed* system, and this means that the product of the endowed operation on the set is part of the set as well. There are different kinds of closure having specific properties in common and the idea with these systems is that they must be capable of describing the complete structure of the study object, which is this case are networks of social relations. What is needed is the closure operation, which is a special type of function on the system that allows us obtaining an algebraic object without proper extensions.

Algebraic Analysis of Social Networks: Models, Methods and Applications using R,
First Edition. J. A. R. Ostoic. Companion website: www.wiley.com/go/ostoic/algebraicanalysis.
© 2021 John Wiley & Sons Ltd. Published 2021 by John Wiley & Sons Ltd.

For a formal definition of the closure operation, consider a collection of binary relations in a set X that characterizes a social network, an *algebraic closure* on X is a function $^{-}:\mathbf{R}_X \to \mathbf{R}_X$, where for every subset $A \in X$ and for each $A, B \subseteq X$, the closure $\overline{A}$ has the following properties:

$$A \subseteq \overline{A} \qquad\qquad\qquad (extensive)$$

$$\overline{(\overline{A})} = \overline{A} \qquad\qquad\qquad (idempotent)$$

$$A \subseteq B \quad \text{implies} \quad \overline{A} \subseteq \overline{B} \qquad (increasing)$$

In this sense, when $A = \overline{A}$ then it means that set A is *algebraically closed* with respect to the function $^{-}$. The extensive property means that the set is included in its closure, whereas the increasing property means that the relation is order preserving, which is a property type that is also known as *isotone*.

The precedent properties defined on relations can be extended to themselves by means of a closure operation as well, and important closure operations for an existing relation R in a defined set are:

$$R^r = R \cup I \quad (reflexive\ closure)$$
$$R^s = R \cup R^T \quad (symmetric\ closure)$$
$$R^t = \cup R^k \quad (transitive\ closure)$$

Thus while reflexive and symmetric closures are achieved with the union of the relation with the identity and with its transpose respectively, transitive closure is the minimal transitive relation that contains R. That is for $k > 0$, $R^1 = R$, and $R^k = R \circ R^{k-1}$.

In addition to that, an *equivalent closure* is obtained through the composition of the reflexive, symmetric, and transitive closures; whereas a *quasi-order closure* is the product of the composition of the reflexive and transitive closures (Boyd, 1991, p. 44).

2.2   Group Structure

A key algebraic object in the study of relational systems is that of a *group*, which is a binary structure made of an element set with an endowed operation

$$\langle\ G,\ \cdot\ \rangle.$$

Groups are particularly characterized by their invertibility. That is, the product of each element with its inverse produces the identity element of the system, and this property gives the group structure a symmetric character.

Besides identity and inversion, the *abstract group* G also satisfies the associativity and closure axioms, which are formally defined for all $a, b, c, e \in G$:

Identity:	$a \cdot e = e \cdot a = a$
Inversion:	$a \cdot a^{-1} = a^{-1} \cdot a = e$
Associativity:	$(a \cdot b) \cdot c = a \cdot (b \cdot c)$
Closure:	$a \cdot b \in G \qquad \text{for all } a, b$

Clearly, e is the identity element in G, and the "inversion" axiom corresponds to the "invertibility" property that relates each element with its "inverse." If the inversion property does not hold, the algebraic structure rather corresponds to a *semigroup*, which is "almost a group," which is typically represented by S.

When a group G acts on object set X, then $|G|$ is the *order* of the group (i.e. its cardinality), and $|X|$ corresponds to its *degree*. In other words, the group degree is the number of elements in the set, whereas the order is the number of elements in the group. There are different types of groups with different order and degree, and a trivial group with order 1 where each element on the finite set remains unchanged is the "identity group."

An example of a group structure is the set of integers $\mathbb{Z}$ under addition + operation. It is easy to see that 0 is the identity element in this case, and the commutative property applies in the arithmetic sum of numbers, which means that this particular algebraic structure is a *commutative group*. With respect to the invertibility property, the "additive inverse" of a number $x \in \mathbb{Z}$ implies that $x + (-x) = 0$; i.e., when added to its negation yields 0, and in this case zero is the additive inverse of itself.

2.2.1 Cayley Graph

The "Cayley color graph" (hereafter Cayley graph) relates graph theory and group theory by encoding the algebraic structure of a given group in a graph format. The structure of "the relationships between relations" in multiplex social networks can be depicted as Cayley graphs, and this representation form allows having a higher level of abstraction of the network configuration than for instance multigraphs do.

Formally, the *Cayley graph* Γ of a group G is defined with respect to a given set of generator relations X:

$$\Gamma = \Gamma(G, X).$$

The set of nodes $\mathcal{N}(\Gamma)$ in Γ is the set of elements in G. A directed edge represents a generator $x \in X$ connecting two nodes $a, b \in G$ whenever $b = xa$. Hence, all pairs of the form $(b, x \cdot a)$ make the edge set in $\mathscr{G}(\Gamma)$.

The choice of the *generating set* of the group structure determines the structure of the Cayley graph as they do for other algebraic systems such as the semigroups of relations. For instance, Figure 2.1 depicts the Cayley graph and table of a group with order of 2 representing a set of integers under addition or $\mathbb{Z}_2$. Here, the graphical representation of the group generated by the elements of $\mathbb{Z}_2$ put the identity element as solid loops, and the relations between the two generators as dashed arcs. This means that the identity element represents the relations from generators e and x to themselves; i.e. by juxtaposition $e = ee$ and $e = xx$, and the other relations are between e and x. In the Cayley table to the left of Figure 2.1, the two kinds of relations of $\mathbb{Z}_2$ are those in the diagonal and in the off-diagonal of the matrix.

Typically, the redundant information in the group structure is deleted from the Cayley graph, and in the case of $\mathbb{Z}_2$ is the identity element that constitutes a redundancy. This is because the relations $x = ex$ and $x = xe$ represent the entire group structure, and therefore the dashed arcs remain in the picture. The direction of reciprocated ties constitute a redundancy as well, and these patterns are usually represented as undirected edges in $\mathscr{G}(\Gamma)$. The Cayley graphs of $\mathbb{Z}_2$ with and without redundancies are depicted to the right of Figure 2.1.

+	e	x
e	e	x
x	x	e

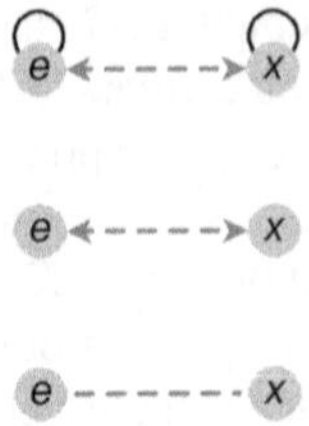

Figure 2.1 Cayley graphs and table of $\mathbb{Z}_2$ under addition. *Left: Cayley table. Top, middle and bottom right: Cayley graph with and without redundancies.*

Cayley graphs serve to represent semigroups as well. Since semigroups tend to be complex in nature, Cayley graphs are appropriate to summarize the information of relatively large algebraic structures because they make it possible to highlight the patterns of relations in a more intuitively manner than a multiplication tables does. Both multiplication tables and semigroups are concepts defined later on in this chapter, but it is worth saying that they serve to model relational structures of multiplex networks.

2.2.2 Permutation Groups

Group theory deals with different possible kinds of symmetries, and a *permutation group* is an algebraic structure that allow us to model symmetric social structures. Permutation groups are collections of permutation operators closed under composition where identity, inversion, and the associativity axiom apply, and they are useful in representing elementary systems in societies such as kinship systems of primitive cultures.

When a permutation group G acts on object set X, then $|G|$ is the *order* of the group (i.e. its cardinality), and $|X|$ corresponds to its *degree*. In other words, the group degree is the number of elements in the set, whereas the order is the number of elements in the group.

One way to look at a permutation group is as a "geometric object" on the Euclidean plane $\mathbb{R}^2$. In this case, a permutation is a bijective function $f : \mathbb{R}^2 \rightarrow \mathbb{R}^2$ that preserves distance between metric spaces X and Y. Such distance-preserving transformation has the generic name of *isometry*, which for two distinct points $x, y \in X$ is defined as:

$$d_X(x, y) = d_Y(f(x), f(y))$$

where d_X and d_Y are the respective "metrics" of the spaces.

The isometries of $\mathbb{R}^2$ form a subgroup of the group of all permutations of $\mathbb{R}^2$, and the isometries of the Euclidean plane that carry any subset of $\mathbb{R}^2$ onto itself forms a subgroup of the group of isometries, known as the *group of symmetries* of the subset in $\mathbb{R}^2$.

Other kinds of isometries that besides distance also preserve orientation are translation and rotation, whereas reflection and glide reflection are transformations that reverse the orientation of the points. Rotations, reflections, and inversions (which is simply the inverse of an isometry) are symmetry-preserving operations, and we will take a closer look at them next.

2.2.2.1 Permutation Operators

A *permutation operator* operates on a sequence of symbols called an *operand* that can be written either as a column or a row vector

$$\begin{bmatrix} 1 \\ 2 \\ 3 \end{bmatrix} \qquad \begin{bmatrix} 1 & 2 & 3 \end{bmatrix}.$$

Each permutation operator is typically represented by a *permutation matrix* having exactly one entry 1 in each row and in each column, and 0 elsewhere

$$\begin{bmatrix} 1 & 0 & 0 \\ 0 & 1 & 0 \\ 0 & 0 & 1 \end{bmatrix}$$

Through matrix multiplication, we evidence that the above operator corresponds to the identity matrix I operating on operand x

$$I_x = \begin{bmatrix} 1 & 0 & 0 \\ 0 & 1 & 0 \\ 0 & 0 & 1 \end{bmatrix} \begin{bmatrix} 1 \\ 2 \\ 3 \end{bmatrix} = \begin{bmatrix} 1 \\ 2 \\ 3 \end{bmatrix} = e$$

Naturally, there exist other options than I operating on x, and for instance the permutation group allows us modeling algebraic structures representing elementary social structures such as the ones made of kinship relations. Other binary operations on permutation groups apart from composition are sum, product, and power group.

A Theorem from Cayley says, "All of group theory can be found in permutations." We are going to analyze group structure by permutations, and more concretely on *permutation symmetry*, but first let us have a look at the formal definition of the group structure.

2.2.3 Presentation of Group Structures

A method of defining the group structure of G is by its *presentation* where the generating set X_G and the set $R_G = R_1, R_2, \ldots, R_n$ of relations or *relators* among the generators in G are given

$$G = \langle X_G \mid R_G \rangle.$$

The elements in R_G "are" the set of relations, which are equations between words in the set of all words over A_G with $A_G = X_G \cup X_G^{-1}$. Typically, the relators are identity relations, commutator elements of the generating set, and in the case of the free group ∅.

For instance, a theorem in group theory says that if a single operator a applies $m < 1$ times producing the identity element e in G, then a forms a *cyclic group* C_m of order m:

$$a^m = e.$$

Hence, the cyclic group of order m has the presentation

$$C_m = \langle a \mid a^m = e \rangle$$

where a constitutes the relator in C_m.

Associations between group structures have presentation as well, and for example the presentation of the direct product of two groups, G and $H = \langle X_H \mid R_H \rangle$ is

$$G \times H = \langle \, X_G \cup X_H \mid R_G \cup R_H \cup R_P \, \rangle$$

where X_G and X_H are disjoint generating sets of G and H, R_G and R_H are defining relations on these groups, and R_P is the set of relations P specifying that every element in X_G commutes with each element of X_H.

2.3   Group of Symmetries: Dihedral Groups

A family of algebraic structures with two generators is found in the *dihedral group* of order $2m$ with the presentation

$$D_{2m} = \langle \, a, b \mid a^m = e, \ b^2 = e, \ (ab)^2 = e \, \rangle$$

where generators a and b represent two types of transformations on the structure.

The *order* of the dihedral group is the number of elements in it, which is twice its *degree*; i.e., $|D_m| = 2m$. Hence, the dihedral group D_m is the group of symmetries of an m-sided regular polygon for $m > 1$.

Next, we take a closer look at two particular cases of symmetric group structure, in order to better understand the notion of groups and more generally of elementary algebraic systems.

2.3.1   Group of Symmetries of the Equilateral Triangle

The composition of series of transformations on the symmetries of the equilateral triangle is a particular type of permutation group known as the *group of symmetries of the equilateral triangle*.

Since the equilateral triangle is a regular polygon, this structure corresponds to the *dihedral group of degree 3*, denoted as D_3, and the order corresponds to the symmetries of the equilateral triangle, which is 6. Hence D_3 constitutes a *finite* group.

Figure 2.2 shows distinct elements of the dihedral group D_3, which correspond to six transformations preserving the vertices of this regular polygon. The two transformations are given on the top of the figure, and they are three *rotations*, and below these transformations are three *reflections*. The three rotations are clockwise about the center of 120° and 240° and where rotation at 0° constitutes the identity e without affecting the structure of the triangle elements. These rotations are denoted by R_{120}, R_{240}, and R_0, respectively.

With respect to the reflections, there is one kind that mirrors the representation through the vertical center axis, and the other kind are "flips" across the two diagonals of the picture. The mirror is denoted by SV, whereas the diagonal flips are as SD and SL, the latter for the bottom-left to upper-right or "left" diagonal.

To arrive to a closed algebraic system, which is one of the properties of the group structure, we perform all possible compositions of these transformations where associativity also holds. In other words, the different types of rotations and reflections in the dihedral group D_3 are functions that combined through the operation of composition yields to a closed system, which constitutes the group structure of D_3.

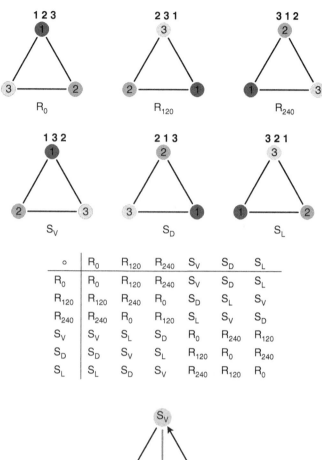

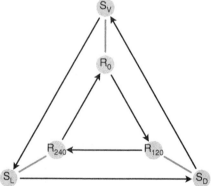

Figure 2.2 Symmetries of the dihedral group of the equilateral triangle D_3 with Cayley graph and table. *Solid directed and undirected edges in the graph represent generators* R_{120} *and* S_V.

The group of symmetries of the equilateral triangle is represented as a Cayley table in the middle of Figure 2.2, and as a Cayley graph to the bottom of the picture. The Cayley table and graph are then two representation forms of the same algebraic structure, and, for example, it is easy to see in the Cayley table the neutral character of the identity element. This is because the composition of R_0 with any other transformation leaves such element invariant. However, the graph representation can have other advantages in highlighting important aspects of the algebraic systems such as cycles and symmetric relations.

In Cayley graphs, undirected edges often represent bi-directional relations and we follow such custom. Since group structures are characterized by the inversion property, the existence of such relations is taken for granted. Figure 2.2 depicts the mirror reflection with a gray color, and this is because in D_3 this generator is always a bi-directional relation. On the other hand, the rotation generator is goes in one direction and therefore solid black arcs are used to represent them.

2.3.2   Group of Symmetries of the Square

The *group of symmetries of the square* is another type of permutation group that concerns with to the composition of series of transformations on the symmetries of the square. Since a square is also a regular polygon, like the equilateral triangle from the previous example, different trans- formations of this structure produce a finite group known as the *dihedral group of degree 4* or D_4. With the square, we are able to evidence the composition of more rotations and reflections than with the equilateral triangle. Because some authors denote permutation groups by its order rather than its degree, this means that D_4 is sometimes expressed as D_8 and it is called an *octic group* (e.g. Miller, 2012).

As with to D_3, the elements in D_4 correspond to rotations where the identity element belongs to, and reflections of different kinds. In this case, however, there are four transformations of each kind rather than six, which makes possible more combinations among these and hence a larger algebraic group structure. Figure 2.3 shows the distinct elements of the dihedral group D_4 where the rotations (also clockwise about the center as before) are in this case of 90°, 180°, and 270°, and 0° that is the identity element of the group. These rotations are denoted by R_{270}, R_{180}, R_{90}, and R_0, respectively. With respect to the reflections, the mirrors are with the square both through the horizontal and vertical center axes; i.e., S_H and S_V, whereas the diagonal flips are S_D and S_L as with D_3.

Notice that with the square the inversion property holds for each element since it constitutes the reciprocal or multiplicative inverse of the transformation, which for the 180° rotation is R_{180}^{-1}; i.e., it implies a back rotation with the same degree. We evidence also that a 270° plus a 90° rotation will produce a 360° rotation that leads to the same arrangement of elements. Such product of functions corresponds to the identity element of the group of transformations, and is expressed as $R_{270} \circ R_{90} = R_0 = e$.

Moreover, a reflection across the diagonal is obtained by mirroring a center axis to the 90° rotation of the group element. For the vertical axis this is expressed as $R_{90} \circ S_V = S_D$, whereas for the horizontal axis S_H it produces the other reflection across the left diagonal; i.e., $R_{90} \circ S_H = S_L$. As a result, all products of the different compositions correspond to one of the elements of D_4 whose group structure is given as a Cayley table and graph in the middle and at the bottom of Figure 2.3. We follow the rule that bi-directional relations are depicted as undirected edges, and different gray colors distinguishes the two types of transformations. Like with the example of the group of symmetries of the equilateral triangle, the graph and table formats have their own advantages and they should complement each other in the analysis of the group structure.

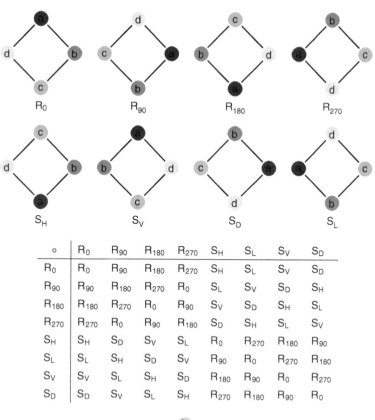

○	R₀	R₉₀	R₁₈₀	R₂₇₀	Sₕ	Sₗ	Sᵥ	S_D
R₀	R₀	R₉₀	R₁₈₀	R₂₇₀	Sₕ	Sₗ	Sᵥ	S_D
R₉₀	R₉₀	R₁₈₀	R₂₇₀	R₀	Sₗ	Sᵥ	S_D	Sₕ
R₁₈₀	R₁₈₀	R₂₇₀	R₀	R₉₀	Sᵥ	S_D	Sₕ	Sₗ
R₂₇₀	R₂₇₀	R₀	R₉₀	R₁₈₀	S_D	Sₕ	Sₗ	Sᵥ
Sₕ	Sₕ	S_D	Sᵥ	Sₗ	R₀	R₂₇₀	R₁₈₀	R₉₀
Sₗ	Sₗ	Sₕ	S_D	Sᵥ	R₉₀	R₀	R₂₇₀	R₁₈₀
Sᵥ	Sᵥ	Sₗ	Sₕ	S_D	R₁₈₀	R₉₀	R₀	R₂₇₀
S_D	S_D	Sᵥ	Sₗ	Sₕ	R₂₇₀	R₁₈₀	R₉₀	R₀

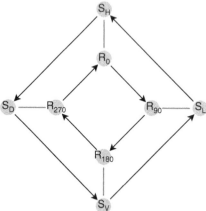

Figure 2.3 Symmetries of the dihedral group of the square D_4. *Generators are* R₉₀ *and* Sₕ *that are represented in the graph by solid directed and undirected edges for reciprocals.*

2.3.3   Generating Set in Symmetric Groups

More generally, rotations and reflections in a dihedral group D are two different types of functions that can be denoted, respectively, by symbols a and b with the special case of the identity relation e. Hence, the *generating set* X of D is made by three elements $X = \{e, a, b\}$ where $a = \pi/n$ and b is any reflection about a line of symmetry.

Even though the algebraic notion of a group is too restrictive to model social networks, it still possible to apply the concept of Group of Symmetries in the modeling of elementary social structures. These structures typically arise from marriage systems among mutually exclusive sections of a society or "clans" in small primitive societies where the distinct classes of families constitute the elements of a *symmetric group* that represent them.

2.4   Semigroup

If we want to characterize the configuration associated with a multiple relational system, the collection of distinct relations in the network serve to define certain algebraic structures. One of such algebraic structure is the **semigroup**, which serves to describe the entailments of the different types of ties in the social system, and hence to represent its relational structure. As the term suggests, a semigroup is "almost a group," which is another mathematical object defined previously, and thus a semigroup has a simpler and more flexible structure for the analysis of complex configurations such as those that arise in multiplex social networks.

There are different types of semigroups, and we start by defining an *abstract semigroup* as a general algebraic structure having a set of elements with an associative operation on it:

$$\langle\, S, *\, \rangle,$$

where S is the underlying or carrier set, and $*$ is a binary operation on an ordered pair $*: S \times S \rightarrow S$, that for all $x, y, z \in S$ satisfies the *associative law*:

$$x * (y * z) = (x * y) * z.$$

Each semigroup S is closed under the operation, and this means that the product of two or more elements in S must be also part of the semigroup.

Two examples of a semigroup are the set of natural numbers $\mathbb{N}$ (i.e. positive integers) either with the multiplication or with the addition as the binary operation, and this is because the product of this class of numbers under the operations is still defined inside the set. However, $\mathbb{N}$ under subtraction is not a semigroup since e.g. the product of one minus two is negative and therefore is not defined in the set. Note that the multiplication and the addition arithmetic operations, besides being associative, are also commutative since $x \cdot (y \cdot z) = (y \cdot x) \cdot z$ assuming that $\{\, x, y, z\, \} \in \mathbb{N}$ and where '$\cdot$' represents the operations.

A more concrete example of a semigroup system related to social networks is given in the representations of the structural balance theory given in Table 7.1 in Chapter 7 Signed Networks. Here the distinct valences in the signed structure constitute the set of elements, and composition of ties is the attached binary operation. In this type of configuration, the ambivalent relation corresponds to the zero element in the matrix semigroup having the absorption property, whereas the positive relationship acts as the identity element in every case.

Types of relations such as the zero and one element are distinguished constituents in the semigroup structure, and we will pay special attention to these elements when it comes to the interpretation of semigroups representing different kinds of social networks. Boyd (1991, p. 70) provides a variety of some significant types of semigroups ordered by the imposition of the commutative axiom, in which a group is regarded as a particular class of semigroup that is placed between the general semigroup object and the one-element semigroup or one-element algebraic structure.

Formally, if an element in a semigroup S under juxtaposition has the property that $s1 = 1s = s$, for all $s \in S$, then 1 is a neutral element called the *identity relation* of S and then the semigroup is termed a *monoid*. On the other hand, if an element satisfies the property $s0 = 0s = 0$, $s \in S$, then 0 is the *zero element* of the semigroup that has the absorbing property.

Each semigroup can have at most one identity element and at most one zero element; however, a semigroup can have more than one *right identity* [resp. *left identity*] elements if $s1 = s$ [resp. $1s = s$], and similarly it can have more than one *right zero* [resp. *left zero*] elements if $s0 = 0$ [resp. $0s = 0$]. If the identity or the zero element are not present in the semigroup S, then it is possible to adjoin these elements to S, and thus a semigroup with identity and zero adjoined are represented by $S^1 = S \cup \{1\}$ if $1 \notin S$, and $S^0 = S \cup \{0\}$ if $0 \notin S$ respectively.

2.4.1   Semigroup of Relations

We know already from abstract semigroups that any set of binary relations R generates a semigroup under the composition operation. Thus for the analysis of a given multiplex network X, the set of unique relations constitutes the underlying set for a *semigroup of relations*, which in formal terms is defined as a *semigroup with generators* $S(R)$ with the triple:

$$\langle\, S,\, \Sigma^*,\, \circ \,\rangle,$$

where $\langle\, S,\, \circ \,\rangle$ is a semigroup having the juxtaposition of strings as its binary operation, and $\Sigma^* \subseteq S$ as the object set. Certainly, the product of the juxtaposition comes from the concatenation of string relations.

In this case, any element x of S is expressed as product of generating relations in the alphabet, $R_i \in \Sigma$ as:

$$x = R_1 \circ R_2 \circ \ \dots \ \circ R_r,$$

and the product of elements in $S(R)$ is described in terms of composition for positive integers n and m as:

$$(x_1, x_2, \ \dots \ , x_n) \circ (y_1, y_2, \ \dots \ , y_m) = (x_1, x_2, \ \dots \ , x_n \circ y_1, y_2, \ \dots \ , y_m).$$

As with the abstract semigroups, it is sometimes convenient to include the identity string among the elements of the alphabet. In consequence, the resultant structure is a semigroup of relations with identity adjoined that is known as a *free monoid over* Σ, and this structure is then represented by $S(R)^1$, where $S(R)^1 = S(R) \cup \{1\}$ and where 1 is the identity string. Certainly, another possibility is to add the zero element to the semigroup of relations that acts as absorbing and get $S(R)^0 = S(R) \cup \{0\}$.

If we include all possible strings of elements in $R_i \in \Sigma^*$ that are the product of the concatenation of ties, it is possible to produce a *free semigroup of relations* generated by Σ. The free semigroup of relations is represented by $FS(R)$, and it is always infinite in length even if the alphabet contains a single element. The importance of the free semigroup lies in the fact that all possible partitions of the semigroup of relations are part of $FS(R)$, and this not only gives a more accurate and complete description of the resultant structure, but it also can be very useful when it comes to compare the configuration of different semigroups.

2.5   Semigroup and Group Properties

2.5.1   Regular Elements

We have just seen in the semigroup of relations for example that the empty string acts as an absorbing element in the structure and an identity relation will be a neutral element with the relational composition operation. There are, however, other types of special elements in a semigroup that serve to distinguish important features of this algebraic structure and which in turn can have significant implications in the analysis of semigroup of relations representing multiplex networks.

In this sense, an element $a \in S$ is regular if there is another element $b \in S$ such that $a = aba$. On the other hand, b is an *inverse* for a if $a = aba$ and $b = bab$. If all the elements of S are regular then S is considered a *regular semigroup*, and when each element of a regular semigroup has at least one inverse then S is called an *inverse semigroup*. A semigroup S having an identity element and where each element in has an inverse is a *group*, and this algebraic structure denoted by G has been subject of a previous section. Formally, the additional conditions for a group means first that $1 \in G$, and second that for all $a \in G$ there exist an element $b \in G$ that is inverse to a such that a $ab = ba = 1$. In this context b is called a *left-* and *right inverse* of a, and it constitutes the *unit* of G.

2.5.2   Subsemigroups and Ideals

A subset H of G is a *subgroup* of G if H also forms a group under the endowed operation, and this means that such subgroup is a subset of a group, i.e. $H \subseteq G$. Every subgroup satisfies the properties of the group, namely closure, associativity, identity, and inverse. In the same way, a subset T of a semigroup S is called a *subsemigroup* of S if is closed under the operation, meaning that $TT \subset T$. Important types of subsemigroups are *ideals* that are subsets closed under the operation even if one factor is outside the subset. The zero element 0 is itself an ideal but there are other elements I such that $0 \subset I \subset S$ that are regarded as *proper* ideals. Thus for a subset I in the semigroup S the expression $IS \subset I$ means that I is an ideal to the right or *right ideal* of S, and $SI \subset I$ means that I is an ideal to the left or *left ideal* of S. Likewise, I is a *two-sided* ideal or just an *ideal* if it is both a right and left ideal, i.e. $SI \subset I, IS \subset I$.

A left ideal I of S is called *principal left ideal* if there exists an element $a \in S$ such that $I = S^1 a$; and the element a is then called a *generator* of I. For a *principal right ideal* generated by an element a the dual expression $I = aS^1$ holds, and for a *principal two-sided ideal* $I = S^1 aS^1$ holds. Semigroups that do not properly contain any two-sided ideal are called *simple* semigroups, and a two-sided ideal is called *minimal* if it does not properly contain another two-sided ideal of

S (Clifford and Preston, 1967). The intersection of all ideals in a semigroup constitutes the minimal ideal that is an associated structure corresponding to the kernel of the semigroup. Each finite semigroup has a kernel that is a simple semigroup and most of the times a number of string equations generates it.

Some of the properties for functions can be also found in a semigroup. For instance if an element $a \in S$ satisfies the property $aa = a$, then *a* is a distinguished element of *S* called an *idempotent*, and it follows that a semigroup in which every element is an idempotent is a regular semigroup known as a *band*. Clearly, any identity and any zero element in the algebraic system is an idempotent, but in general idempotents are crucial in the search for subgroups of a semigroup (Boyd, 1991, p. 66). Its importance stems from the fact that each idempotent is the identity element for a unique *maximal subgroup*, i.e. a subgroup of *S* not properly contained in another subgroup of *S* (Clifford and Preston, 1967, p. 22). Maximal subgroups are disjoint to each other, and Clifford and Preston suggest visualizing them as "islands in a sea" where an idempotent element is the "peak" of an island.

2.6   Ring and Semiring

Now is turn to look at even more complex structures than the semigroup. Recall that both groups and semigroups are algebraic systems made of a set of elements and an operation associated with the set. There are also systems where there is more than one operation involved, and we introduce next some of these varieties of algebraic structures.

A *ring* is a set *R* that is defined in formal terms as the quintuple:

$$\langle\ R, +,\ \cdot,\ 0,\ 1\ \rangle$$

where *R* is a nonempty set associated with the operations named addition '+' and multiplication '·' together with a pair of special elements, 0 and 1, which are the neutral elements under addition and multiplication respectively, and where 0 acts as an absorbing element under multiplication as well.

As with the group, the ring is a closed algebraic structure where both operations are associative, and the elements under addition are commutative as well. There is also a rule that combines both operations and the multiplication distributes over addition; this means that for all $x, y, z \in R$:

$$x \cdot (y + z) = (x \cdot y) + (x \cdot z) \qquad \text{and} \qquad (x + y) \cdot z = (x \cdot z) + (y \cdot z).$$

Besides, each element $x \in R$ has an additive inverse $-x$ in *R* such that

$$x + (-x) = 0.$$

For example, the set of all integers, $\mathbb{Z}$, consisting of positive and negative numbers constitutes a ring where the familiar properties for addition and multiplication of integers apply.

2.6.1   Semiring

An algebraic structure that is more complex system than the ring and the semigroup is found in semirings. This is not only because the semiring structure combines two different kinds of operations with a single underlying set, whereas a semigroup associates just a single operation, but also because the lack of additive inverses of the elements in the system makes it prone to asymmetries.

In formal terms, a *semiring* is defined as the quintuple:

$$\langle\, Q, +, \cdot, 0, 1 \,\rangle$$

where Q is a nonempty set associated with the operations named addition '+' and multiplication '·' together with a pair of special elements, 0 and 1, which are the neutral elements under addition and multiplication respectively, and where 0 acts as an absorbing element under multiplication as well. In this sense, the semiring is the combination of two algebraic objects, namely an abstract semigroup with identity under multiplication and a commutative monoid under addition.

As with the semigroup, the semiring is a closed algebraic structure where both operations are associative, and the elements under addition are commutative as well. There is also a rule that combine both operations and the multiplication distributes over addition; this means that for all $x, y, z \in Q$:

$$x \cdot (y + z) = (x \cdot y) + (x \cdot z) \qquad \text{and} \qquad (x + y) \cdot z = (x \cdot z) + (y \cdot z).$$

In other words, multiplication has a higher precedence than addition.

The term "semiring" suggests that this structure is "almost a ring," which is another type of algebraic object defined above. The difference between a semiring and a ring is that the elements in the semiring do not necessarily have an additive inverse, whereas for a ring the addition yields in a commutative group, i.e. for $a, b \in G$, $ab = ba$.

There are different types of semirings that are useful for the analysis of networks. For instance, the set of natural numbers $\mathbb{N}$ with the usual arithmetic addition and multiplication operations make the *combinatorial semiring*, $\langle\, \mathbb{N}, +, \cdot, 0, 1 \,\rangle$. Similarly, a *max-plus algebra* constitutes a semiring structure over the union of real numbers, $\mathbb{R}$, when the **max** operator replaces addition, and the **plus** operator stands for multiplication; now if $\mathbb{R}$ is closed, the **min** operator stands for addition, and the neutral elements under addition and multiplication are ∞ and 0, respectively, then the algebraic structure constitutes the *shortest paths semiring*, $\langle\, \mathbb{R}_0^+, \textbf{min}, +, \infty, 0 \,\rangle$ (Batagelj and Praprotnik, 2016).

Closure in semirings is obtained if for each $x \in Q$:

$$\bar{x} = x \cdot (1 + (x^* \cdot x)) = x \cdot (1 + (x \cdot x^*))$$

where $x^* = 1 + x + x^2 + \cdots + x^n + \dots$ for $n > 0$.

2.7   Lattice Structure

An algebraic structure based on a partial ordered set where any two elements $x, y \in L$ has a *supremum* (or least upper bound of the join) $x \vee y$, and a unique *infimum* (or greatest lower bound of the meet) $x \wedge y$ is a lattice structure, and which plays a significant role in the analysis of complex network structures. Lattice structures are also known as *trellises*, and they have two binary operations, $\wedge$ and $\vee$, which denote the union and intersection of the pair of elements of the lattice.

A *lattice* is defined in formal terms as a triple:

$$\langle\, L,\ \wedge,\ \vee\, \rangle$$

where the two operators $\wedge$ and $\vee$ are called *join* and *meet*, respectively. We note that $x \vee y$ can be represented as $\sup\{x, y\}$ as well, and $x \wedge y$ as $\inf\{x, y\}$.

For a set of related elements, in order to make a lattice structure the join and meet operations need to satisfy important axioms. These axioms are formally defined for any $x, y, z \in L$ as:

$x \vee y = y \vee x$ and $x \wedge y = y \wedge x$	(commutativity)
$x \vee (y \vee z) = (x \vee y) \vee z$ and $x \wedge (y \wedge z) = (x \wedge y) \wedge z$	(associativity)
$x \vee x = x$ and $x \wedge x = x$	(idempotence)
$x \vee (x \vee y) = x$ and $x \wedge (x \wedge y) = x$	(absorption)

Each axiom in the definition of a lattice has two statements that are the opposite or *dual* of each other. Such principle of duality can also be applied to both the supremum and the infimum when they, together with L, produce two *semilattices*, $\langle L, \vee \rangle$ and $\langle L, \wedge \rangle$. As a result, a lattice is viewed either as a partial ordering structure or else as a set with two algebraic configurations, namely "meet" and "join," where these two last structures are called *join-* and *meet-semilattice*, respectively.

A *complete lattice* is a partially ordered set in which any subset L_i has both a join and a meet. The greatest element in a complete lattice is called *suprema* and is represented by 1, whereas the least element or *infima* is represented by 0, where in a lattice of sets, for example, $\{1\}$ represents the entire set and 0 represents $\{\varnothing\}$.

The partial ordering of the elements on the lattice is given by

$$x \leq y \ \text{ if and only if }\ x = x \wedge y \text{ or } y = x \vee y$$

where there is a *covering relation* in which y covers x, or $x < y$.

Furthermore, two equivalence relations $\equiv_1$ and $\equiv_2$ are represented by a lattice where $\equiv_1 \cap \equiv_2$ corresponds to its infimum, and $(\equiv_1 \cup \equiv_2)^t$ corresponds to its supremum, and we notice that the supremum is product of transitive closure. Since equivalence relations on a finite set or fixed domain are partially ordered by set inclusion, this means that $\equiv_1 \leq \equiv_2$ if and only if $\equiv_1 \subseteq \equiv_2$. In such case $\equiv_1$ is said to be *finer* than $\equiv_2$, or conversely the equivalence $\equiv_2$ is *coarser* than the equivalence $\equiv_1$.

2.7.1   Congruence Lattice

Another significant lattice structure that is useful for the analysis of social networks is known as a *congruence lattice*. This particular configuration type corresponds to the lattice of equivalent elements in an algebraic structure, which allows us to recognize whenever a particular semigroup may be represented as the subdirect product of smaller semigroups (Boyd, 1991, p. 63).

For a formal treatment, we start with an equivalence relation Θ on a lattice L that is called a *congruence of L* if and only if $x \equiv w(\Theta)$ and $y \equiv z(\Theta)$ imply the two *substitution properties* for join and meet.

The two substitution properties in the congruence lattice are defined as

$$x \wedge w \equiv y \wedge z(\Theta)$$
$$x \vee w \equiv y \vee z(\Theta).$$

Congruence lattices are important for the decomposition or "reduction" of the algebraic structures because the partition relations of a decomposed algebraic system lie in this lattice structure. We will see more of congruence lattice structures with applications to empirical networks in Chapter 6 where congruences of L serve to represent the set of all equivalences on a semigroup made of collective relations among actors in a given multiplex social network.

2.7.2   Modular and Distributive Lattice

Two important families of lattices that play a significant role in the factorization of algebraic structures (cf. Chapter 6 Decomposition of Role Structures) are the *modular* and the *distributive* lattices . These two types of lattices result when the carrier set, L, satisfies the *distributivity* and *modularity* conditions, respectively, which are formally defined for any $x, y, z \in L$ as:

$$x \vee (y \wedge z) = (x \vee y) \wedge (x \vee z) \text{ and } x \wedge (y \vee z) = (x \wedge y) \vee (x \wedge z) \quad \text{(distributivity)}$$
$$x \leq z \text{ implies that } x \vee (y \wedge z) = (x \vee y) \wedge z \quad \text{(modularity)}$$

Any distributive lattice is modular but the converse is not necessarily true, and any lattice made of equivalent elements is isomorphic to a finite distributive lattice (Dilworth's Theorem, cf. e.g. Grätzer, 2005).

Examples of two significant types of lattices are given in Figure 2.4, which depicts nondistributive lattice structures with 5-elements. The lattice to the left of the picture is called M_3 and

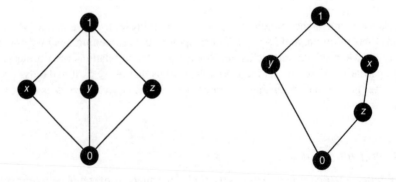

Figure 2.4   Five-element nondistributive lattices. *Left: Modular, M_3, right: Nonmodular, N_5.*

the structure to the right is known as N_5, and where N_5 is M_3 with an addition. There are two covering relations, which determines the ordering structure in N_5: $0 < x < z < 1$ and $0 < y < 1$, whereas M_3 has three covering relations; $0 < a < 1$, $0 < b < 1$, and $0 < c < 1$. M_3 "fails" the distributivity property because $x \vee (y \wedge z) = x$ and $(x \vee y) \wedge (x \vee z) = 1$, which implies that these relations are not equal.

As with congruence lattices, nondistributive lattices are important in the decomposition of algebraic structures as well. For example, a decomposition method that we are going to see in Chapter 6 Decomposition of Role Structures is factorization, which uses congruence lattices of equivalence relations for the partition of the algebraic system. The existence of modular lattice structures such as M_3 implies, roughly speaking, that the algebraic structure to decrease does not have a unique representation of the reduced system.

2.8   Algebraic Structures: Summary

In this Chapter, we constructed elementary and complex algebraic structures with properties, functions and operations from the previous Chapter. The algebraic structures, though, are still mathematical constructions of an abstract character, and they complement the basic notions introduced so far. Hence, the analysis of multiplex social networks–which is the subject of interest–and their relational arrangements was made by applying these mathematical concepts associated with algebraic systems such as groups, semigroups, semirings, and lattice structures among others.

In common of all these algebraic configurations is that they consist of a set of elements and a set of operations on the elements set. For social networks, simple and compound ties make the element set of a group or semigroup structures where the associated operation corresponds to the concatenation of these string relations. Like semigroups, we have seen structures such as semirings that combine two different operations, and these will serve for the analysis of particular types of social networks. Furthermore, for representing inclusion relations that yield hierarchies in the elements set, another significant kind of algebraic system is that of a lattice diagram. In the next Chapters, we are going to use all these algebraic structures for the analysis of diverse types of multiplex networks.

2.9   Learning Algebraic Structures by Doing

2.9.1   *Dihedral Group of the Equilateral Triangle D₃*

```
# find all possible combinations of three elements
R> pm <- expand.grid(rep(list(seq_len(3)),3))

# construct a data frame to record the unique combinations in 'pm'
R> pmd <- data.frame(matrix(NA, ncol = 3, nrow = 0))
R> for(i in rev(seq_len(nrow(pm)))) {
+     pmv <- as.vector(unlist(pm[i,]))
# make sure that each element appears just once
+     if(isTRUE(sum(pmv)==6L)==TRUE && length(unique(pmv))==ncol(pm)) {
+         pmd[nrow(pmd)+1L,] <- as.vector(pm[i,]) }
+     }
```

```
# the combinations for the dihedral group D₃
R> pmd

   X1 X2 X3
22  1  2  3
20  2  1  3
16  1  3  2
12  3  1  2
8   2  3  1
6   3  2  1
```

```
# object 'tri' represents the equilateral triangle
R> tri <- transf(c("1, 2", "1, 3", "2, 3"))

# define scope of node / edge / graph characteristics
R> scpt <- list(directed = FALSE, cex = 16, lwd = 12, ecol = 1, pos = 0, cex.main = 4)

# Fig. 2.2  symmetries in D₃
R> for(i in seq_len(nrow(pmd))) {
# cluster vector from a single combination
+     clu <- as.vector(unlist(pmd[i,]))
# permuted matrix and title according to 'clu'
+     ptri <- perm(tri, clu = clu)
+     titl <- jnt(clu, sep = " ", unique = TRUE)
# vertex color
+     colv <- gray(2:4/5)[as.vector(as.numeric(attr(ptri,"dimnames")[[1]]))]
# plot graph with the respective transformation
+     multigraph(ptri, scope = scpt, vcol = colv, main = titl)
+     }
```

```
# Cayley table of D₃

# the generator relations
R> SD3 <- transf(list(R = c("2, 1", "3, 2", "1, 3"),
+     S = c("1, 1", "3, 2", "2, 3")), type = "toarray", sort = TRUE)

, , R

  1 2 3
1 0 0 1
2 1 0 0
3 0 1 0

, , S

  1 2 3
1 1 0 0
2 0 0 1
3 0 1 0
```

```
# find the group structure of 'SD3'
R> SD3S <- semigroup(SD3)

# for the rearrangement of 'SD3S', find strings and the set of equations
R> strings(SD3, equat = TRUE, k = 3)

...

$st
[1] "R"   "S"   "RR"  "RS"  "SR"  "SS"
...
$equat
$equat$R
[1] "R"   "SSR" "RSS"

$equat$S
[1] "S"   "SSS" "RSR"

$equat$RR
[1] "RR"  "SRS"

$equat$RS
[1] "RS"  "SRR"

$equat$SR
[1] "SR"  "RRS"

$equat$SS
[1] "SS"  "RRR"


$equate
$equate$e
[1] "e"   "SS"  "RRR"
```

```
# the rearrangement of 'SD3S': first identity, then generators...
R> SD3S <- perm(SD3S$S, clu = c(2, 4, 3, 5, 6, 1))
```

```
# customized relabel 'SD3S' in a (semi)group structure with generators
R> SD3S <- as.semigroup(SD3S, gens = c(2, 4),
+    lbs = c("R0", "R120", "R240", "SV", "SD", "SL"))

$ord
[1] 6

$st
[1] "R0"    "R120" "R240" "SV"    "SD"    "SL"

$gens
[1] "R120" "SV"

$s
      R0 R120 R240    SV    SD    SL
R0    R0 R120 R240    SV    SD    SL
R120 R120 R240   R0    SD    SL    SV
R240 R240   R0 R120    SL    SV    SD
SV    SV   SL   SD    R0 R240 R120
SD    SD   SV   SL R120   R0 R240
SL    SL   SD   SV R240 R120   R0

attr(,"class")
[1] "Semigroup" "symbolic"
```

```
# define scope of node / edge characteristics
R> scp0 <- list(cex = 7, lty = 1, lwd = 3, pos = 0, vcol = 8, fsize = 16)

# Fig. 2.2 Cayley graph of D_3 with concentric layout and 2 radii
R> ccgraph(SD3S, conc = TRUE, nr = 2, scope = scp0, undRecip = TRUE)
```

2.9.2   Dihedral Group of the Square D_4

```
# matrix construct for the square
R> sqr <- transf(c("a, b", "b, c", "c, d", "d, a"))

# define scope of node / edge / graph characteristics
R> scps <- list(directed = FALSE, cex = 16, vcol = gray(2:5/6), clu = 1:4, lwd = 8,
+    ecol = 1, pos = 0, fsize = 60)

## Fig. 2.3  elements in D_4
# Identity (no action)
```

```
R> multigraph(sqr, scope = scps)
# Rotation clockwise 90 degrees about the center
R> multigraph(sqr, scope = scps, rot = 90)
# Rotation clockwise 180 degrees about the center
R> multigraph(sqr, scope = scps, rot = 180)
# Rotation clockwise 270 degrees about the center
R> multigraph(sqr, scope = scps, rot = 270)
# Reflection through the horizontal center line
R> multigraph(sqr, scope = scps, mirrorH = TRUE)
# Reflection through the vertical center line
R> multigraph(sqr, scope = scps, mirrorV = TRUE)
# Reflection across diagonal Y = X
R> multigraph(sqr, scope = scps, mirrorD = TRUE)
# Reflection across diagonal Y = -X
R> multigraph(sqr, scope = scps, mirrorL = TRUE)
```

```
# Cayley table of D₄

# the generator relations
R> SD4 <- transf(list(R = c("2, 1", "3, 2", "4, 3", "1, 4"),
+      S = c("1, 3", "2, 2", "3, 1", "4, 4")), type = "toarray", sort = TRUE)

, , R

  1 2 3 4
1 0 0 0 1
2 1 0 0 0
3 0 1 0 0
4 0 0 1 0

, , S

  1 2 3 4
1 0 0 1 0
2 0 1 0 0
3 1 0 0 0
4 0 0 0 1
```

```
# group of 'SD4' is a semigroup class object with a symbolic format
R> SD4S <- semigroup(SD4, type = "symbolic")
```

```
# rearrangement of semigroup in'SD4S': identity first, then cycle (sub)group R³
R> SD4S <- perm(SD4S$S, clu = c(2, 5, 3, 6, 8, 1, 4, 7))
```

```
        SS   R   RR  RRR    S   RS  RRS    SR
SS      SS   R   RR  RRR    S   RS  RRS    SR
R        R  RR  RRR   SS   RS  RRS   SR     S
RR      RR RRR   SS    R  RRS   SR    S    RS
RRR    RRR  SS    R   RR   SR    S   RS   RRS
S        S  SR  RRS   RS   SS  RRR   RR     R
RS      RS   S   SR  RRS    R   SS  RRR    RR
RRS    RRS  RS    S   SR   RR    R   SS   RRR
SR      SR RRS   RS    S  RRR   RR    R    SS
```

```
# relabel 'SD4S'
R> SD4S <- as.semigroup(SD4S, gens = c(2, 5),
+    lbs = c("R0", "R90", "R180", "R270", "SH", "SL", "SV", "SD"))
```

```
# Fig. 2.3  Cayley graph of D₄ with concentric layout
R> ccgraph(SD4S, conc = TRUE, nr = 2, scope = scp0, undRecip = TRUE)
```

2.9.3   Modular and Nonmodular Lattices

```
# modular (nondistributive) lattice M₃
R> M3 <- transf(c("1, 1","2, 1","2, 2","3, 1","3, 3","4, 1","4, 4",
+    "5, 1","5, 2","5, 3","5, 4","5, 5"), type = "toarray")
```

```
   1 2 3 4 5
1  1 0 0 0 0
2  1 1 0 0 0
3  1 0 1 0 0
4  1 0 0 1 0
5  1 1 1 1 1
```

```
# Fig. 2.4 lattice M₃ with customized shapes
R> diagram(M3, shape = "circle", col = 1, lwd = 4)
```

```
# nomodular (nondistributive) lattice N₅
R> N5 <- transf(c("1, 1","2, 1","2, 2","3, 1","3, 3","4, 1","4, 2","4, 4",
+    "5, 1","5, 2","5, 3","5, 4","5, 5"), type = "toarray")

  1 2 3 4 5
1 1 0 0 0 0
2 1 1 0 0 0
3 1 0 1 0 0
4 1 1 0 1 0
5 1 1 1 1 1
```

```
# Fig. 2.4  lattice N₅ with customized shapes
R> diagram(N5, shape = "circle", col = 1, lwd = 4)
```

3

Multiplex Network Configurations

3.1 Multiple Networks

In this Chapter, we take a look at different multiplex network configurations produced by interactions in human societies, and this means that we begin to apply the algebraic structures introduced in Chapter 2 in an abstract manner to real-life social networks. Before the applications, however, we briefly look at multiple networks terminology.

Recall from Chapter 1 that the term "multiplex network" was defined as a system having more than one type of relation. Such definition was made in opposition to a "simple" network, which are those systems having just one kind of relation. However, even if the social system is not a simple network, and it is of a multiple character, the term *multiplex* can be defined either in opposition to a *monoplex* structure or else in opposition to a *uniplex* edge.

When the "multiplex" notion is in opposition to monoplex then it refers to the graph structure, whereas when this concept is stated in opposition to uniplex then multiplex refers to the edge of the graph. Putting in other words, in case we refer to the network structure rather than to the links in the configuration, then multiplex represents single or collapsed levels in the set of relations, whereas if we refer just to the edges then the multiplex term represents single or collapsed levels in the relationships among network members.

It is worth mentioning that other terms used for social systems made of multiple relations such as "multimodal network" has the same meaning than multiplex systems, but it is most used with flows or transportation modes. The term "multilevel network" represents a structure made of individuals and groups; that is, affiliation networks where both the domain and co-domain or entity levels are between related and they can be within related as well. Both affiliation networks and multilevel structures are subject of Chapters 8 and 10, and we will see some algebraic procedures of the analysis.

Moreover, the term "multilayer network" represents a cascade structure with multiple subsystems and layers of connectivity, which means that the entire system has more than one multiplex (or monoplex) structures that can be interrelated with each other. Some of this terminology still, however, under current development within the social network analysis discipline, and therefore it is important to have a formal definition of the network, particularly when is made of different domains and types of relationships between and across them.

Algebraic Analysis of Social Networks: Models, Methods and Applications using R,
First Edition. J. A. R. Ostoic. Companion website: www.wiley.com/go/ostoic/algebraicanalysis.
© 2021 John Wiley & Sons Ltd. Published 2021 by John Wiley & Sons Ltd.

Table 3.1 Typology of multiple network structures.

Simple networks:
  - *(Simple) graphs, matrices* → for relations between actors

Multiplex networks:
  - *Multigraphs, arrays* → for (types of) relations between actors
  - *Cayley graphs, tables* → for relationships between relations

(a) Representations for simple and multiplex networks

Type of structure	Algebraic system
Elementary	*Group*
Complex	*Semigroup, Semiring, Lattice*, etc.

(b) Algebraic systems representing multiplex networks

3.1.1   Types of Multiple Networks

A typology of multiple network structures where we can look at different aspects of multiple network structures is given in Table 3.1, with representations for simple and multiplex networks, and algebraic systems representing multiplex networks as well. In this typology, we distinguish multiplex networks from simple configurations first by their nature in the associations, and then there are two different kinds of multiplex structures. Hence, simple social networks are configurations for relations between actors, which are represented by adjacency matrices and graphs, and when the social tie has no directionality, then the adjacency matrix is symmetric and the graph is undirected.

With multiplex networks, there are other representation forms of the different types of relations among the network members. One are multigraphs having parallel edges and another one are three-dimensional arrays that is actor by actor by ties. Moreover, there is also the possibility to model the relationships between the relations themselves with multiplex networks, and a table made by ties only represents the resultant configuration, which is graphically depicted as a Cayley color graph. Mapping relationships between relations in multiplex networks implies a higher level of abstraction in the analysis, and algebraic structures allow us to represent such types of association.

Table 3.1 also provides algebraic systems representing multiplex networks, and in this case there is a distinction of algebraic groups that are termed as *elementary* structures, whereas algebraic systems with the same properties of the groups except for inversion property are called *complex* structures. Complex structures include different types of algebraic systems, in which semigroups, semirings, and lattice structures are most commonly used for the analysis of multiplex social networks.

If we concentrate on multiplex configurations, one important aspect to notice with the group structure is that groups represent symmetric systems of related elements such as the Dihedral groups from Chapter 2 Algebraic Structures. This is due to the inversion axiom applied to the algebraic object that distinguish groups from semigroups. This means that the group structure does not have implicit within a hierarchy of relations, but algebraic structures without the inversion property such as the semigroup do have some ordering among their elements, which implies an additional *algebraic constraint* for the analysis. Algebraic constraints are some sort of restrictions in the network relational structure that can guide us in the substantial interpretation of multiplex network structures, and we will look at them more closely in Chapter 11 Comparing Relational Structures.

The existence of elementary structures in human societies seems to be difficult, and this is because their symmetrical character that typifies elementary structures. There are, however, types of social relations that are undirected in nature and one of these corresponds to kinship relations. In fact, some kinship systems from primitive human societies can yet effectively be represented by algebraic group structures where the term *primitive* means that is "a first of its class."

In the next section, there is a classic algebraic analysis of a network made of different types of kinship relations that characterizes an elementary structure. It is an elementary structure despite the system has different types of ties, and later on, there are analyses of complex structures of multiplex networks that correspond to most–if not all–social systems in today's societies where asymmetry occurs since not all relations are reciprocated.

3.2   Kinship Networks and Group Structure

In accordance with the work of Weil (Lévi-Strauss, 1949, appendix by A. Weil), permutation groups allow interpreting algebraically certain types of kinship structures with marriage rules governed by prohibition. In particular, kinship systems based on a "prescriptive" marriage type correspond to an *elementary structure* represented by the algebraic group object. In this context, elementary structures are opposed to "complex systems" where individual preferences run the marriage rules of the society (White, 1963; Korn, 1973). However, we use the term "complex" rather for algebraic arrangements where the inversion property does not hold, and hence are not group structures even if such complex systems are established in a deterministic manner.

Following this line of thought, Kemeny, Snell and Thompson (1974) proposed an integrated set of axioms that serve to investigate the properties of kinship systems in primitive societies based on ideal types of prescribed marriage. The first axiom for a kinship group states

Axiom 1 *The entire population is partitioned into n > 0 clans*

Then there are two axioms where the permutation operation applies and which correspond to marriage rules

Axiom 2 *Two individuals can marry only if they are in the same clan marriage type*

Axiom 3 *A man from two different clans cannot marry a woman of the same clan*

For the descent rules, there is a pair of axioms as well where the permutation operation apply and that—loosely speaking—state

Axiom 4 *All descendants of a couple are assigned to a single clan determined by the clans of the father and mother*

Axiom 5 *Descendants of a married couple from different clans must be in different clans*

White (1963, cf. also Kim and Roush (1983)) also provides rules for kinship systems where some permutation matrices do not apply. However, what is important right now is to distinguish two kinds of rules in kinship structures, one for marriage and one for descent that are modeled with permutation matrices.

Permutation matrices are square arrays having 1 in each row and in each column, and 0 in all other entries. Both Korn (1973) and Kemeny, Snell and Thompson (1974) use permutation matrices to model "elementary" structures where the logical true represent a transformation of parent's marriage type into son's and daughter's marriage type. Permutation matrices were employed as well by Bush (White, 1963, appendix by R. R. Bush) as "operators" F and G, whereas White (1963) builds kinship structures based on W and C, which are the two types of matrices that correspond to marriage and descent rules, respectively.

3.2.1   Marriage Types in Kinship Systems

Since algebraic structures in kinship networks are based on marriage arrangements, we start by looking at significant marriage types that may be allowed or not in a given society. For instance, Figure 3.1 gives the *matrilineal* and *patrilineal* first cousins marriage types in the general kinship systems with the equations above each diagram. The kinship diagrams have square and circle identifiers for male and female, respectively, and the letters F and G that correspond to son's and daughter's marriage type. Such relationship results from a right multiplication product like other compound relations of descendant like FG, GF, etc., and this is when we trace the descendant's line in the kinship diagram "upwards".

There are important aspects to notice with the categorization of these kinship diagrams. First, the difference between "patrilineal" and "matrilineal" first cousins lies in whether the parent of

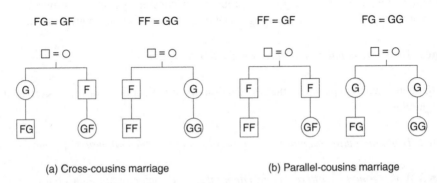

Figure 3.1   Matrilineal and patrilineal first cousins marriage types in general kinship systems. F *and* G *stand for rules of marriage and descent, respectively. Equations above are for cross-cousins marriages.*

the ego male who is sibling to the parent of his wife is male or female, whereas *parallel-* differs from *cross-* in the sense that in the former the parents of the first cousins are of the same sex, whereas the parents of the first cousins are of different sex in the latter case.

The equations above each kinship diagram depicted in Figure 3.1 allow us to detect permitted marriage types by commutation in the given society. Hence, in the matrilineal cross-cousin marriage, for example, a male is allowed to marry his father's sister's daughter. Similarly, a female is allowed to marry her mother's brother's son. The same kind of reading applies to the rest of first cousins marriage types in the general kinship system.

3.3   Rules for Marriage and Descent in the Kariera Society

Now is time to see how algebraic procedures can help us to perform an analysis of elementary structures, and we are going to look at a kinship system with ideal types of *prescriptive* marriage. This corresponds to a "primitive" society of the **Kariera** people that lived as such in Australia in the early period of the last century. We bear in mind, however, that such kinds of marriage constitute ideal types of structure and, as White pointed out the question should be "to what extent [the society] conforms to one or to some mixture of ideal types of prescribed marriage systems" (White, 1963, p. 148).

In the case of the Kariera society, the kinship network has four different clans with specific rules of marriage and descent: *Banaka, Burung, Karimera*, and *Palyeri* (or Palyori). Figure 3.2 shows the distinct marriage and descent rules in this society both as kinship diagrams and as multigraphs. As before, the kinship diagrams have square and circle identifiers for male and female, whereas shadowed nodes in the multigraph at the bottom of the picture serve to depict the four clans. Hence, the kinship diagrams together with the multigraph serve to represent the four types of descent rules of the Kariera where the multigraph unifies all rules for this society where solid and dashed arcs stand for the rules of marriage and descent, respectively.

The kinship diagram at the top of Figure 3.2 illustrates the rules of marriage and descent for a Banaka and Palyeri men, whereas the kinship diagram below is for the rules of marriage and descent for a Burung and Karimera husbands. The graphs to the right of each kinship diagram represent this information with vertices and edges of different shape for the clans and the two kinds of rules for this people.

As a result, the rules of marriage for the four Kariera clans are

A Ba(Pa) *man marries a* Bu(Ka) *woman, while a* Bu(Ka) *man marries a* Ba(Pa) *woman.*

While the rules of descent among the Kariera are

The child of a Ba(Pa) *man is a* Pa(Ba), *while the child of a* Bu(Ka) *man is a* Ka(Bu).

Next, we are going to perform an algebraic analysis of these rules by applying the principles of the group structure.

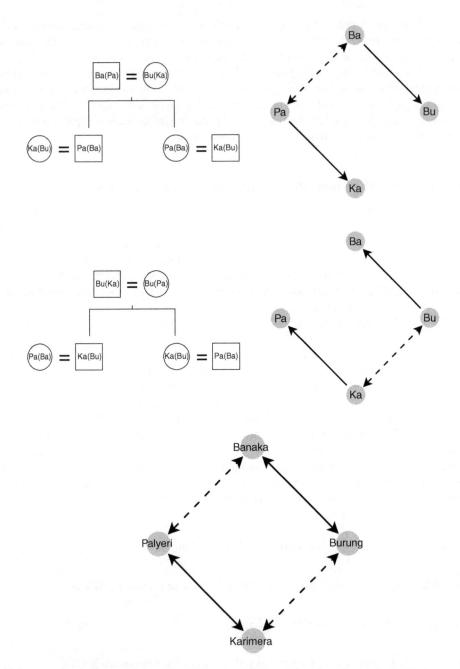

Figure 3.2   Kinship diagrams and multigraph with rules for marriage and descent in the Kariera society. *Banaka (Ba), Burung (Bu), Karimera (Ka), Palyeri (Pa), and male as Ego. Solid and dashed arcs represent marriage and descent types of relationships.*

3.3.1   Group Structure and Set of Equations

As already pointed out, despite its symmetry, algebraic groups can model human societies. In order to construct the group structure of the Kariera society, the marriage and descent relations will constitute the generators that eventually lead to the group structure of the clan system.

The two structures representing the rules for marriage and descent among the Kariera are expressed by permutation matrices

$$\begin{bmatrix} 0 & 1 & 0 & 0 \\ 1 & 0 & 0 & 0 \\ 0 & 0 & 0 & 1 \\ 0 & 0 & 1 & 0 \end{bmatrix} \quad \begin{bmatrix} 0 & 0 & 0 & 1 \\ 0 & 0 & 1 & 0 \\ 0 & 1 & 0 & 0 \\ 1 & 0 & 0 & 0 \end{bmatrix}$$

The anti-diagonal character of the matrix for the descent relations is evident, which allows a systematic "change" in the clan denomination among the descendants. Since the product of two anti-diagonal matrices is a diagonal matrix, then

$$GG = e,$$

which means that "a daughter is in the same class as her mother's mother." In fact this also true for the son, and this is because of the equation

$$FF = e$$

which also applies among the Kariera people. Hence, "a son is in the same class as his father's father." Naturally, these two statements also apply for the grandchildren.

Since both FF and GG equal the identity element, this means that these composite relations equal to each other. In fact, the two most significant equation among the Kariera are:

$$FF = GG$$
$$FG = GF.$$

And this is because according to these two equations both cross-cousins marriages are permitted in the Kariera society as we evidence from the general matrilineal and patrilineal first cousins marriage types in the general kinship systems in Figure 3.1.

From the *set of equations* we are able to construct the group structure of the Kariera society, which is given in Figure 3.3. In this case, a table of relations or a multiplication table represents the group structure of this particular kinship system where the two operators make compound kinship relations through composition.

Besides, the Cayley graph in Figure 3.3 represents the group structure in a graphical format and, as with the multiplication table, we evidence the symmetric character of the symmetric character of the algebraic group structure representing the Kariera society. The Cayley graph is actually isomorphic to the multigraph, representing the marriage and descent rules with the four clans given in Figure 3.2, and this is because the order of the group structure is similar to the number of dimensions, which is the amount of clans among the Kariera.

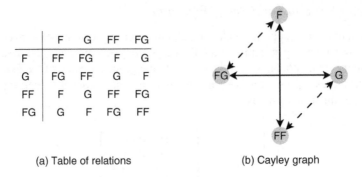

(a) Table of relations   (b) Cayley graph

Figure 3.3   Group structure of the Kariera represented in two formats.

3.4   Algebraic Constraints

Algebraic constraints are structural constrictions in the network relational or role system that are useful in the analysis of multiplex networks in substantial terms. For instance, there are two kinds of algebraic constraints in the analysis of the Kariera society from the previous section. One refers to the set of equations among relations, in which representative strings of the structure are created. A second algebraic constraint in the Kariera group structure constitutes the *structure* itself where different relations are associated through composition.

Unlike the Kariera people, most social networks from today and past societies are, however, not symmetric. With a non-symmetric algebraic system where the inverse property does not hold, there is another algebraic constraint that is of the *set of inclusions*, and that is associated with the hierarchy of ties in the relational structure.

The set with the different equations among string relations in the social system allows finding, or rather "choosing", the representative elements in the multiplication table that represents the system structure. That is why the multiplication table results being a consequence of the set of equations. On the other hand, the inclusions among tie strings produce an ordered structure that is expressed in the partial order table that is related to another table, the multiplication table.

In the schema given in Table 3.2 there are the names of three types of algebraic constraints with the involved operation and the way these constrictions are represented. Hence, the set of equations provides the representative strings of the network relational structure, where these representatives are related by the composition operation. The relational structure of the network then is represented by a multiplication table where all the representative strings are related through composition.

Table 3.2   Algebraic constraints with involved operation and representation form.

Algebraic constraint	Operation in relation	Representation form
Set of **equations**	$R_1 = R_2$	representative *string*
Set of **inclusions**	$R_1 \leq R_2$	*partial order table*
Relational **structure**	$R_1 \circ R_2$	*multiplication table*

On the other hand, in the partial order table represents the hierarchy of the string relations occurring in the network relational structure, which are related through a "weak" partial order relation where the reflexive property holds. Hence, set of inclusions makes the relational structure an ordered system, which is typically represented by the partially ordered semigroup in the case of multiplex social networks.

In the case of elementary systems such as the group structures from the Kariera society, there are no inclusions among the different representative string relations, and the set of inclusions involves only reflexive relations. The partial order table in symmetric structures makes no sense, and this is because it corresponds to the identity matrix without any hierarchy among the ties.

3.5 Link Generalizations and Complex Structures

The Kariera kinship network we have just seen in the previous section represents a group structure of a symmetric character; however, if the network structure is not symmetric it means that there is an ordering among the network relations. This "ordering" constitutes an additional constraint that makes the algebraic structure not "elementary" anymore, but rather a *complex* structure. Complex structures play a central role in the analysis of multiplex social networks and this is because complex structures lack of symmetry, and the lack of symmetry in social relations is what typically characterizes human societies from today.

To analyze multiple network structures, we need to differentiate the *social structure* of a given social network that is a system of relations among actors in the system, from the network relational structure, which is the structure produced by the relations among the relationships themselves. Hence, the relational structure of the network represents the intertwining of ties in the system that are of different kind, and it constitutes a characteristic exclusively in multiplex network structures.

The analysis and substantial interpretation of the network relational structure and the social system in particular often constitute a great challenge because of the complexity inherent in the network relational structure, and certainly one approach to deal with such complexity is to reduce the network structure. There are diverse techniques within the field of blockmodeling that produce classes of structurally related actors; however, most of the blockmodeling methods are based on the embeddedness of the actors in the network, which is typically grounded on a single type of relationship (see Lorrain and White, 1971; Doreian, et al., 2004, for examples).

For multiplex networks, there is a significant loss of structural information if the condensed system fails to reflect the multiplicity of the ties. Hence, it is desirable to produce a simpler and single structure that integrates the different types of relations, and at the same time provides valuable insights into the whole system for its substantial interpretation.

Integrating relations in multiplex networks implies a *generalization* of the links between pairs of actors in the system, and we recognize the multiplicity of the ties in the social structure either at the different levels in the relationship or by considering chains of several kinds of relations that influence the structuring process of the social system. If we account for the levels in the ties, there is a *wide* extension of the pairwise link, whereas the generalization with chains of ties constitutes a *long* extension of the link. The combination of both extensions can naturally occur in the multiplex network.

The wide extension of the ties is expressed in different classes of configurations that occur at the dyadic level in the network structure and we call them *bundles*. Such categories include the well-known patterns for single networks like the *null, asymmetric,* and *reciprocal* (or mutual) dyads (Holland and Leinhardt, 1970, 1976), and other arrangements that have the multiplexity property, which generalize these kinds of dyadic relations.

The entrainment and the exchange of ties in directed networks are fundamental patterns where at least two types of tie are involved and, although there are "bundle configurations" where these patterns are mixed as well, these two configurations represent entirely different realities with consequences in the structural analyses of social phenomena in multiplex network structures.

With respect to the long extension of the link, this generalization form corresponds to the interrelations among the ties that produce chains and paths of relations. In this sense, simple ties such as social interactions, flow of information, co-occurrence, etc. constitute primitive relations in the system whereas the concatenation of these create compound relations.

Strings is a generic name for both primitives and compounds, and string relations are also known metaphorically as "words" composed by letters or primitive relations from an "alphabet." Even though it is likely to have an infinite amount of string relations in a multiplex network structure, most of them will connect precisely the same individuals. Hence, there are a limited number of isomorphic strings in the closed system that constitute the building blocks of the network relational structure.

3.6   Bundle Patterns

For multiplex network structures, the wide component of the link is expressed in *bundle patterns*, which are different types of configurations at the dyadic level that encode simultaneity of various kinds of ties. Two fundamental classes of bundle patterns with the multiplexity property are the entrainment and the exchange of ties, which are depicted as graphs in Figure 3.4. As the pictures illustrate, these bundle patterns are extensions of the asymmetric and reciprocal dyad, respectively.

A formal definition of these classes are given below for an ordered pair of actors $(i,j) \in R_r$ as:

$$B_{ij}^E, \text{ Tie Entrainment} : (i,j) \in R_{1,\ldots,s} \wedge (j,i) \notin R_{1,\ldots,s}, \quad \text{for } 1 < s \leq r$$

$$B_{ij}^X, \text{ Tie Exchange} : (i,j) \in R_p \wedge (j,i) \in R_q, \quad \text{for } R_p \neq R_q$$

Hence, while *tie entrainment* has an asymmetric character, the exchange of ties of different type or *tie exchange* implies—as the reciprocal dyad—that the bundle pattern has a mutual

Figure 3.4   Multiplex patterns: tie entrainment and tie exchange.

character. The tie exchange bundle class is termed as "mixed reciprocity" by Snijders (2017) when specifying dependencies in multivariate or multiplex networks at the dyadic level. Besides, the mixture of an asymmetric and a reciprocal dyad represents another bundle class presumably with a mutual character, but that will depend on the relational content of the ties.

The definition of the asymmetric and reciprocal dyads implies in the case of multiplex networks that the remaining levels in the relationship (if any) are null.

$$B_{ij}^N, \text{Null:} \qquad (i,j) \notin R_{1,\dots,r} \ \wedge \ (j,i) \notin R_{1,\dots,r}$$

$$B_{ij}^A, \text{Asymmetric:} \qquad (\ (i,j) \in R_p \ \wedge \ (j,i) \notin R_p\ )$$

$$(\ (i,j) \notin R_{r-p} \ \wedge \ (j,i) \notin R_{r-p}\ )$$

$$B_{ij}^R, \text{Reciprocal:} \qquad (\ (i,j) \in R_p \ \wedge \ (j,i) \in R_p\ )$$

$$(\ (i,j) \notin R_{r-p} \ \wedge \ (j,i) \notin R_{r-p}\ )$$

And the other bundle classes with the multiplexity property are:

$$B_{ij}^M, \text{Mixed:} \qquad B_{ij}^A \ \wedge \ B_{ij}^R, \qquad\qquad r = 2$$

$$B_{ij}^A \ \wedge \ B_{ij}^X, \qquad\qquad r > 2$$

$$B_{ij}^F, \text{Full:} \qquad (i,j) \in R_{1,\dots,r} \ \wedge \ (j,i) \in R_{1,\dots,r}$$

That is, as the simplest structure the mixed bundle is the union or the *co-occurrence* of the asymmetric dyad with the reciprocal with two types of relations, or else an asymmetric with a tie exchange pattern for more types of tie. On the other hand, the *full* pattern is a special case of a mixed bundle that is the complement of the null dyad.

Bundle patterns with a mutual character are said to be *strong bonds* while asymmetric bundles represent *weak bonds*, which are two generalizations of the Strength of Weak Ties theory for multiplex networks. We are going to review briefly the Strength of Weak Ties theory in this chapter, but we look first at some properties of the different bundle class patterns.

3.6.1 Bundle Class Properties

In addition to the three dyadic patterns found in simple networks, we have with multiplex networks four configuration classes with ordered pair of actors. For instance, the full bundle is a class where all other bundles are included except for the null dyad, and to some extent, it represents the universal element of the bundle classes. On the other extreme, the null dyad is an empty pattern where none of the other properties exists, but in the context of multiple relations the null property as such is often occurring in the rest of the patterns except for the full bundle.

This means that for multiplex networks, pairs of actors having either an asymmetric or a mutual tie alone constitute configurations having the null property as well. However, these "multiplex" patterns are still considered as asymmetric and reciprocal dyads, since the null property acts as a neutral element and does not have an influence on the distinctiveness of these or on the other properties. This means that in a multiplex network an asymmetric dyad—for example—is a bundle class where the remainder types of relations are kept null.

Although patterns with the multiplexity property occurs in undirected networks as well, both tie entrainment and tie exchange are properties exclusively applied to networks having directed edges on it. This is because when the direction of the ties is ignored in an undirected network, then the bundles become isomorphic pairs of unlabeled nodes with no structural distinction between them.

This means that the total number of potential patterns to be found in undirected networks is less than the potential patterns in directed ones. For a directed network, the total number of potential bundles is

$$h = 2^{2r},$$

while for an undirected network the number of bundles is just

$$2^r.$$

3.6.2 Bundle Isomorphic Classes

Table 3.3 provides the amount of potential isomorphic patterns in each bundle class involving a single, two-, three-, and a generalization for r-types of relationships. The bundle classes with the multiplexity property are only allowed to occur when $r > 1$, and we can see that there still a single isomorphic pattern for both the null and full bundles regardless the number of relational types involved.

If we look at the bundle isomorphic patterns, we verify that two possible forms for the asymmetric dyads that can arise when the tie direction change in single directed and labelled network. On the other hand, we have already seen that a changing the direction of the ties in a reciprocal dyad produces the same isomorphic pattern. In fact, for every type of relationship added to the system, there are two extra asymmetric patterns while there is just one extra reciprocal pattern possible to occur.

Table 3.3 Bundle isomorphic patterns in dyads with single, 2-, 3-, and r-types of tie.

Bundle Class	single	2-types	3-types	r-types
Null, N	1	1	1	1
Asymmetric, A	2	4	6	$2r$
Reciprocal, R	1	2	3	r
Tie Entrainment, E	–	2	8	$r(r+1) - 4$
Tie Exchange, X	–	2	6	$r(r-1)$
Mixed, M	–	4	39	$h - (r(2r+3) - 2)$
Full, F	–	1	1	1
Total	4	16	64	h

Pair	Pattern and bundle class															
$R_1(i,j)$	0	1	0	0	0	1	0	1	0	1	0	1	1	1	0	1
$R_1(j,i)$	0	0	1	0	0	1	0	0	1	0	1	1	1	0	1	1
$R_2(i,j)$	0	0	0	1	0	0	1	1	0	0	1	1	0	1	1	1
$R_2(j,i)$	0	0	0	0	1	0	1	0	1	1	0	0	1	1	1	1
	N	A	A	A	A	R	R	E	E	X	X	M	M	M	M	F

Figure 3.5 Patterns for a dyad with $r = 2$

The difference between the amount of the tie entrainment and tie exchange bundles is because the former occurs by just having parallel edges with the same direction, whereas a "pure" tie exchange will occur only with a pair of ties of different types between the actors. With respect to the amount of bundle patterns within the mixed class, this is simply the difference between the total number of possible bundles in the network and the sum of the pairs having exclusively the rest of the properties.

Figure 3.5 shows the 16 patterns for a dyad and two types of relations, i.e., $r = 2$, where ones are zeroes are for the presence or absence of a link as usually, and at the bottom of the table is the respective bundle class of the pattern. We can see that the amount of classes correspond the the ones already given in Table 3.3; that is, asymmetric and mixed bundles occur most, whereas for the rest of the bundle classes there are just two combinations of the pattern. This is naturally excluding the null and the full bundle class pattern that has one isomorphism no matter the amount of types of relations involved.

3.6.3   Statistical Approach to Bundle Patterns

Bundle patterns allow us to model structural features of multiplex networks in statistical terms. For instance, Wasserman (1980) proposed a simple stochastic model for measuring both the level of "cohesion" and "reciprocity" in a simple network that is based on the three dyadic parameters studied by Holland and Leinhardt (1976).

The maximum likelihood estimate for *group cohesion* is the proportion of asymmetric dyads in the network, A, to twice the amount of null dyads, N (cf. also Proctor and Loomis, 1951). That is,

$$\hat{\gamma}_{\text{cohesion}} = \frac{A}{2 \cdot N},$$

where $A = (i,j) \in R \wedge (j,i) \notin R$, and $N = (i,j) \notin R \wedge (j,i) \notin R$, for $i > j$ and $r = 1$.

The *reciprocity* level in the network is defined by the log odds of the ratio of *group coherence*, which is the proportion of twice the mutual dyads M and asymmetric patterns, to the score for group cohesion. This is the log odds ratio

$$\log \left(\frac{\hat{\gamma}_{\text{coherence}}}{\hat{\gamma}_{\text{cohesion}}} \right)$$

where

$$\hat{\gamma}_{\text{coherence}} = \frac{2 \cdot M}{A}$$

and $M = (i, j) \in R \wedge (j, i) \in R$.

Since weak and strong bonds generalize asymmetric and reciprocal dyads, respectively, the estimation of group cohesion and the reciprocity level in multiplex network structures results straightforward by counting the amount of strong and weak bonds in the bundle census:

$$\text{Cohesion} = \frac{\# \text{ weak bonds}}{2 \cdot \# \text{ null bundles}}$$

$$\text{Reciprocity} = \log \left(\frac{\frac{2 \cdot \# \text{ strong bonds}}{\# \text{ weak bonds}}}{\text{Cohesion}} \right),$$

where the amount of null bundles in the proportion for cohesion for a network of n actors equals

$$\# \text{ null bundles} = \binom{n}{2} - \# \text{ strong and weak bonds}.$$

3.7   Co-occurrence of Ties Model

What typifies a multiplex social network is the relational interlocking of the system. That is, the fact that actors are linked with ties having different relational content where it is possible to associate several types of ties to characterize the entire relational structure. However, while the relational interlocking occurs by alternating many types of tie in a sequence among actors and hence forming compound relations, there are also specific patterns between a pair of subjects. In this latter case, there is a co-occurrence of different types of ties that constitute a particular relational bundle that characterizes the bond between a pair of actors.

Recall that bundles with multiplex character are obtained through the patterns found in simple networks, which for directed pairwise relations are the asymmetric, reciprocal, and the null dyad. For instance, if we consider in a given dyad just the asymmetric relations then there are two possible patterns based on the direction of the tie. From the standpoint of an individual actor the asymmetry consist on sending or by receiving a tie, and we represent such options as A_s and A_r, respectively.

If we adjoin to the asymmetric dyad another unequal relation that corresponds to a different relational content, then two different patterns can arise from the co-occurrence of these ties. One possibility is that the two relations are in the same direction (whether sending or receiving from an actor's point of view), and the other option is that the ties have opposite direction. These two possibilities correspond to a tie entrainment bundle pattern matching the direction of the ties, and to a tie exchange bundle pattern for ties with reverse direction.

Through an algebraic structure as below where the co-occurrence operation is represented by the symbol $/\!/$, we can obtain *rules of association* among asymmetric and other types of dyads or bundle patterns where the bundle classes constitute the set of elements or generators of the system whereas the co-occurrence of ties acts as the endowed operation. For instance, the multiplication table below associates A_s and A_r, and the product of such co-occurrences leads to other kinds of patterns in the cells, which are bundles representing the sending and receiving entrainment of links (E_s and E_r) are in the diagonal, whereas the off diagonal of the table is populated by the tie exchange pattern X.

//	A_s	A_r
A_s	E_s	X
A_r	X	E_r

A mixed pattern is obtained by the associations of the two asymmetric dyads with the bundles with a multiple character, except when the relations coincide in a unilateral direction.

//	A_s	A_r	E_s	E_r	X
A_s	E_s	X	E_s	M	M
A_r	X	E_r	M	E_r	M
E_s	E_s	M	E_s	M	M
E_r	M	E_r	M	E_r	M
X	M	M	M	M	M

However, the two above structures remain open, and this is because all associations between the implied elements are not yet achieved. In other words, there are some patterns in the algebraic structure not belonging to its elements set. To make the structure a closed system, we also need to consider the products of the mixed bundle with the other patterns in the array structure.

//	A_s	A_r	E_s	E_r	X	M
A_s	E_s	X	E_s	M	M	M
A_r	X	E_r	M	E_r	M	M
E_s	E_s	M	E_s	M	M	M
E_r	M	E_r	M	E_r	M	M
X	M	M	M	M	M	M
M	M	M	M	M	M	M

It is clear that a mixed bundle is the absorbing element in the system since the product of any other element with this pattern gives M, and the same circumstance is valid for tie exchange as well. This leads us to classify the bundle patterns in two categories, which are the asymmetric and tie entrainment patterns in one hand, and the tie exchange and mixed bundle classes in the other hand. The first category is characterized by the asymmetry in the relations, whereas the second category has patterns with a mutual character.

Yet there are still two types of tie patterns that belong to simple networks that are not considered in the co-occurrence model, namely the reciprocal and null dyads. Clearly a reciprocal dyad, R, has a mutual character since ties are in both directions at the same level, whereas the null dyad, N, represents the absence of a tie. In the context of multiple networks, the absence of a link does not affect the pattern made of a tie or ties that are occurring simultaneously between a pair of actors. As a result, the produced algebraic structure is closed under the co-occurrence operation with a finite set of 8 elements, and M and N are respectively the "null" and the "identity" elements of the system.

$/\!/$	N	A_s	A_r	R	E_s	E_r	X	M
N	N	A_s	A_r	R	E_s	E_r	X	M
A_s	A_s	E_s	X	M	E_s	M	M	M
A_r	A_r	X	E_r	M	M	E_r	M	M
R	R	M	M	M	M	M	M	M
E_s	E_s	E_s	M	M	E_s	M	M	M
E_r	E_r	M	E_r	M	M	E_r	M	M
X	X	M	M	M	M	M	M	M
M	M	M	M	M	M	M	M	M

Hence, it is clear that there is a congruence between the products of the reciprocal and the tie exchange bundle pattern, i.e. $R \cong X \cong M$.

Whether a bond between a pair of actors has an asymmetric or a mutual character has important social implications. For instance, cohesion influence is achieved mainly through mutual relationships among social actors, whereas hierarchical systems are primary made of asymmetric tie patterns. However, in the end, whether or not a bundle pattern has a mutual character or not depends on the relational content of the ties.

3.8   Relational Structure

The algebraic structure produced with the co-occurrence operation corresponds to a *semigroup*, and now we are going to apply the semigroup structure directly to all the elements involved in the multiple networks. That is, not only to the different kinds of relationships among the actors, but also the different characteristics of the actors. This implies that the relations and the attributes of the actors are no longer going to be considered as separated entities that are subject to a correlation as before, but they will be constituent elements of an integrated system of relations, which means that the attributes of the actors are going to be treated in a relational manner.

In this sense, when considering different types of relationships in the analysis, there are two significant aspects to take into account. Clearly, the first one is the modeling of the actors, even though this is done in the context of a multiple network; a second more intriguing aspect, however, is to consider the "interrelations between relations" occurring in the social system. Modelling ties rather than modeling actors implies a higher level of abstraction in the analysis, and we can benefit from algebraic models to perform such task.

We have seen that the relations among the actors in the network, which can occur at different levels, determine the structure of the social system. On the other hand, it is possible to represent the relationship among the different types of ties in a multiple network by using specific algebraic structures with particular axioms and other characteristics. In this sense, we define the relational structure of the entire network as a system made by the different rules of association between the unique social relations existing in the network. What "unique" exactly means in this context is a matter of the chapter, but what is important now is to realize that the relational structure plays a central role in uncovering different types of structural constraints, which serve to explain diverse types of behavior of the actors in the networks.

3.8.1 Strength of Weak Ties Model as Relational Structure

The theory of the Strength of Weak Ties formulated by Granovetter (1973) posits that transitivity is a function of the strength of the ties rather than be a general feature of the social structure (p. 1377), where the strength of a tie depends on a mixture of intensity and extent of the tie. In particular, strong ties tend to transitivity and—since this type of tie is part of transitive triples—they cannot act as bridges between different components of the system. Thus only weak ties have the power to link the individual actor to a broader social circle, which implies that they are source of non-redundant information and also significant channels for the diffusion of novel information (Valente, 2010; Granovetter, 1982).

Although Granovetter's Strength of Weak Ties original argument (1973) was formulated for the analysis of single networks, he in fact mentions the multiplexity of relations in the definition of strong ties suggesting that "most of the multiplex ties [are] strong but also allow for other possibilities" (p. 1361, fn. 3). In this sense, it is possible to generalize the notions of strong and weak ties to multiple networks where the qualification of relational bundles based on the mutuality of relations permits the tie entrainment bundle to be distinguished as a weak pattern from the rest of the bundles with a multiple character that are considered as strong patterns.

Within this line of thought, it has been hypothesized that strong and weak ties are interrelated in a different way with the other network relations. While strong ties exhibit inbreeding, the compound paths constructed from weak ties tend to be increasingly weaker as the path length increases (Pattison, 1993, pp. 257–58; cf. also Breiger and Pattison 1986). By considering these premises, then we can represent such strong-weak structure as different forms for relational composition. If we account for the transitive closure of these types of tie, then the different combinations are expressed in the equations below, where S and W represent concrete strong and weak relations respectively:

$$(S \circ S)^t = S$$

$$(S \circ W)^t = W$$

$$(W \circ S)^t = W$$

$$(W \circ W)^t = W.$$

Even though a significant implication of the Strength of Weak Ties theory for the structure of the network is that the triadic configurations with only two strong ties are logically likely to become transitive over time, and for this reason, such kind of arrangement is called as a "forbidden triad" (Granovetter, 1973, p. 1363). Granovetter did not specify in his original argument what sort of tie would come out from the closure of a triad. The only assertion on this respect is that "the [closure] tie is always present (*whether weak or strong*), given the other two strong ties" (1973, p. 1316, emphasis added).

In this sense, it is plausible that the transitivity effect of a pair of strong ties may be the underlying cause for a weak tie (cf. Fig. 4.2 in Borgatti and Lopez-Kidwell, 2011), and it can be related e.g. to phenomena like social distance. For instance, Bourgeois and Friedkin (2001) point out that "strong ties may be difficult to form and maintain over large distances in social space" (p. 247) and it is not hard to imagine for example that some relatives of one's non-sibling family members have weak ties with oneself.

As a result, the transitivity effect of two strong ties can be interpreted in some particular circumstances with the closure of a weak tie, and this implies that in a number of cases the first equation in this section can be rather represented as:

$$(S \circ S)^t = W.$$

Now this latest situation suggests that the concatenation of two weak ties—if there exists—tends to lack of transitivity and consequently it needs to be an "even weaker" type of tie. Then we can interpret the product of two weak ties in terms of closure as an undefined relation:

$$(W \circ W)^t = \emptyset.$$

Table 3.4 presents the two versions of relational structures that correspond to the Strength of Weak Ties theory where the transitive closure operation is given in terms of composition of strong and weak ties in the system. Hence, while the classic representation has only an identity and a zero element in the structure, the alternative version lacks the identity relation in the multiplication table.

3.8.2   Graph Representation of the Strength of Weak Ties

Figure 3.6 provides a visual representation of the Strength of Weak Ties theory in the form of graphs where solid and dashed arcs stand for strong and weak relations, respectively. Here the transitive closure is a strong, weak, or an empty path that results from the combination of strengths in the indirect relationship between nodes a and c. , and the absorbing character of the empty tie applies to both a strong and a weak type of tie. The forbidden triad at the bottom of the picture means that transitive closure is always possible from a two-path relationship of a strong character.

Table 3.4   Two relational structures of the Strength of Weak Ties theory. *The Alternative version has a higher order than the Classic.*

$\circ^t$	S	W
S	S	W
W	W	W

$\circ^t$	S	W	$\emptyset$
S	W	W	$\emptyset$
W	W	$\emptyset$	$\emptyset$
$\emptyset$	$\emptyset$	$\emptyset$	$\emptyset$

(a) Classic   (b) Alternative

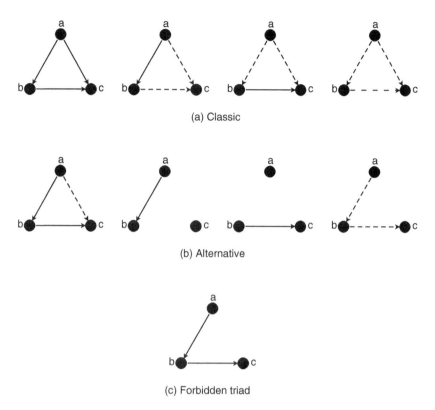

(a) Classic

(b) Alternative

(c) Forbidden triad

Figure 3.6 Graph representation of the Strength of Weak Ties theory. *Solid and dashed arcs represent strong and weak ties, respectively.*

Whether one of these versions of the Strength of Weak Ties theory—the classic or the alternative version—is more adequate requires empirical support, and probably it depends on the relational content of the ties. Moreover, in the case of multiple networks it also needs to take into account the relational structure of the entire system as well. The important thing right now is that the characterization of the Strength of Weak Ties hypothesis and the system of rules made in Equations in section 3.8.1 characterizes a theoretical representation of a relational model.

The fact that there are different versions of the association between strong and weak ties means that the resulting relational structures differ as Table 3.4 evidences. The relational structures of the two versions of the Strength of Weak Ties hypothesis are in fact closed algebraic structures. These systems are abstract semigroups, where the underlying set is made by elements W, S, and ∅ while the associated binary operation is the transitive closure of the composition of these types of tie, which is expressed by ∘^{t}.

The algebraic representation of the relational structure has the advantage that the system is closed under the defined operation, and in addition, we are able to identify important properties in the relations. As mentioned before, the semigroup structure representing the first hypothesis has S representing the identity element and W is the zero element of this particular semigroup structure. On the other hand, the semigroup representing the second hypothesis has no identity, and ∅ is the zero element in the alternative version.

Given that the first hypothesis of the Strength of Weak Ties is only made by the identity and zero elements, the structure representing the classical version of the theory is an example of a *commutative semigroup* given that the product of its elements commute, i.e. SW = WS, etc. Boyd (1991, p. 64) characterizes a commutative semigroup as "self-dual semigroup" or a semigroup with "left-right" duality where for example xS substitute Sx.

3.9   Semigroup of Relations in Multiplex Networks

Recall from Chapter 2 Algebraic Structures that like the abstract semigroups, there are other types of algebraic systems aimed to represent relational structures, and that are flexible enough to include simultaneously all types of relations considered in the study, including the attributes of the actors as relations. However, the possible combinations of the different types of ties in multiplex networks can easily result in complex and quite large structures, and thus a great part of the analysis takes the reduction of the relational structure in theoretical terms.

To understand the rationale behind the methods to pursue an algebraic analysis of complete network systems, we introduce next other types of semigroups that serve to represent the relational structure of empirical multiplex networks. Afterwards, we will look at some types of functions useful for the analysis of these algebras and also to strategies especially designed for the reduction of the complexity of the resultant relational structures.

To illustrate the semigroup of relations and its properties in multiplex network structures, we take a "small configuration" network represented by adjacency matrices in Table 3.5. We denote this network as $\mathscr{X}_Z$, and the structure—and the relational content as well—is isomorphic to one of the components of the Incubator network $\mathscr{X}_C$ depicted at the beginning of Chapter 11 Comparing Relational Structures. Each actor in $\mathscr{X}_Z$ has an identifier that is a lower case letter, whereas C and F represent two distinct ties that stand for collaboration and friendship relations among the actors.

In order to make an algebraic representation of the relational structure of $\mathscr{X}_Z$, then these string letters constitute the alphabet of the semigroup with generators for this network. We illustrate next the different steps in constructing the semigroup of relations for the small configuration network, and then we apply such principles to empirical multiplex network structures that are a bit larger.

Table 3.5   Adjacency matrices for the small configuration $\mathscr{X}_Z$ with two types of relations.

	a	b	c	d	e			a	b	c	d	e
a	0	0	0	0	0		a	0	1	0	0	0
b	0	0	1	0	0		b	0	0	1	0	0
c	0	0	0	0	0		c	1	0	0	0	0
d	0	0	0	0	0		d	1	1	0	0	0
e	0	0	0	0	0		e	0	0	0	1	0

C   F

3.9.1   Partial Order Relations and the Axiom of Quality

There are a number of implications to be made with the information provided by Table 3.5. For instance, we can see that the only C relation occurring in this system is repeated in F, which means that the set of relations F in this case includes the set of relations C. Such aspect suggests that there is a partial order between these two types of relations with the form

$$C \leq F.$$

In formal terms, two strings, say R_1 and R_2 in X, define a *partial order relation*

$$R_1 \leq R_2$$

whenever for all $(i,j) \in X$,

$$(i,j) \in R_1 \text{ implies } (i,j) \in R_2,$$

and this means that relation R_2 "contains" relation R_1.

It is, however, possible that R_1 and R_2 link precisely the same pair of elements in the set and the two relation types are alike; that is $R_1 = R_2$. According to the *Axiom of Quality* (Boorman and White, 1976, also Pattison 1993) any two relations defining the same set of connections among pairs of individuals are to be equated, and this is defined as

$$R_1 = R_2 \text{ implies } R_1 \leq R_2 \text{ and } R_2 \leq R_1.$$

Besides the partial ordering of the two primitive relationships, it is also possible to construct a set of compound ties from these letters through the operation of composition with precisely the same terms presented in Figure 1.2 in Chapter 1. In this way, the different paths of relations make together a relational structure of compound ties in a multiple network that is suited for instance for the analysis of the different channels of influence existing in the system.

As a result, if we take only $R_1 = C$ from the example of Table 3.5, then it is clear that it is not possible to have a compound relation made just of collaboration ties in this small system. This means that the product of two collaboration relations is empty, or else $CC = \emptyset$. Now this implies that the compound of two collaboration ties acts as an absorbing element and all products made with this two-path cycle will be empty. From the Axiom of Quality then we have that $CC = CCC = CCF = \ldots$ etc. $\ldots = \emptyset$.

However, if we consider the friendship relations in the analysis, then it is possible to have several types of compound relations that are not empty. This certainly brings more significant characteristics to the relational structure, and Figure 3.7 provides in a graph format some of the compound relations occurring in this system, which involve one or the two types of these primitives. Such relations are then the product of composition of relations, and they represent paths of relations where information and other resources flow among the members of the network. Naturally, it is possible to have even longer paths of relations than those represented in the figure and the semigroup of relations serves to associate such paths in a meaningful structure.

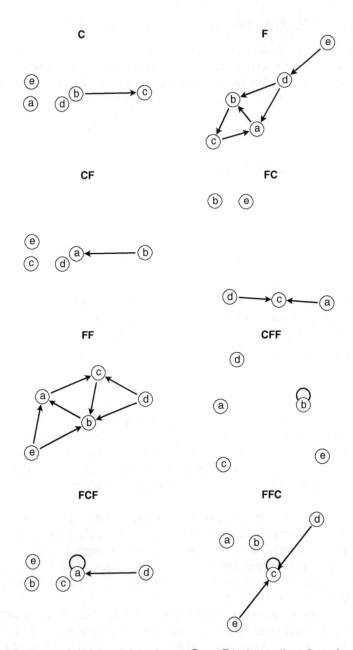

Figure 3.7   Different compounds involving relations C and F in the small configuration $\mathcal{X}_Z$. *Adjacency matrices are given in Table 3.5.*

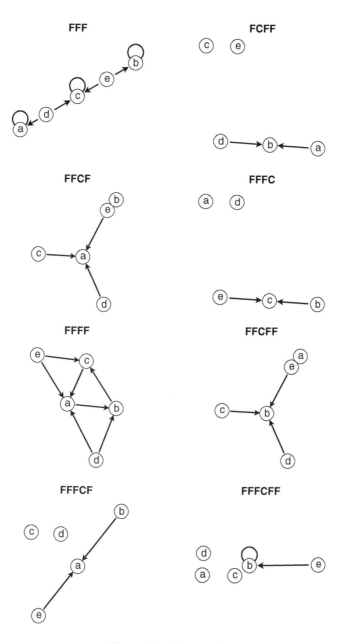

Figure 3.7 (*Continued*)

3.9.2 Multiplication Table

We already know that each compound relation having CC in the system example of Table 3.5 is empty and it is meant to be equated with this compound. However, there are other paths of relations that are not empty and they happen to coincide as well. This means that according to the Axiom of Quality such compound relations are also going to be equated, and thus at a certain point of time there will be a finite set of unique equations that represent the relational structure of the entire system.

In the case of $\mathscr{X}_Z$, besides the two primitive relations $1 = C$, $2 = F$, the set of unique compound relations is the following:

3 = CC	6 = FF	9 = FFC	12 = FFCF	15 = FFCFF
4 = CF	7 = CFF	10 = FFF	13 = FFFC	16 = FFFCF
5 = FC	8 = FCF	11 = FCFF	14 = FFFF	17 = FFFCFF

Consequently, there are 17 distinct relationships in this example that represent unique configurations made of different types of relational paths, and each class is represented by the shortest string of relations indexed by an identification number. The set of relations in terms of composition between these unique strings then makes a representation of the whole relation structure, and each one of the longer compound relations are derived from one of these strings. This means that any of such compounds in principle could be the representatives in the network relational structure for its interpretation.

In this sense, the semigroup with generators $S(R)$ serves to record such distinct relations in the form of a particular table called a *multiplication table* of the semigroup, which is also known as the "Boolean matrix semigroup." This table has a *dimension* that equals the number of primitives, and an *order* that equals the amount of unique string of relations. As a result, while the semigroup of relations is a pure algebraic interpretation of the structure of the network, the multiplication table is a matrix representation of the structure summarizing all the information concerning the associations between the unique relations of the relational system.

Table 3.6 gives the multiplication table of the system component example both in a *symbolic* form and in a *numerical* format. The first table is made of the strings relations with concatenation, whereas the second table uses the identification number to denote the class string as given above.

Thus we can see in the multiplication tables that, just with having a very small system with a couple of primitive ties such as $\mathscr{X}_Z$, the structure of the semigroup gets quite large and complex. The fact that the semigroup of relations becomes very quickly intricate and large implies that one of the main tasks in the analysis of the semigroup with generators is to ease the arrangement of the multiplication table to be able to proceed with the interpretation of the relational structure in more substantial terms. There are some methods devoted to this aspect, and a simplification of semigroup structures is especially needed where we can incorporate simultaneously the attributes of the actors in the relational system.

A reduction of the multiplication table requires capturing the essential structure of the system, and for that we need to obtain all relevant information about the related elements. In this sense, we introduce next another type of algebraic structure that complements the semigroup of relations by defining the ordering of its elements, and which also plays an important role in the establishment of the different constraints existing in the system.

Table 3.6 Matrix multiplication tables for $\mathscr{D}_Z$.

o	C	F	CC	CF	FC	FF	CFF	FCF	FFC	FFF	FCFF	FFCF	FFFC	FFFF	FFCFF	FFFCF	FFFCFF
C	CC	CF	CC	CC	CC	CFF	CC	CC	C	F	CC	CF	FC	FF	CFF	FCF	FCFF
F	FC	FF	CC	FCF	FFC	FFF	FCFF	FFCF	FFFC	FFFF	FF	FFFCF	FC	FF	FFFCFF	FCF	FCFF
CC	CC	CC	CC	CC	CC	CC	CC	CC	CC	CC	CC	CC	CC	CC	CC	CC	CC
CF	CC	CFF	CC	CC	C	F	CC	CF	FC	FF	CFF	FCF	FFC	FFF	FCFF	FFCF	FFCFF
FC	CC	FCF	CC	CC	CC	FCFF	CC	CC	FC	FF	CC	FCF	FFC	FFF	FCFF	FFCF	FFCFF
FF	FFC	FFF	CC	FFCF	FFFC	FFFF	FFCFF	FFFCF	FC	FF	FFF	FCF	FFC	FFF	FCFF	FFCF	FFCFF
CFF	C	F	CC	CF	FC	FF	CFF	FCF	FFC	FFF	FCFF	FFCF	FFFC	FFFF	FFCFF	FFFCF	FFFCFF
FCF	CC	FCFF	CC	CC	FC	FF	CC	FCF	FFC	FFF	FCFF	FFCF	FFFC	FFFF	FFCFF	FFFCF	FFFCFF
FFC	CC	FFCF	CC	CC	CC	FFCFF	CC	CC	FFC	FFF	CC	FFCF	FFFC	FFFF	FFCFF	FFFCF	FFFCFF
FFF	FFFC	FFFF	CC	FFFCF	FC	FF	FFFCFF	FCF	FFC	FFF	FFFF	FFCF	FFFC	FFFF	FFCFF	FFFCF	FFFCFF
FCFF	FC	FF	CC	FCF	FFC	FFF	FCFF	FFCF	FFFC	FFFF	FF	FFFCF	FC	FF	FFFCFF	FCF	FCFF
FFCF	CC	FFCFF	CC	CC	FFC	FFF	CC	FFCF	FFFC	FFFF	FFCFF	FFFCF	FC	FF	FFFCFF	FCF	FCFF
FFFC	CC	FFFCF	CC	CC	CC	FFFCFF	CC	CC	FFFC	FFFF	CC	FFFCF	FC	FF	FFFCFF	FCF	FCFF
FFFF	FC	FF	CC	FCF	FFC	FFF	FCFF	FFCF	FFFC	FFFF	FF	FFFCF	FC	FF	FFFCFF	FCF	FCFF
FFCFF	FFC	FFF	CC	FFCF	FFFC	FFFF	FFCFF	FFFCF	FC	FF	FFF	FCF	FFC	FFF	FCFF	FFCF	FFCFF
FFFCF	CC	FFFCFF	CC	CC	FFFC	FFFF	CC	FFFCF	FC	FF	FFFCFF	FCF	FFC	FFF	FCFF	FFCF	FFCFF
FFFCFF	FFFC	FFFF	CC	FFFCF	FC	FF	FFFCFF	FCF	FFC	FFF	FFFF	FFCF	FFFC	FFFF	FFCFF	FFFCF	FFFCFF

(a) Symbolic representation

Table 3.6 (*cont.*)

o	1	2	3	4	5	6	7	8	9	10	11	12	13	14	15	16	17
1	3	4	3	3	3	7	3	3	1	1	3	4	3	4	7	3	3
2	5	6	3	8	9	10	11	12	13	14	15	16	5	6	17	8	11
3	3	3	3	3	3	3	3	3	3	3	3	3	3	3	3	3	3
4	3	7	3	3	1	1	3	4	3	4	7	3	3	7	3	3	3
5	3	8	3	3	3	11	3	3	5	5	3	8	3	8	11	3	3
6	9	10	3	12	13	14	15	16	5	6	17	8	9	10	11	12	15
7	1	1	3	4	3	4	7	3	3	7	3	3	1	1	3	4	7
8	3	11	3	3	5	5	3	8	3	8	11	3	3	11	3	3	3
9	3	12	3	3	3	15	3	3	9	9	3	12	3	12	15	3	3
10	13	14	3	16	5	6	17	8	9	10	11	12	13	14	15	16	17
11	5	5	3	8	3	8	11	3	3	11	3	3	5	5	3	8	11
12	3	15	3	3	9	9	3	12	3	12	15	3	3	15	3	3	3
13	3	16	3	3	3	17	3	3	13	13	3	16	3	16	17	3	3
14	5	6	3	8	9	10	11	12	13	14	15	16	5	6	17	8	11
15	9	9	3	12	3	12	15	3	3	15	3	3	9	9	3	12	15
16	3	17	3	3	13	13	3	16	3	16	17	3	3	17	3	3	3
17	13	13	3	16	3	16	17	3	3	17	3	3	13	13	3	16	17

(b) Numerical representation

3.10   Partially Ordered Semigroup

The Axiom of Quality allows us to count with a finite set of elements in the structure, and we already seen that there is possible to have a partial ordering on the elements in $S(R)$. In this sense, the set of inclusions among the unique strings of relations produce another algebraic structure that complements the semigroup of relations. This sort of structure is called a *partially ordered semigroup* or simply *ordered semigroup*, which is defined as the triple:

$$\langle\ S,\ \circ, \leq\rangle,$$

where S represents either an abstract semigroup or the semigroup of relations, $\circ$ is the operation of composition, and $\leq$ represents an inclusion ordering among the semigroup elements.

The partially ordered semigroup is represented by the multiplication matrix semigroup S together with a corresponding *partial order table*, which is a square array matrix that gives the partial ordering of the semigroup elements that corresponds to the inclusions among the unique strings in the semigroup of relations.

Elements a, b in a partial order table for S are defined as

$$s_{a,b}^{\leq} = \begin{cases} 1 & \text{iff relation } a \text{ is contained in relation } b \\ 0 & \text{otherwise.} \end{cases}$$

It is certainly possible in the definition to switch the assignments in the partial order table, where $s^{\leq}_{a,b} = 1$ if relation a contains relation b, and 0 otherwise (e.g. Pattison, 1993). In such case, the structure of the partial order remains intact, and it equals to the transposition to the partial order table representation given here.

3.10.1 Partial Ordering in $\mathscr{X}_Z$

We have seen for instance in $\mathscr{X}_Z$ that one of the primitive relationships is included in the other generator tie. Similarly, the different compound relations given in Figure 3.7 can be partially ordered as well, and we can see for instance that all ties that are present in the configuration for FC, are also occurring in the configuration for FF, which means that FC $\leq$ FF. The same situation occurs for CF that is contained in FF but not in FC in this case. Also, note that FCFF is not included in FF given that the inclusion of elements considers the direction of the ties, and the system considered here is directed.

Table 3.7 gives the partial order table for the entire semigroup of relations with a numerical labelling of its elements and according to the definition of the partial order given in terms of containment. We note in this table that the character of this type of structure is reflexive, and that agrees to the partial order definition given in the first chapter. In this sense, the information encoded in this arrangement complements the multiplication table, and together both constitute the partially ordered semigroup for the relational system.

Table 3.7 Partial order table for $S(R)$ in $\mathscr{X}_Z$.

$\leq$	1	2	3	4	5	6	7	8	9	10	11	12	13	14	15	16	17
1	1	1	0	0	0	0	0	0	0	0	0	0	1	1	0	0	0
2	0	1	0	0	0	0	0	0	0	0	0	0	0	0	0	0	0
3	1	1	1	1	1	1	1	1	1	1	1	1	1	1	1	1	1
4	0	0	0	1	0	1	0	0	0	0	0	0	0	0	0	1	0
5	0	0	0	0	1	1	0	0	0	0	0	0	0	0	0	0	0
6	0	0	0	0	0	1	0	0	0	0	0	0	0	0	0	0	0
7	0	0	0	0	0	0	1	0	0	1	0	0	0	0	0	0	1
8	0	0	0	0	0	0	0	1	0	1	0	0	0	0	0	0	0
9	0	0	0	0	0	0	0	0	1	1	0	0	0	0	0	0	0
10	0	0	0	0	0	0	0	0	0	1	0	0	0	0	0	0	0
11	0	1	0	0	0	0	0	0	0	0	1	0	0	1	0	0	0
12	0	0	0	0	0	0	0	0	0	0	0	1	0	1	0	0	0
13	0	0	0	0	0	0	0	0	0	0	0	0	1	1	0	0	0
14	0	0	0	0	0	0	0	0	0	0	0	0	0	1	0	0	0
15	0	0	0	0	0	1	0	0	0	0	0	0	0	0	1	0	0
16	0	0	0	0	0	1	0	0	0	0	0	0	0	0	0	1	0
17	0	0	0	0	0	0	0	0	0	1	0	0	0	0	0	0	1

3.11   Word and Edge Tables

Although the multiplication table has such large order, the essential information of the semi-group of relations structure can be summarized in the form of two tables called the *Word table*, and the *Edge table* (Pattison, 1993, also Cannon 1971). The Word table gives a list of indexed elements in the complete semigroup, and two columns that represent the nodes and generators respectively. The former gives the edges that record the result of post-multiplying the compound relations by the generators, and the latter column are the collection of unique relations from where the paths begin.

On the other hand, the Edge table is a rectangular array where each column represents a generator relation and there is a row for every element in the semigroup. In consequence, the Edge table constitutes the right multiplication table that encodes the different relationships of the unique generators.

Table 3.8 presents the Word and the Edge tables for $\mathscr{X}_Z$ with the attributes, and we evidence that both tables have the same dimension and order. The order of the semigroup is 17 and the longest compound has a length of 7. The relations between the different strings of relations are given in the Edge table, and the rest of the semigroup elements can be deduced from the edge table by association. For instance, the product of relations 1 and 3 results in $1 \cdot (1 \cdot 1)$, which is 3; Similarly, $1 \cdot 4 = 1 \cdot (1 \cdot 2)$ or by the associativity property equals $(1 \cdot 1) \cdot 2 = 3$, etc. In this sense, the Edge table has the capability to encode all the necessary information to represent the relational structure of the system.

Table 3.8   Semigroup of relations from Small configuration $\mathscr{X}_Z$ as Word and Edge tables. *Only compound relations have Node and Generator string numerical identifiers.*

Semigroup Element	Word Table			Edge Table	
	Word	Node	Generator	1	2
1	C	–	–	3	4
2	F	–	–	5	6
3	CC	1	1	3	3
4	CF	1	2	3	7
5	FC	2	1	3	8
6	FF	2	2	9	10
7	CFF	4	2	1	1
8	FCF	5	2	3	11
9	FFC	6	1	3	12
10	FFF	6	2	13	14
11	FCFF	8	2	5	5
12	FFCF	9	2	3	15
13	FFFC	10	1	3	16
14	FFFF	10	2	5	6
15	FFCFF	12	2	9	9
16	FFFCF	13	2	3	17
17	FFFCFF	16	2	13	13

If we consider words of length 3 or less, the only equation in the semigroup structure corresponds to strings to the composition of a pair of C relations.

$$CC = CCC = CCF = FCC = CFC$$

Naturally, there will be many more equations if we consider words with a larger length. In this sense, besides the length of the word other criteria in choosing the labelling for the semigroup representation are that the attribute relation prevails over the other types of tie, and that the relation itself prevails over its transpose. Accordingly, we "choose" CC over CCC and CCF over CFC, and this is both to keep simplicity in the interpretation of the relational structure and also to highlight the attribute relation when feasible. Nevertheless, the compound made of C ties in this case might serve to explain certain features of the relational structure when it comes to the substantial interpretation of it.

In order to continue with the analysis of networks such as $\mathscr{X}_Z$, we need to address the size and complexity of the semigroup of relations resulting for the systems, and this is a crucial aspect to tackle in the study of almost any multiplex network structure. Besides, by looking at the partial ordering of the elements, one way to handle this is to detect ideal structures inside the system by looking at significant properties of the semigroup and the elements involved.

Ideal structures usually have a more clear interpretation and predictive power than the "ordinary" structure and therefore it is important to identify them. Another way to reduce the complexity of the relational system is to model both the actors and their relations of a given multiplex network in order to capture the essential structure of the system, and this latter strategy is the chosen in most of the analyses made in the following chapters.

3.12   Multiplex Network Configurations: Summary

After a brief preamble with notions associated with relational structures in the previous two Chapters, there are some applications of algebraic systems to real-world social networks and their relational structures. For instance, some primitive societies like the Kariera have ideal types of prescriptive marriage types and descent, and their kinship systems follow the rules of a group structure having a symmetric character. However, it is more likely that cultures from today follow patterns of preferential marriage types, and the kinship networks for most–if not all–modern societies produce rather complex structures.

Although the representation of symmetric configurations through algebraic group structures resulted a fine illustration, the modeling of most multiplex social networks requires alternative approaches. The distinction of bundle class patterns allows making distinction of particular systems inside the network structure, and a co-occurrence of ties model represents relations between these bundle classes. A widely used model in the analysis of social networks with multiplex patterns comes from the Strength of Weak Ties theory, and the relational structure with the two types of tie was represented in a graph and table format.

There has also been a treatment to different representation forms of semigroup of relations. Semigroups typically stand for the relational and role structures of multiplex networks, and by applying an ordering on the relations, the semigroup object becomes a partially ordered structure. Word and Edge tables abbreviate the multiplication table, and the partial order table characterizes the partially ordered semigroup. From the different structures presented, three algebraic constrains arise: equality with a set of equations, hierarchy of strings represented by a set of inclusions, and semigroup tables that connected relationships or role relations with each other.

3.13   Learning Multiplex Networks by Doing

3.13.1   Kariera Kinship Network

```
# construct two permutation matrices for marriage and descent rules
R> kks <- transf(list(F = c("1, 2", "3, 4", "2, 1", "4, 3"),
+    G = c("1, 4", "2, 3", "3, 2", "4, 1")), type = "toarray")


, , F

  1 2 3 4
1 0 1 0 0
2 1 0 0 0
3 0 0 0 1
4 0 0 1 0


, , G

  1 2 3 4
1 0 0 0 1
2 0 0 1 0
3 0 1 0 0
4 1 0 0 0
```

3.13.1.1   Figure 3.3 Kariera Society

```
# redefine node / edge / graph characteristics
R> scpKS <- list(cex = 10, lwd = 5, pos = 0, vcol = 8, ecol = 1, collRecip = TRUE)

# Kariera 1
R> multigraph(transf(list(M = c("Ba, Bu", "Pa, Ka"), D = c("Ba, Pa", "Pa, Ba")),
+    lbs = c("Ba","Bu","Ka","Pa")), scope = scpKS)
# Kariera 2
R> multigraph(transf(list(M = c("Bu, Ba", "Ka, Pa"), D = c("Bu, Ka", "Ka, Bu")),
+    lbs = c("Ba","Bu","Ka","Pa")), scope = scpKS)
# Kariera 3
R> multigraph(kks, lbs = c("Banaka", "Burung", "Karimera", "Palyeri"),
+    scope = scpKS, fsize = 20)


# string equations in the relational structure
R> strings(kks, equat = TRUE)
```

```
...
$equat
$equat$FF
[1] "FF" "GG"

$equat$FG
[1] "FG" "GF"

$equate
$equate$e
[1] "e"   "FF" "GG"
```

```
# multiplication table of the Kariera group structure
R> Skks <- semigroup(kks, type = "symbolic")

...
$S
    F  G FF FG
F  FF FG  F  G
G  FG FF  G  F
FF  F  G FF FG
FG  G  F FG FF

attr(,"class")
[1] "Semigroup" "symbolic"
```

```
# define graph scope
R> scpKSS <- list(cex = 10, lwd = 6, pos = 0, vcol = 8, collRecip = TRUE, fsize = 28)

# plot Cayley graph of the Kariera group structure
R> ccgraph(Skks, scope = scpKSS, conc = TRUE)
```

3.13.2   Multiplex Networks

3.13.2.1   Figure 3.4 Multiplex Patterns

```
# redefine node / edge / graph characteristics
R> scp <- list(cex = 16, lwd = 16, vcol = 0, vcol0 = 1, ecol = 1, showLbs = FALSE,
+    rot = -90)

# Fig. 3.4.a  plot tie entrainment with customized format (default layout)
R> multigraph(list("1, 2", "1, 2"), scope = scp)

# Fig. 3.4.b  plot tie exchange as above but swap tie levels
R> multigraph(list("1, 2", "2, 1"), scope = scp, swp = TRUE)
```

3.13.3   Strength of Weak Ties

3.13.3.1   Figure 3.6 Visualization of Strength of Weak Ties Theory

```
# redefine node / edge / graph characteristics
R> scp <- list(cex = 12, vcol = 1, lwd = 9, ecol = 1, mirrorX = TRUE)

# Fig. 3.6.a  plot classic Strength of Weak Ties
R> multigraph(c("a, b", "b, c", "a, c"), scope = scp)
# a list with two levels for each type of tie
R> multigraph(list("a, b", c("b, c", "a, c")), scope = scp)
R> multigraph(list("b, c", c("a, b", "a, c")), scope = scp)
# first tie level is empty
R> multigraph(list(NULL, c("a, b", "b, c", "a, c")), scope = scp)

# Fig. 3.6.b  plot alternative hypotheses of Strength of Weak Ties
R> multigraph(list(c("b, c", "a, b"), "a, c"), scope = scp)
# add isolates to the plot  with different labels
R> multigraph("a, b", add = "c", scope = scp)
R> multigraph("b, c", add = "a", scope = scp)
# plot only weak ties
R> multigraph(list(NULL, c("a, b", "b, c")), scope = scp)


# Fig. 3.6.c  plot the forbidden triad
R> multigraph(c("a, b", "b, c"), scope = scp)
```

3.13.4   Relational Structure

3.13.4.1   Table 3.5 Small Configuration with Two Types of Relations

```
# create a small multiplex network with ties "C" and "F" and labeled actors
R> mnet <- transf(list(C = "b, c", F = c("a, b", "e, d", "c, a", "d, a",
+      "d, b", "b, c")), sort = TRUE)

R> mnet

  a b c d e
a 0 0 0 0 0
b 0 0 1 0 0
c 0 0 0 0 0
d 0 0 0 0 0
e 0 0 0 0 0

, , F

  a b c d e
a 0 1 0 0 0
b 0 0 1 0 0
c 1 0 0 0 0
d 1 1 0 0 0
e 0 0 0 1 0
```

3.13.4.2   Figure 3.7 Compound Relations in $\mathscr{X}_Z$

```
# find the string relations in "mnet"
R> mnets <- strings(mnet)
```

```
# representatives of the 17 unique string relations
R> mnets$st

 [1] "C"      "F"      "CC"     "CF"     "FC"     "FF"     "CFF"    "FCF"
 [9] "FFC"    "FFF"    "FCFF"   "FFCF"   "FFFC"   "FFFF"   "FFCFF"  "FFFCF"
[17] "FFFCFF"
```

```
# redefine node / edge / graph characteristics
R> scp <- list(cex = 9, vcol = "#FFFFFF", vcol0 = "#000000", fsize = 20, pos = 0,
+    lwd = 4, ecol = 1, cex.main = 3)
```

```
# string relations
R> for(i in seq_len(dim(mnets$wt)[3])) {
# string name
+    sname <- mnets$st[i]
# avoid empty strings
+    if(isTRUE(sum(mnets$wt[,,i])==0)==TRUE)
# plot compound strings in "mnets" with force layout algortihm
+    multigraph(mnets$wt[,,i], layout = "force", seed = 3, scope = scp,
+    loops = TRUE, main = sname)
+    }
```

4

Positional Analysis and Role Structure

4.1 Roles and Positions

The reduction of a social structure into a configuration made of classes of structurally equivalent actors constitutes a research method known as *positional analysis*, which is an important part in the study of social networks. Positional analysis performs on simple network or in configurations made of multiple types of tie. One outcome of the positional analysis process is a *positional system*, denoted by $\mathscr{S}$ that is a reduced version of a given network $\mathscr{X}$. The positional system hence is made of collective roles representing the ties, and "positions" or the different classes of actors related in the social system.

To talk about of roles and positions we start with Lévi-Strauss (1967), who pointed out that a society is not merely an ensemble of the existing relations, but a further order over and above the one implicit and that interrelates these relationships. In addition, according to Nadel (1957), the ordered arrangement of parts in a social system, which we call the social structure, "is arrived through abstracting from the concrete population and it behavior the pattern or network (or system) of relationships obtaining between actors and their capacity of playing roles relative to one another" (p. 12).

Such statement suggests that the patterns of relations among the actors reflects the social structure that is not made simply by a collection of individuals but by categories of agents with similar social *roles* in the system. As a result, the different types of equivalence we have been looking at so far reflect different forms for social differentiation among the actors in terms of relational patterns. In this sense, the notion of social roles proposed by Nadel serves as a basis for a theory of social structure where a role is an intermediary between society and the individual that operates when a particular behavior becomes a social conduct.

A significant background in Nadel's theory of roles comes from the study on social elites made by Pareto (1916), since he was among the first to realize the sociological significance of *labels* in order to indicate a person's proper place at the various classes of people. Societies classify its population in accordance with the jobs, offices, or functions that individuals assume with a linguistic notice of the differential parts the people are expected or briefed to play. Such

Algebraic Analysis of Social Networks: Models, Methods and Applications using R,
First Edition. J. A. R. Ostoic. Companion website: www.wiley.com/go/ostoic/algebraicanalysis.
© 2021 John Wiley & Sons Ltd. Published 2021 by John Wiley & Sons Ltd.

classification has been to turn into an analytical tool where the linguistic notice denote a label role that is made either by a simple or by a compound relation.

For instance, label roles such as "father" or "boss" indicate a single relation, whereas a *label role* such as "uncle" is made of a compound relation that is one parent's brother or a parent's sister's husband. Although there are a number of label roles that have more than a single word in absence of a single linguistic label, not all social roles, whether they are simple or composed, have a linguistic label and yet they are also important and interesting to study, as Wasserman and Faust (1994) pointed out.

Social roles, then, suggest the presence of classes of individuals that constitute social positions in the system, and within the social structure context, this concept goes back at least to the social differentiation studies made by Linton (1936). In these studies, *status* and role constitute a duality between the patterns of behaviors and relations with other actors, and the rights and duties defined by the pattern. Status is then the "polar position" in the patterns of relations, whereas the role puts the status into effect by the rights and duties of the actors.

Although Nadel restricted the notion of status to a quasi-role and limited to a hierarchical position (Nadel, 1957, pp. 28–29), status as such has become a synonymous of position in the social network analysis discipline, where a social role represents the dynamic aspect of status. In a similar way, Parsons (1991) asserted that role is status translated into action and it constitutes the processual aspect of status, whereas status is the positional aspect of the social role.

On the other hand, Merton (1957) distinguished between status and the relationship between statuses, where a status is a named position that may have a role relationship between it and other statuses by virtue of rights and duties. All the role relationships that a status enters into are called the *role set* and all statuses that an individual inhabits are the *status set*.

Actors occupying the same status set are expected to perform a similar role set with other status sets, and, for instance, "doctors"—a status set—relates with "patients" and "nurses"—other status sets—in a different ways, which constitute the role sets. As a result, a hierarchy of power exists in such relations that are disclosed by the pattern of their relations, and to unfold such kind of structure constitutes one of the main goals of the positional analysis, which we will see in Chapter 5 Role Structure in Multiplex Networks with concrete examples.

To found roles and positions of related actors in multiplex networks in terms of their relational interlocking, it is important first to make a definition of the notion of "equivalence" among the network members. We look for types of structural equivalence that are capable to handle multiplex network structures and preserving the multiplicity in the ties.

The definition of equivalence in networks is closely related to the notion of "graph-" or "network homomorphism," and we start by looking at the abstraction that involves a pair of functions, one for the graph nodes and another for the edges in the graph. Later on, we will see the patterns between equivalent actors that can include different types of relations as both primitive and compound ties, and then we look at some strategies to make possible the inclusion of actor attributes in the positional analysis in a relational manner.

4.2   Network Homomorphism

Recall from Chapter 1 Structural Analysis with Algebra that networks made with different types of ties are relational structures represented by multigraphs of the form

$$\mathcal{G}^+ = \langle\, \mathcal{N}\,,\, \mathcal{R}\,\rangle = \mathcal{X}^+\,.$$

There exists a network homomorphism that serves to simplify the sets of ties in a network, which contemplates a pair of functions, $h = \langle h_1, h_2 \rangle$ where $h_1 : \mathcal{N} \to \mathcal{M}$ and $h_2 : \mathcal{R} \to \mathcal{T}$ are onto. That is, the first function is defined between the actors and the second one is defined between the ties of the network, and where $\mathcal{M}$ and $\mathcal{T}$ are the elements of another domain that in this context is called the *graph image* or simply *image*.

There are different forms of network homomorphism, and in the case of reducing a social network, the basic idea is that the configuration obtained in the image after performing the homomorphic reduction preserves the essential structure of the original network. In this sense, a *full network homomorphism* is obtained when each tie in the domain is mapped onto some tie in the graph image domain.

White and Reitz (1983) points out that the full network homomorphism induces a full graph homomorphism from $\langle \mathcal{N}, \mathcal{R} \rangle$ to $\langle \mathcal{M}, h_2(\mathcal{R}) \rangle$, where a graph homomorphism is a structure preserving mapping between two graphs, even though these functions are surjective. Since graphs are abstract representations of networks, graph homomorphisms will be taken from now on.

An essential point to take into account is that the reduced-form representation of actors and their relations is a product of a graph homomorphism and it is achieved by means of a technique known as blockmodeling (Lorrain and White, 1971; White et al., 1976; Doreian et al., 2004). Blockmodeling is based on the equivalences among the actors in the network, whereas the semigroup homomorphism produces the reduced-form representation of the algebra of the actors' relations (Breiger and Pattison, 1986, p. 216; cf. also Bonacich (1982)).

4.2.1   Weak and Strong Graph Homomorphisms

Now is time to define in formal terms different types of graph homomorphism that can be applied to social networks. We start with the cases within simple networks and then we will see homomorphisms for multiple network structures.

A function $h : \mathcal{G} \to \mathcal{H}$ is a *graph homomorphism* if nodes i and j are mapped correspondingly to $h_1(i)$ and $h_1(j)$ under the mapping h_2, and provided that if (i,j) is a line in $\mathcal{G}$; then the line $h_1(i,j) \in h_2(R)$ must be the image of $(i,j) \in R$ under h_2 of $\mathcal{H}$.

The mappings h_1 and h_2 determine a *weak graph homomorphism* whenever for each $i, j \in \mathcal{N}$ and $R \in \mathcal{R}$:

$$\langle i,j \rangle \in R \quad \text{implies} \quad h_1\langle i,j \rangle \in h_2(R); \qquad \text{(WGH1)}$$

and additionally, for every $h_1(i,j) \in \mathcal{M}$ and $h_2(R) \in \mathcal{T}$:

$h_1\langle i,j \rangle \in h_2(R)$ implies there exists $k, l \in \mathcal{N}$ such that

$$h_1(k) = h_1(i), h_1(l) = h_1(j) \text{ and } \langle k, l \rangle \in R. \qquad \text{(WGH2)}$$

These conditions assure first that any relation in $\mathcal{G}$ is carried over into the image graph $\mathcal{H}$, and second that each relation in $\mathcal{H}$ is the image of some relation in $\mathcal{G}$. In other words, if i and j are nodes in $\mathcal{G}$, then any edge between them there must correspond to the same type of relation between their images $h_1(i)$ and $h_1(j)$ in $\mathcal{H}$ (Bonacich 1982, p. 276; White and Reitz 1983, p. 204; Reitz and White 1992, p. 442).

Such type of graph homomorphism is called "weak" because it does not account for the direction of the edges. When the mapping considers the direction of the edges, then the function becomes a *strong graph homomorphism*, meaning that:

$$(i,j) \in R \quad \text{implies} \quad h_1(i,j) \in h_2(R); \quad \text{and} \qquad \text{(SGH1)}$$

$h_1(i,j) \in h_2(R)$ implies there exists $k, l \in \mathcal{N}$ such that

$$h_1(k) = h_1(i), h_1(l) = h_1(j) \text{ and } (k,l) \in R. \qquad \text{(SGH2)}$$

As a result, in the strong network homomorphism each edge in the graph image is induced by an edge for every different pairs of points in the preimage, and not just by some other edge.

4.2.2   Juncture Graph Homomorphism

To capture the global structure of multiple relational systems, White and Reitz (1983) propose a special type of homomorphism for graphs having different edge sets. This type of homomorphism is based on the notion of "relational bundles," which has been introduced in Chapter 3 Multiplex Network Configurations, and we will see how these patterns can be useful for the modeling of the network relational structure.

In this sense, when an ordered pair of actors (i,j) is associated with a determinate bundle pattern **B**, and if a second ordered pair (k,l) shares the same bundle with the first one, then we have $B_{ij}^{\mathbf{B}} = B_{kl}^{\mathbf{B}}$. Thus for every bundle type there is a *multiplex relation* R_B induced by $\mathcal{G}^+$ where $(i,j) \in R_B$ if and only if $B_{ij} = B$, and a unique bundle associated with each pair of nodes (i,j) that can be any class of **B**. As a result each pair of multiplex relations R_{B_i}, R_{B_j} are disjoint, which means that either $R_{B_i} \cap R_{B_j} = \varnothing$ or $R_{B_i} = R_{B_j}$ holds.

A special case of bundle homomorphism is termed *juncture graph homomorphism* by White and Reitz (1983), which in this case is adapted for the bundle class patterns already introduced. The juncture graph homomorphism is defined as a mapping $h : \mathcal{G} \to \mathcal{H}$ iff:

$h_1(i) = h_1(k)$ and $h_1(j) = h_1(l)$ implies

$$B_{ij} = B_{kl}, \; B_{ij} = B_{ij}^N \text{ or } \; B_{kl} = B_{kl}^N \qquad \text{(JGH)}$$

This means that juncture graph homomorphism requires that two pairs of nodes either shares the same bundle type or that the edge set of one of the pairs belong to the null bundle pattern.

White and Reitz (1983) point out that every strong graph homomorphism is also a juncture homomorphism, and they also claim that juncture graph homomorphism preserves the properties necessary to the description of roles in multiple networks; such conditions are multiplexity, the semigroup of relations, and the composition of relations as well. This means first that the juncture homomorphism h induces a map from $\mathcal{G}^+$ to a multigraph $\mathcal{H}^+$ where $\mathcal{G}^+ = \langle \mathcal{N}, \mathcal{R} \rangle$, $\mathcal{H}^+ = \langle \mathcal{M}, \mathcal{T} \rangle$, and $h = \langle h_1, h_2 \rangle$; and second that if $S = \langle \mathcal{R}, \circ \rangle$ is a semigroup with $\circ$ as a relation composition, then $h_2 : \langle \mathcal{R}, \circ \rangle \to \langle \mathcal{T}, \circ \rangle$ is a isomorphism (p. 210).

In other words, if $S(\mathcal{G}^+)$ and $T(\mathcal{H}^+)$ are the semigroups of $\mathcal{G}^+$ and $\mathcal{H}^+$, and are isomorphic, then the composition of relations is unaffected under the juncture graph homomorphism. However, Everett and Borgatti (1991) show that $T(\mathcal{H}^+)$ and its image are not isomorphic to

$S(\mathcal{G}^+)$ but just homomorphic, since any juncture homomorphism is *regular*—a type of equivalence defined in the next section—and because the network itself usually is not the semigroup generated by $\mathcal{R}$.

*
* *

After defining useful homomorphisms for systems made of relations, it is time now to look at some important types of equivalences for social networks. As already noticed, each graph homomorphism induces an equivalence relation on the domain, and every *type* of graph homomorphism induces a particular kind of equivalence (White and Reitz, 1983, p. 199). This means that the elements in the graph image are partitioned according to the mathematical properties of the homomorphism in use.

Likewise, recall that any equivalence relation induces a partition on the set. As a result of this, by applying an equivalence relation on a social network we create different classes of actors in the system according to the chosen equivalence type. Hence the set of equivalence classes constitutes a partition of the system that in case of multiplex social networks is compatible with $\mathcal{R}$.

The partition of the network in different classes serves to reduce the complexity of the network. However, another essential motivation to apply an equivalence relation on a social network is to be able to operationalize the concept of social role, which is presumed to be played by social positions representing classes of actors. We will develop further the sociological theory of social roles later on, but before that, we are going to introduce some types of equivalences that are defined by different types of homomorphisms, and which are useful for the analysis of social networks. The main concern with the different types of correspondence among the actors is to try to express their role equivalence that is based on their patterns of relationships in the networks.

In the definition of classes of equivalent actors in the network, there are two perspectives to take into account. One perspective considers the overall structure, while the other perspective takes the point of view of individual actors. We will see later on that is possible to combine the two perspectives, but first in the next points we take a look at different types of equivalences from both standpoints.

4.3   Global Equivalences

We start by looking at important types of equivalences for social networks taking into account the perspective of all actors simultaneously within a social system, which means that such kinds of correspondences have a *global perspective*. However, since any collection of equivalent actors in a social network is meant to play a particular role in the system, the global perspective does not necessarily mean that these roles are "perceived" directly by the actors in the network; it only means that the equivalence definition is made by considering the elements of the system as a whole.

In this sense, several types of equivalence have been defined based on different conditions, which try to capture and operationalize the notion of role in a system of relations. The order in the presentation of the equivalence types follows a historical development in the Positional analysis of social networks, and most importantly relaxes the level of stringency in the definition of classes of equivalent actors.

4.3.1 Structural Equivalence

Perhaps the earliest characterization of an equivalence type applied to social networks is *Structural equivalence* (Lorrain and White, 1971) where corresponding nodes are related in identical way to all other nodes in the graph.

A formal definition of Structural equivalence means that two nodes, say i and j in $\mathcal{N}$, are structurally equivalent, which is denoted by $i \overset{SE}{\equiv} j$, if for any $k \in \mathcal{N}$, such $k \neq i, j$, implies:

$$
\begin{array}{rcl}
(i,k) \in R & \Leftrightarrow & (j,k) \in R \\
(k,i) \in R & \Leftrightarrow & (k,j) \in R \\
(i,j) \in R & \Leftrightarrow & (j,i) \in R \\
(i,i) \in R & \Leftrightarrow & (j,j) \in R
\end{array}
\tag{SE}
$$

According to Doreian et al., (2004) these statements represent the "in-neighborhood" and the "out-neighborhood" conditions of the nodes with loops when self-relations are also considered in the network. This means that nodes i and j must have identical adjacent nodes regarding both the incoming and the outgoing edges in order to be structurally equivalent in the system.

White and Reitz (1983, p. 200) (cf. also Reitz and White (1992, p. 438)) interpret the last statement concerning the loops as $(i,i) \in R \Leftrightarrow (i,j) \in R$, which means that the conditions for Structural equivalence is for nodes that are adjacent to each other. On the other hand, Wasserman and Faust (1994, p. 678) consider only the first two statements for Structural equivalence for two nodes that are not directly connected.

The notion of Structural equivalence provided by Lorrain and White was a fundamental seed that initiated the positional analysis of social networks. However, the conditions imposed by the Structural equivalence concept are too stringent to effectively apply to empirical social networks. It is unlikely that two actors have identical neighbors in real social networks for a various reasons. We human beings need to construct an identity of our own, and in addition we experience a diversity of circumstances in our social life, which make us to have different social circles. This means that the conditions imposed by the definition of Structural equivalence need to be relaxed in order to be effectively applicable to the analysis of human social networks.

In this sense, we present next important generalizations of Structural equivalence that are more suitable to represent patterns of social relations among sets of actors with structurally equivalent locations in a given network of relations.

4.3.2 Automorphic Equivalence

Winship (1988) argued in year 1974 that nodes are equivalent as well when they relate to the same *kinds* of nodes, rather than to exactly the same ones. Thus based on this idea, a fundamental abstraction of Structural equivalence is found in *Automorphic equivalence* (Winship and Mandel, 1983; Everett, 1985), where the graph nodes are relabeled to form an isomorphic graph.

Automorphic equivalent nodes then can be interchanged and still sharing the properties that are label independent. Since Automorphic equivalence is defined in terms of graph isomorphism, this type of equivalence is also known as *structural isomorphism*, and it is also considered a natural generalization of weak structural equivalence by Everett and Borgatti (1994).

Formally, nodes i and j are automorphic equivalent, or $i \overset{AE}{\equiv} j$, if the mapping $h : \mathscr{G} \to \mathscr{H}$ is a graph isomorphism for all $i, j \in \mathscr{N}$ where:

$$(i, j) \in R \quad \text{implies} \quad (h(i), h(j)) \in R \tag{AE}$$

An alternative definition considers nodes i and j as automorphic equivalent if h is an automorphism of $\mathscr{G}$ with

$$h(i) = j. \tag{AE}$$

As a result, i and j belong to the same *orbit* of $\mathscr{G}$, and these nodes are undistinguished in terms of graph structure when they are unlabeled. Naturally, all structurally equivalent nodes are also automorphically equivalent.

Borgatti, Boyd and Everett (1989), and also Pattison (1993) noticed that Automorphic equivalence relations are automorphic irreducible in the sense that the only homomorphism is the identity mapping and automorphic equivalent are "primary" substitutable. However, it is possible to generalize this type of correspondence by applying it recursively to the set of Automorphic equivalence classes. If these nodes are also substitutable, they are "secondary" substitutable and they induce to an *extended* or *iterated* version of Automorphic equivalence, which is not sensitive to the nodes degree.

The notion of Automorphic equivalence has a nice and clear mathematical definition, and it has been pointed out that the classes of automorphic nodes are useful representations of role similarity (Everett, 1985), where the set of orbits of a graph represents a partition into role equivalent actors (Everett et al., 1990). However, the asymmetry of ties and other irregularities found in social networks make this type of equivalence hard to be applicable to detect social roles in the system. This means that it is required even more generalization in the definition of the equivalence; one that considers the direction of the ties and which is sufficiently flexible to capture the patterns of relations found in empirical social networks.

4.3.3 Regular Equivalence

A further generalization of both Structural and Automorphic equivalence is found in *Regular equivalence*. The concept of Regular equivalence was first suggested informally by Boyd as "structural relatedness" (cf. Sailer, 1978, p. 78), and when Sailer 1978 wrote a critique of structural equivalence; they noticed that the joint occupancy of social position or performance of social role derives from the fact that the occupants have structurally similar worlds rather than identical ones. Since Regular equivalence appears to express more accurate the notion of a social role in the system, Everett and Borgatti (1991) term this kind of equivalence as "role assignment."

Formally, nodes i and j are regular equivalent (White and Reitz, 1983), or $i \overset{RE}{\equiv} j$, if for any k and some $l \in \mathscr{N}$:

$$(i, k) \in R \quad \text{implies} \quad (j, l) \in R, \text{ and } l \overset{RE}{\equiv} k$$
$$(j, k) \in R \quad \text{implies} \quad (i, l) \in R, \text{ and } l \overset{RE}{\equiv} k \tag{RE}$$

As with to the definition of Structural equivalence, these are conditions of both "in-neighborhood" and "out-neighborhood" but now in the Regular equivalence context, which means that this time nodes are required just to be related in a similar way to matching pairs of nodes rather than identical ones.

It is important to mention, however, that there is not always a unique optimal partition when applying Regular equivalence in the graph. All structurally equivalent actors are also regular equivalent, but—as noted by Borgatti and Everett (1989), Borgatti et al. (1989)—since any Regular equivalence induces a "regular partition," there is a variety of regular equivalences, and one might consider the "best" model for the analysis. This decision certainty tries disregarding trivial partitions such as the complete partition and the equality partition, and in the case of digraphs having just those partitions, it is perhaps better to have a local perspective as the only reasonable technique in extracting information of roles and positions (Everett et al., 1990, p. 168).

4.3.4 Generalized Equivalence

Batagelj et al. (1992) and Doreian et al. (2004) noticed that both Structural equivalence and Regular equivalence induce special "types of connections" between sets of equivalent nodes, which they call *clusters*.[1] Such connection types are described by the predicates that have to be satisfied by a *block* $(X, Y) \in R$, which specifies the relation between two clusters of equivalent nodes. For instance, if $(X, Y) \notin R$ then the block is termed null, and in the case that for all $x \in X$, $y \in Y$, and $x \neq y$ implies that $(x, y) \in R$, then the block is regarded as complete or com.

In this sense, a *clustering* is a special type of partition of the network that is *structural* if and only if each of its blocks is either null or a complete type. However, one can relax this circumstance and the clustering can be regarded as *regular* when reg blocks are also permitted. A regular block has at least a tie in each row and column, and this formally means that for all $x \in X$, and for some $y \in Y$ implies that $(x, y) \in R$ and $(x, y) \in R^{-1}$, which are the "row-regular" and the "column-regular" conditions respectively.

The mentioned authors propose other kinds of block types to describe the relation between clusters, but for most of the purposes it is sufficient to consider a set **C** with three block types: null, com, and reg. As a result, a *Generalized equivalence* is a correspondence over $\mathscr{G}$ that is compatible with **C** if and only if for all $x, y \in \mathscr{N}$ there are some $C \in \mathbf{C}$ such that:

$$([x], [y]) \in C \tag{GE}$$

where [x] and [y] are clusters of equivalent units where Generalized equivalence is underlying the partition, and C is a block type defined in **C**.

Boyd (2002) considers the Generalized equivalence as "approximately" Regular equivalence, and a more accessible equivalence type for social networks. Empirically, this sort of equivalence is one of the most used in recent analysis of social networks. Batagelj et al. (1992) and Doreian et al. (2004) propose a special type of homomorphism aimed to perform a positional analysis in networks as the "generalized blockmodeling" procedure that is based on selecting the block types.

---

[1] We do not have to confuse this term with the cluster semiring, which is the subject of Chapter 7. Although both procedures have the same purposes, namely to find classes of structurally equivalent actors, they emphasize different aspects of the relational structure.

4.4 Global Equivalences Applied

We illustrate the global equivalence types introduced above in Figure 4.1, and we take again the small configuration example from Table 3.5 in Chapter 3 Multiplex Network Configurations, and which was called $\mathscr{X}_Z$. For the sake of simplicity, we start by just considering one type of relation occurring in this system that is F and we also apply a symmetric closure to the ties in order to obtain an undirected graph representation. This latter aspect is important because both Structural and Automorphic equivalences have very strict conditions and there are no actors in this network having identical ties with matching neighbors. This means that the type of homomorphism is in this case "weak," which is in contrast with the strong homomorphism applied to digraphs or directed networks.

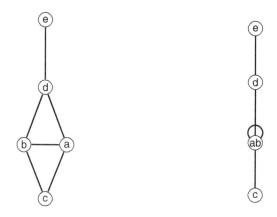

(a) Structural and Automorphic equivalences

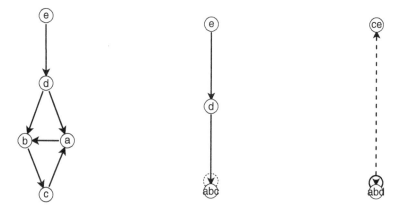

(b) Regular and Generalized equivalences

Figure 4.1 Global equivalence types in network $\mathscr{X}_Z$. *The stricter versions of equivalence* SE *and* AE *are given as weak graph homomorphisms, while* RE *and* GE *are given as strong graph homomorphisms where solid and dotted edges stand for complete and regular blocks.*

In case we consider a weak graph homomorphism on this configuration with no distinction on the direction of the ties as in Figure 4.1a, we can see that elements a and b have both identical edges with the same nodes. This means that these elements are structurally equivalent and they belong to a same class. Similarly, it is also possible to interchange the labels of these elements and yet get a graph isomorphism, which means that these nodes are automorphic equivalent as well. In this sense, a reduced model of the configuration in terms of both Structural and Automorphic equivalence is given to the right of the graph. This image structure has a chain of relations among the classes of elements and a self-relation for the internal tie existing between the two equivalent elements a and b in the class.

The conditions for Regular equivalence are more relaxed, and in this case it is possible to find equivalent elements in the directed graph. Figure 4.1b presents the directed graph of the example and considers two possible images: an image with three classes of nodes in the digraph, plus another image having only two classes of regular equivalent nodes. The labelling in the edges of the images indicates in both cases the type of "block" that exists between the classes in the Generalized equivalence and, as we have seen before, there are also self-relations among some of the classes of approximately regular nodes.

The Tables in 4.1 present the permuted and partitioned matrices for the system just depicted for the homomorphisms that correspond to the global equivalence types considered for this small network. In the case of structural equivalence, this model only permits one to have blocks of relations with either a perfect fit or a zeroblock, which in the network analysis literature are called as *fat-fit* and *lean-fit*, respectively. However, blocks of relations among regular equivalent

Table 4.1 Global equivalence types with Image matrices of network $\mathscr{X}_Z$. *There are two partitions of the network with a pair of equivalence types each.*

	a	b	c	d	e
a	0	1	1	1	0
b	1	0	1	1	0
c	1	1	0	0	0
d	1	1	0	0	1
e	0	0	0	1	0

	a	b	c	d	e
a	0	1	0	0	0
b	0	0	1	0	0
c	1	0	0	0	0
d	1	1	0	0	0
e	0	0	0	1	0

1	1	1	0
1	0	0	0
1	0	0	1
0	0	1	0

reg	null	null
com	null	null
null	com	null

(a) Structural and Automorphic
equivalences

(b) Regular and Generalized
equivalences (3 clusters)

elements do not require perfect fit but just the presence of ties in all the rows and columns inside the block relation.

The partition of the network then expresses the categorization of equivalent actors in terms of the defined correspondence. There is also an image matrix below each table that represent the existing relations in a system of classes of equivalent nodes in each graph, configurations express blocks of relations among the different classes of actors in the system. Since the self-relations are ignored in this case, these reduced models include two types of blocks of relations for structural and automorphic equivalence, and three types of block relations for regular and generalized equivalence.

If we take the Generalized equivalence from Table 4.1b, for instance, we can see that the model with three classes of actors has an incongruent tie that is where the block is considered as com, whereas there is no incongruence with the model with only two classes (whose matrix is not shown here). However, one might choose a model based on substantial reasons, even if such model has a small margin of error, since that form of representation reflects in a more accurate way the relational structure of the network. In this sense, the possibility of having different models in the regular equivalence representation could constitute a potential source of ambiguity in the analysis, and different kinds of criteria play in the decision to choose the best representation of the network of relations.

One of the main goals with the different equivalence types is to find positions in the network made of actor who are meant to play a similar role, and therefore the equivalence types try to mimic the role equivalence. As the smallest image in Figure 4.1 shows, non-adjacent nodes can belong to the same class, and this is because of the particular way they are connected to the rest of the system. Yet in order to have a full representation of the structure for a multiple network, we need to account for different types of relations, and this means that the analysis of $\mathscr{X}_Z$ must include also the existing collaboration tie C in the modeling.

A possibility to account for the multiplexity of ties could be to collapse the two types of relations in the blockmodeling of the example, and thus regard the entrainment of the two types of ties as a single edge, which in this case would lead to the same results. However, there are at least two difficulties with such decision: One is the ambiguity of the ties when there is an exchange of different types of tie, and the other aspect is that there is a neglecting of the multiplexity of the relational bonds by collapsing the levels in the relational structure. Certainly, there must be other options by performing separately the modeling of the multiple ties, and this is in order to consider the concordance in the different bundle class patterns, or else to generalize in a certain way the global equivalence types just reviewed.

From Chapter 3, we know that it is possible to obtain an integrated system with the semigroup structure, which is not only made by the different types of ties, but also by the combination of the diverse paths of relations. Moreover, an algebraic analysis permit one to include the attributes of the actors in a relational way and it is possible to perform a more complete analysis of the empirical networks and the attributes of the actors in the form of innovation adoption.

In this sense, another approach to the role equivalence among actors is by looking at the algebraic models associated with the system of relations. By accounting simultaneously for the chain of relations that an individual actor has in the network it is possible to characterize the manner in which that the actor is embedded in the system, where those actors having similar set of ties—whether they are direct or compound—are meant to be playing a similar structural role in the network. Therefore in the following point we take an algebraic approach to the analysis of roles and positions in a multirelational system.

4.5 Local Equivalences

Local equivalences differ from the global correspondence types in the sense that this time the definition of the equivalence is made from the perspective of the individual actors in the network rather than the general structure of the social system. However, we will see that it is possible to combine both angles and come up with an intermediary type of equivalence among the actors.

4.5.1 Relation-Box $\mathbf{R}(\mathbf{W})$

To recognize local equivalences among the actors within the algebraic framework, we rely on a three-dimensional array similar to the adjacency matrices where the primitive ties of the network and compounds relations—whether they are unique or not—are "stacked together". A partially ordered structure by increasing relation similar to the array shown in Figure 4.2 has been proposed by Winship and Mandel (1983) for the definition of local equivalences, and they called this device a *Relation-Box*, which here is denoted as $\mathbf{R}$ (cf. also Mandel, 1983; Reitz and White, 1992). Hence, the Relation-Box is a device that is built up from the set of adjacency matrices of the generating ties together with the indirect relation matrices product of composition of these generators and their string products. As a result, $\mathbf{R}$ resembles to $\mathbf{A}(\mathscr{R})$ of a network $\mathscr{X}$ but now with the set of compound relations as well.

In principle, the set of compound relations can be infinite in length; however, since the number of primitive relations is finite, much of the compound relations will result identical in form. Then it is possible to apply the Axiom of Quality by equating the matching compound relations, and

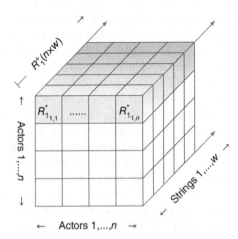

Figure 4.2 Relation-Box $\mathbf{R}(\mathbf{W})$ with an emphasized Relation Plane R_i^+, and its Role Relations $R_{i_{x,j}}^*$. *The unrestricted version of* $\mathbf{R}(\mathbf{W})$ *has dimensions* $n \times n \times \infty$.

have in this way a Relation-Box of finite size with the set of unique relations. However, even with relatively small multiplex networks with three or more types of relations, the total number of unique compound relations that results is still relative large and as well both the size of the Relation-Box and the length of the Relation Plane.

One mode to tackle this "issue" is by using the argument that humans are not aware of very long chain of relations, and such long indirect ties are not capable of having a practical influence in the network either. This is perhaps true in societies where the extended family is a strong institution some long kinship relations may have an effect on the individual person. In this sense, Winship and Mandel (1983) argued in their analysis of local role algebra that it is probably not necessary to account for chain relations that are no longer than three or four. However, the term "local role algebra" implies that the constituent structure corresponds to an individual actor of the network, and there is a type of structural correspondence between actor that is based on these structures and is defined later on in this chapter.

The argument of Winship and Mandel means that one can restrict the analysis on a *truncated* version of the Relation-Box $\mathbf{R}(\mathbf{W}_k)$ with ties up to a certain length k imposed on substantial reasons. This system is in opposition of the *unrestricted* version of the Relation-Box $\mathbf{R}(\mathbf{W})$ where the analysis of multirelational structures is without a cut off value on the length of the chain of relations.

In this sense, a Relation-Box $\mathbf{R}(X)$, being $\mathscr{S}_X$ the relational structure of a network X, and n the number of actors in X,

$$w = \text{number of matrices in } \mathbf{R}(X)$$

or w represents the length of the third dimension of $\mathbf{R}(X)$.

In the truncated version of the Relation-Box $\mathbf{R}(X_k)$,

$$w = \text{number of primitive and compound relations in } \mathscr{S}$$

and in the unrestricted version of the Relation-Box $\mathbf{R}(X)$,

$$w = \infty.$$

4.5.2 Relation Plane and Role Relations in $\mathbf{R}(\mathbf{W})$

The Relation-Box provides essential information about the individual roles and the chains of relations in the network, and there are important characteristics of this configuration that serves as the basis for the structural analysis of multiple networks. For instance, in Figure 4.2 the shadowed horizontal "slice" across the string relations for the outgoing ties of a single actor (the first one in this case) which reflects the actor's activity linked to the rest of the members through the different string relations, both primitives and compounds, that are occurring in the network.

Each horizontal slice in the Relation-Box is called a *Relation Plane*, denoted as R^+l, and encodes the distinct primitive and compound relations that a single actor l has with the rest of the network members. There are n Relation Planes R^+ in total of size $n \times w$ in a Relation-Box where the value of w depends on whether $\mathbf{R}(X)$ is unrestricted or truncated.

Each row in the Relation Plane is a *Relation Vector*, which represents the collection of distinct primitive and compound relations for a distinguished individual actor. On the other hand, the column of R^+ corresponds to the actor ties of a particular kind to the rest of the actors in the network, and it represents the *Role Relation* of the actor with size $1 \times w$ and that is denoted by R^*. For each network member l in a given network X, there is a Role Relation, represented by $R^*_{l_{x,j}}$, with actor j on relation x. An individual actor has a total of n Role Relations in its Relation Plane including the relation to itself, and since there are n Relational Planes in $R(X)$ there are in total n^2 Role Relations in the Relation-Box.

Role Relations characterize the direct and indirect ties existing between a pair of actors, and they result from the roles that they occupy with respect to each other in the network. However, for each individual actor in the social structure some Role Relations are identical, and it is possible to extract the set of non-identical Role Relations. In this sense, the collection of n distinct relations for a particular actor when duplicates are removed will constitute the *Role Set* R^*_l of actor l (Winship and Mandel, 1983; Wasserman and Faust, 1994).

The distinct Role Sets of the individual actors play a significant part in the establishment of equivalence types from a local perspective, and the definition of this concept is very similar to the concept of "Role Set" used by Merton (1957). Some of the consequences of the distinct Role Sets in multiplex network structures are expanded later on in this chapter, and also in Chapter 5; however, we first continue with the definition of some equivalence types that are from a "local perspective" this time.

4.5.3   Local Role Equivalence

A direct consequence of the Relation-Box leads to one of the earliest definitions of equivalence in social networks, and that is from an algebraic perspective. This type of equivalence takes into account two significant elements of $R(W)$: the Role Relation of an individual actor i in a network X, and its respective Role Set.

One kind of equivalence that is a direct consequence of the Role Relation and Role Set in $R(W)$ is *Local Role equivalence* (Winship and Mandel, 1983; Mandel, 1983) or $\overset{\text{LRE}}{\equiv}$. This local correspondence type is formally defined for nodes i and j that are meant to be *Local Role equivalent*, or $i \overset{\text{LRE}}{\equiv} j$ on relation $x \in X$, if and only if:

$$
\begin{aligned}
&\text{there exists some } R^*_{j_{x,l}} \in R^*_j \text{ such that } R^*_{i_{x,k}} = R^*_{j_{x,l}}, \text{ for all } R^*_{i_{x,k}} \in R^*_i \\
&\text{there exists some } R^*_{i_{x,k}} \in R^*_i \text{ such that } R^*_{j_{x,l}} = R^*_{i_{x,k}}, \text{ for all } R^*_{j_{x,l}} \in R^*_j
\end{aligned}
\tag{LRE}
$$

That is, for every Role Relation related to each individual actor in the network, there is at least one (but usually more) identical Role Relations associated with the other individual, and the Role Sets of role equivalent nodes contain the same Role Relation (Borgatti and Everett, 1989, p. 81).

This actually means that such nodes are Local Role equivalent if the pair of nodes (i, k) has the same Role Relation with a pair (j, l). As a result, for all $R^*_{i_{x,k}} \in R^*_i$ and some $R^*_{j_{x,l}} \in R^*_j$ nodes i and j are Local Role equivalent iff:

$$
R^*_{i_{x,k}} = R^*_{j_{x,l}}
\tag{LRE}
$$

Or even more generally, nodes i and j are equivalent iff they have identical Role Sets:

$$
R^*_i = R^*_j
\tag{LRE}
$$

Local Role equivalence is also a way to characterize social roles in incomplete or ego-centerd networks while preserving the distinction of diverse types of relationships. Winship and Mandel (1983) point out that Local Role equivalence is a generalization of Automorphic equivalence (cf. AE), given that both kinds of equivalence involve the same types of role relations. However, the Automorphic equivalence condition in addition requires the same number of role relations, and hence the same Role Sets and local role algebras (Pattison, 1993).

4.6   Compositional Equivalence

Following the idea of the individuals' Role Relation and Role Set, another type of equivalence that is suitable for chains of relations of different kinds is *Compositional equivalence* (Breiger and Pattison, 1986, also Mandel 1978). Compositional equivalence or $\overset{CE}{\equiv}$ has some important characteristics to take into account. First, it is built on the algebra of relational structures and more particularly on local role algebras, and second is that Compositional equivalence works both at the "local" and at the "global" level.

Breiger and Pattison (1986) point out that this equivalence type "shifts from the global analysis of relational interlock to see how the algebra impinges on the perspective of individual actors who are embedded in a multiple network of relations" (p. 219). Because there is an algebraic emphasis in the modeling, Compositional equivalence is also called as "ego algebra" (Wasserman and Faust, 1994), and this name is also given because the establishment of structurally equivalent actors is based on the network local role algebras (Breiger and Pattison, 1986).

With Compositional equivalence, the analysis of local roles takes the information expressed in the different Relation Planes of the Relation-Box corresponding to particular network members, and whose rows and columns represent—according to Breiger and Pattison—the *dual structure* of the actors and their relations. While such aspect characterizes the local role equivalence type, there is a step forward from a local perspective in the equivalence definition since all the information from particular role relations is generalized to the entire network structure.

The fact that Compositional equivalence generalizes local roles to the entire system implies that this type of correspondence works both at the local and at a "global" level. That is, the establishment of roles and positions in the network are from the perspectives of individual actors, whereas the characterization itself of equivalence is made by considering the relational features that are common to all members in the network. This last feature, though, works better with middle size networks, and hence Compositional equivalence can be regarded as a local to "middle-range" type of correspondence.

The local portion of Compositional equivalence lies in the particular views of the actors in the system, which is in terms of inclusions among the Role Relations of the actors' immediate neighbors. Recall that the Role Relations are recorded in the columns of the individual Relation Planes, which means that there are in total n^2 of these vectors in the network, one for each actor in every Relation Plane of size $w \times n$ with string relations up to length k.

Isolated actors in networks are unable to "see" any type of relationships among other actors through the defined links. This implies that Role Relations for isolates are empty no matter the type of tie or its length, and that any Role Relations in the Relation Plane are blank as well. However, connected actors have a different perspective about whether or not there is an inclusion between the other actors. The collection of inclusions (or lack of them) for each actor or class is reflected in the actor Person Hierarchy, which is a square array size $n \times n$ belonging to this entity and is denoted by **H**.

However, substantially the awareness of compound relations on the actors becomes less as the different paths increase in length. In this sense, paths of relations that are longer than three should not have much effect on the relational structure, particularly within an economic context for example. As a result, the analysis of the Relation-Box typically starts with compounds whose length is no longer than three or $k = 3$ in $\mathbf{R}(\mathbf{W}_k)$, and this should be enough to express the entire relational or role structure of a given multiplex network.

4.6.1   Formal Definition of Compositional Equivalence

In order to define the Compositional equivalence in formal terms, we still use the Relation-Box device $\mathbf{R}(\mathbf{W})$ as the common underlying structure for a network partition. However, since the individual perspectives of the actors are also considered in the establishment of the equivalence type, it is also required in this case to take a reference node from the network, which is called as the *ego* in this context.

As a result, formally nodes j and k that a given multiplex structure $\mathscr{G}^+$ are *Compositional equivalent with respect to* the ith ego; that is $j \overset{CE}{\equiv} k \bmod i$ for a relation $x \in \mathscr{X}$, if and only if for each Role Relation R^*:

$$R^*_{i_{xj}} \text{ implies } R^*_{i_{xk}}, \text{ and } R^*_{i_{xk}} \text{ implies } R^*_{i_{xj}},$$

$$\text{and there exist at least one } R^*_{i_{xj}} \tag{CE}$$

This means that both the equivalences among relations and the composition of ties are equal from the standpoint of each actor.

Hence two relation types R_1 and $R_2 \in \mathscr{R}$ are indistinguishable from the perspective of actor i if and only if both types of relations coincide as incoming and outgoing ties to a given actor. That is iff for $j = 1, 2, \ldots, n$:

$$(i,j) \in R_1 \text{ iff } (i,j) \in R_2 \text{ and } (j,i) \in R_1 \text{ and iff } (j,i) \in R_2. \tag{CE}$$

Yet a more algebraic approach to the definition of Compositional equivalence is based on both the network role structure and on the role algebra of the actors. Although these topics are subjects in the next section, we note that while the local role algebra characterizes the composition of relations from the perspective of the actor, the role structure describes how relations are associated in the network independently of the particular individuals involved. The representation of the local role structure is based on the right multiplication table that contains all of the relationship types, whereas the partially ordered semigroup represents the network role structure.

Since each possible direct and indirect tie in the relational system is contained in the free semigroup of relations, then one would expect that the number of relations from i's perspective is less (or maybe equal) to the total relations present in the semigroup of relations $S(R)$. Most importantly, it means that all relations in S that are identical from the perspective of an actor i is assigned to the same class in $S(R)$.

Subsequently, we get a partition of the semigroup based on i's perspective, which is represented by $S(R)_i$ and is termed the *ego algebra* of i. If the partition of $S(R)$ by actors j and k produces identical right multiplication tables, we can consider these two actors as correspondent in terms of Compositional equivalence as well.

As a result, nodes j and k are Compositional equivalent or $j \overset{CE}{\equiv} k$ if and only if they have both equal ego algebra and identical local role algebra; that is iff:

$$S(R)_j = S(R)_k \tag{CE}$$

*
* *

Although both Local Role equivalence and Compositional equivalence search for an abstract characterization of local roles, the notion of Compositional equivalence generalizes the Structural equivalence (cf. (SE)) concept but still working with specific individuals (Breiger and Pattison, 1986, p. 254). This makes the Compositional equivalence a less abstract type of correspondence than for instance the Regular equivalence concept (cf. RE), and a useful alternative in the search of equivalences that is valid at the middle and global level that considers at the same time local structural characteristics of the network. In the next section, we are going to apply the Compositional equivalence notion to reduce multiplex network structures in different ways depending on the characteristics of the social system.

4.7   Positional Analysis with Compositional Equivalence

A key constituent in establishing a system of collective action of the network is the Role Set R^*, and a fundamental step in capturing the essential structure of a multirelational system is the ability to group the network members according to their patterns of relationships having a similar Role Set. As a result, the main reason to choose the notion of Compositional equivalence for the analysis is because this type of correspondence not only permits the modeling of relational ties at different levels, but it also allows including paths of relations in the modeling process.

In order to perform a positional analysis of multiplex networks in terms of Compositional equivalence, we need to define the set of primitive relations of the network, and we will see that there are some strategies to capture this information in the modeling when dealing with complex network structures with actor attributes. However, we look first in the next point at a partial order structure that is linked to the Relation-Box of Figure 4.2, and which plays a crucial part in the operationalization of Compositional equivalence.

4.7.1   *Cumulated Person Hierarchy, $\mathcal{H}$*

The Relation Plane in a Relation-Box $\mathbf{R}(X)$ carries significant information in the definition of Compositional equivalence, particularly the notions of Role Relation and Role Set, which are part of the Relation Plane R^+. For instance, Breiger and Pattison (1986) point out that the Relation Plane encodes in the rows and its columns the dual structure of actors and relations. The set of all row inclusions forms a partial order that is the ego's *Relation Hierarchy*, whereas the set of ego or person's inclusions as a partial order constitutes the *Person Hierarchy* of the ego (p. 228).

The Compositional equivalence procedure relies on the existing Person Hierarchies of the social system, which we denote as **H**, which means that by means of Compositional equivalence we seek to classify the actors in the multiplex network rather than the relations occurring between them. A further step in the analysis will look at the structure made of links between the resulting classes of actors, and at this phase of the analysis we are most concerned most with the structure resulting from the different types of relations among the actors. Hence, the hierarchy among actors is most relevant for the Compositional equivalence procedure that is obtained by searching equations in the different Person Hierarchies of the network.

In more formal terms, from the standpoint of a given actor l, actor i is "contained within" actor j whenever there is a string x between l and i, there is a same type string between l and j (cf. Breiger and Pattison (1986, p. 229)). Furthermore, the collection of all perceived inclusions in R_l^+ represents the Person Hierarchy $\mathbf{H}_l$, which is defined for actors $l, i, j \in X$ and relation x as:

$$
\mathbf{H}_{l_{ij}} = \begin{cases} 1 & \text{iff } R_{l_{xi}}^* \leq R_{l_{xj}}^* \\ 0 & \text{iff } R_{l_{xi}}^* \nleq R_{l_{xj}}^* \\ 0 & \text{iff } \sum R_{l_{xi}}^* = 0 \end{cases}
$$

The last proposition implies that there is no inclusion between actors i and j in the Person Hierarchy of l, and this is simply because actor i has an empty Role Set. Notice as well that there is a perceived containment among actors in a given relational plane if their role relations are identical. That is, $\mathbf{H}_{l_{ij}} = 1$ if and only if $R_{l_{xi}}^* = R_{l_{xj}}^*$.

The global part of Compositional equivalence occurs with the union of the different personal hierarchies into a *Cumulated Person Hierarchy* across actors, which is denoted by $\mathscr{H}$. This means that $\mathscr{H}$ is represented by a single square matrix of size $n \times n$ having the properties of a partially ordered structure, namely reflexive, antisymmetric and transitive. The structural information in the Cumulated Person Hierarchy lays the foundations for categorizing the actors and performing a reduction of the network that—as Breiger and Pattison pointed out—comes from the *zeroes* or the absence of inclusions among the different actors.

The partition of the network itself is then a product of a global type of equivalence that is performed on the Cumulated Person Hierarchy $\mathscr{H}$. However, we should bear in mind that the matrix $\mathscr{H}$ does not represent social ties as in the adjacency matrix **A**, but constitutes a partial order structure indicating the lack of containments among the network members. Hence, we assess classes of actors in the network according to their placement in such graded system that can be visualized through a lattice structure that is aimed for partially ordered sets.

As a result, while elements in **A** disconnected to the rest of the members in the adjacency matrix represent isolated actors in the social network, the actors in $\mathscr{H}$ that are not related to any other actors in this partial order structure except for themselves are said to be *incomparable* elements in the Cumulated Person Hierarchy. One consequence of this is that incomparable actors in $\mathscr{H}$ make a class of structurally equivalent actors.

4.7.2   Set of Generators in Complex Networks

When dealing with complex network structures, most of the times we need to account for the system essential structure while we try to reduce the complexity of the network through the employment of different procedures. For instance, recall that one procedure to reduce the complexity of a given relational system while highlighting its essential structure is positional analysis. With positional analysis, we try to locate classes of structurally equivalent actors who are meant to perform a similar role in the network from a common structural position, and this is done even if the actors are linked with different types of relations.

Another strategy for the analysis of complex directed network structures that can be used alongside the positional analysis is by creating some sort of "symmetry" in the relationships. This is because symmetric objects are easier to interpret than asymmetric ones, and complex structures product of directed multiplex networks are typically asymmetric. To take care of this asymmetry in the relations we employ a *relational contrast* to the directed ties since it serves to counteract the asymmetries in the relational structure. The relational contrast is achieved by the converse operation applied to the ties. Hence, the converse of a relation type R is obtained by the transposition of the relation set, i.e. R^T that is most conveniently expressed for the entire network by the transposition of the adjacency matrix $\mathbf{A}(R)^T$.

For instance, to perform a positional analysis and the modeling of a multiplex network we take the small configuration $\mathscr{X}_Z$ in Table 3.5, which has been used as example in Chapter 3 Multiplex Network Configurations. Recall that we count in this case with a set of primitive relations made of relations C and F representing two distinct ties that stand for collaboration and friendship relations among the actors. If we account for the tie transpose on the set of primitive relations, which is a common practice for, and we end up with an alphabet set that is made of four elements, which are:

$$\Sigma_Z = \{ \ \mathsf{C}, \ \mathsf{F}, \ \mathsf{C}^{-1}, \mathsf{F}^{-1} \ \}.$$

As a result, the set of primitive relations for the modeling of the members in $\mathscr{X}_Z$ by means of Compositional equivalence with relational contrast is:

$$\Sigma_Z = \{ \ \mathsf{C}, \ \mathsf{F}, \ \mathsf{D}, \ \mathsf{G} \ \}.$$

As an illustration, we take the two types of relationships occurring in the small configuration $\mathscr{X}_Z$. Thus from collaboration and friendship represented by C and F, we obtain relations D and G that are given in matrix form in Table 4.2 for this small example.

Consequently, in the tie converse representation for relation C in Table 4.2, actor b "receives collaboration" *from* actor c, and the relational content in this case is equal than its transpose. Similarly, actor a "is pointed as a friend" by actors d and c, and the same situation occurs for the other actors, except for e for this particular type of tie. In this sense, by transposing the entire adjacency matrices for C, F in $\mathscr{X}_Z$, we get the relational contrast for the application of the Compositional equivalence to this multiplex network.

Compositional equivalence allows constructing a positional system of the multiplex network, which can be represented by a role structure where social roles are related with each other.

Table 4.2 Converse of the ties C and F for network $\mathscr{X}_2$. *Tie converse is an operation made to achieve relational contrast in directed systems.*

	a	b	c	d	e			a	b	c	d	e
a	0	0	0	0	0		a	0	0	1	1	0
b	0	0	0	0	0		b	1	0	0	1	0
c	0	1	0	0	0		c	0	1	0	0	0
d	0	0	0	0	0		d	0	0	0	0	1
e	0	0	0	0	0		e	0	0	0	0	0

| D | G |

When constructing the role structure of a given directed multiplex network by a partially ordered semigroup structure, Boyd (2000) points out that the tie converses ensure that the semigroup of relations is closed with respect to the inverse (cf. also Boyd and Everett, 1999).

4.7.3   Incorporating Actor Attributes

An important aspect when dealing with social networks, whether they are made of different types of relations or not, is the ability to incorporate actor attributes in the modeling of the network relational structure. In this case, the actor characteristics do not need to have a structural character, which means that the traits that are inherent to the actors do not depend directly on their embedment in the network such as individual centrality measures, dyadic covariates, etc. On the other hand, although actor attributes can be independent variables, they are not ascribed to the actors in the same way as age, gender or other demographic information, but it they are governed by the action of the actors themselves. The belief is that such kind of actor attributes should be part of the modeling of the network positional system and also of the establishment of role structure when the attribute has a structural effect, which are all cases when individual attributes are neither shared by all network members nor by just none of the actors. In other words, an actor attribute has a structural effect when is not represented by the identity or the null matrix that are two extreme cases.

The incorporation of the changing attributes of the actors in the relational structure of the network implies that subjects sharing a characteristic constitute a subset of self-reflexive ties associated with the social system. In formal terms, actor attributes are to be represented by the elements of a *diagonal matrix* A^α where each value is defined as:

$$a_{ij}^\alpha = c_i \delta_{ij}.$$

Accordingly, for a given attribute defined in α, and for $i = x_1, x_2, ..., x_n$, the possible values of the first variable in the right hand expression are:

$$c_i = \begin{cases} 1 & \text{if the corresponding attribute is tied to actor } i \\ 0 & \text{otherwise.} \end{cases}$$

On the other hand, δ_{ij} is defined for nodes $i, j = x_1, x_2, ..., x_n$ in $\mathbf{X}$ by the delta function or Kronecker delta as:

$$\delta_{ij} = \begin{cases} 1 & \text{for } i = j \\ 0 & \text{for } i \neq j. \end{cases}$$

Hence, the general representation of A^α constitutes a diagonal matrix with the form:

$$\begin{pmatrix} c_1 & 0 & \cdots & 0 \\ 0 & c_2 & \cdots & 0 \\ \vdots & \vdots & \ddots & \vdots \\ 0 & 0 & \cdots & c_n \end{pmatrix}$$

that records as self-relationships the attributes of the total number of actors in the system. In other words, the "possession" of the attribute produces a reflexive closure in the respective element of the system.

The establishment of the indexed diagonal matrix implies that each type of attribute considered for the actors in the network is represented by its own array, and it constitutes an additional generator to the relational structure. In case that all network members share a given attribute, the result is the identity matrix without any structural effect, whereas in the case where none of the actors possesses the characteristic, the representation is the null matrix with an annihilating effect, where no composition is possible.

Clearly, we are mainly interested in the differentiation of the actors who share an attribute as opposed to those who do not share the trait, and we avoid having the identity and the zero matrices as generator relations for the attribute-based information. This is because the resulting matrix that is neither a neutral nor an absorbing element in the algebraic structure has structuring consequences in the network relational system.

It is important to note as well that with a matrix form for representation of actor attributes, a social tie many times cannot be contained within an attribute relation. The reason is evident that low dimensional arrays with empty off-diagonal cells can eventually be contained in an indexed diagonal matrix representing actor attributes. This means that there are no ties occurring between the actors in the ordering of the elements of the relational system.

As a result, the alphabet corresponding to network $\mathscr{X}_Z$ with the social relations and actor attributes, which is represented by $\mathbf{A}$, has the following primitive relations:

$$\Sigma_Z = \{ \text{ C, F, A, D, G } \}.$$

These five string relations then constitute the generators to produce the relational structure for this particular network $\mathscr{X}_Z$ with attribute-based information where the combination of these strings with the Compositional equivalence notion constitute as well part of the Relation-Box structure. Both the distinct Person Hierarchies and the Cumulated Person Hierarchy, provide the basis of the network partition, and they are built from the structure inherent in the Relation-Box device.

In the next chapter, we take a more closely look at the whole process of establishing positional systems in empirical social networks made of multiplex kinds of tie. This form for network reduction is going to be done by means of Compositional equivalence, which is a correspondence type that allows us to incorporate the multiplicity of the ties in the modeling and even actor attributes. Then we construct the aggregated relational structure of the network where the distinct social roles among positions are related with each other. Such type of structures can then be aggregated as well where the aim of such reduction process is to decode the different constraints in the network role structure. At the end, we seek to make a substantive interpretation of the complexity in the network structure.

4.8   Positional Analysis and Role Structure: Summary

Multiple networks typically yield large and complex relational structures, and the ability to reduce configurations made of different types of ties is a crucial aspect in the analysis. In this Chapter, we reviewed some algebraic procedures to reduce multiplex networks, and we looked at different kinds of equivalence relations that help us to group structurally correspondent actors in the network. The product is a system as positions made of classes of actors, which are meant to occupy social positions and play specific role relations at different levels.

Equivalence types can have a global perspective where the criterion to establish correspondence among actors considers the overall network structure. However, an equivalence type can also have a local perspective where the modeling takes the individual points of view of the actors in the system for the categorization. An equivalence type from a local perspective that is suitable for performing a positional analysis of multiplex networks without losing the multiplicity of the ties is Compositional equivalence that is based on a device called Relation-Box where primitive and compound relations are stacked together.

From the Relation-Box, Compositional equivalence searches for equations in the existing Person Hierarchies of the system and construct a matrix of cumulative inclusion person relations. On the other hand, the structural implications of Relation Hierarchies from multiplex networks remain to date an almost unexplored topic of research in the algebraic analysis of social networks. Compositional equivalence also allows incorporating actor attributes in a relational manner by using diagonal matrices to represent such features and at the end of the Chapter, there was an exemplification of a positional analysis with Composition equivalence for a multiplex network structure with actor attributes.

4.9   Learning Positional Analysis and Role Structure by Doing

Object `mnet` records the small multiplex network with ties C and F with labeled actors defined in Chapter 3.

4.9.1   Equivalence Relations

4.9.1.1   Figure 4.1 Global Equivalence Types

```
# choose the 'F' relations from Fig. 3.5
R> net <- mnet[,,2]

# symmetrise the adjacency matrix
R> mnplx(net, directed = FALSE)


  a b c d e
a 0 1 1 1 0
b 1 0 1 1 0
c 1 1 0 0 0
d 1 1 0 0 1
e 0 0 0 1 0
```

```
# class automorhic equivalent nodes with weak homomorphism
R> net2 <- reduc(net, clu = c(1,1,2,3,4), lbs = c("ab","c","d","e"))

R> net2

   ab c d e
ab  1 1 0 0
c   1 0 0 0
d   1 0 0 0
e   0 0 1 0
```

```
# class regular equivalent nodes (cf. Table 4.1.b)
R> reduc(net, clu = c(1,1,1,2,3), lbs = c("abc","d","e"))

    abc d e
abc   1 0 0
d     1 0 0
e     0 1 0
```

```
# class of regular equivalent nodes with two clusters
R> reduc(net, clu = c(1,1,1,2,2), lbs = c("abd","ce"))


     abd ce
abd   1  1
ce    1  0
```

```
# separate complete and regular three clusters
R> net3 <- transf(list(com = c("d, abc", "e, d"), reg = "abc, abc"))

, , com

    d abc e
d   0  1 0
abc 0  0 0
e   1  0 0


, , reg

    d abc e
d   0  0 0
abc 0  1 0
e   0  0 0
```

```
# separate complete and regular two clusters
R> net4 <- transf(list(com = "abd, abd", reg = c("abd, ce", "ce, abd")))

, , com

    abd ce
abd   1  0
ce    0  0

, , reg

    abd ce
abd   0  1
ce    1  0
```

```
# redefine node / edge / graph characteristics
R> scp <- list(layout = "force", cex = 6, vcol = "#FFFFFF", vcol0 = "#000000",
+    fsize = 20, pos = 0, lwd = 3, ecol = 1)
```

```
# Fig. 4.1.a  structural and automorphic equivalences
R> multigraph(net, directed = FALSE, seed = 3, scope = scp, rot = -25)

R> multigraph(net2, directed = FALSE, seed = 3, scope = scp, rot = 150, loops = T)
```

```
# Fig. 4.1.b  regular and generalized equivalences with strong homomorphisms
R> multigraph(net, seed = 3, scope = scp, rot = -25)

R> multigraph(net3, seed = 6, scope = scp, rot = -90, loops = TRUE)

R> multigraph(net4, seed = 2, scope = scp, rot = -90, loops = TRUE, collRecip = T)
```

5

Role Structure in Multiplex Networks

It is now time to perform applications of Compositional equivalence to empirical network structures into systems of classes of structurally equivalent actors. In this case, the "structurally" conception arises from the definition of Compositional equivalence made in Chapter 4 Positional Analysis and Role Structure, and the framework where the network reduction takes place is positional analysis, which allows performing network reduction for both directed and undirected relations of different types occurring in a given network. With a positional analysis, it is possible to model attribute-based information, and an extended description on how to integrate this type of information into the positional analysis framework is a matter of this Chapter.

We start by performing a positional analysis to a directed multiplex network structure, and then we continue with the positional analysis of an undirected network that is made of relations of different types. The two network structures have also different kinds of actor attributes that the network relational structure incorporates them. This means that we treat the actor attributes in a relational manner in the sense that these actor traits represent additional kinds of tie in the network structure.

For the positional analysis of a directed structure, the empirical case is a directed network made with three types of relations among Danish entrepreneurial firms and two kinds of measured attributes of the firms. This empirical network, called **Incubator network A** is denoted by $\mathscr{X}_A$, and $\mathscr{S}_A$ represents its reduced positional system. For details of the data collection and for more information about the relational content of this network refer to Ostoic (2013).

The second study within the positional analysis approach is for the case of an undirected multiplex network structure. For this second example, we take the classic network dataset of the **Florentine families** from the 15th century with Business and Marriage ties (Padgett and Ansell, 1993). $\mathscr{X}_F$ and $\mathscr{S}_F$ denotes the Florentine families network and its reduced configuration, respectively, and there are attributes of the actors as well.

Since with $\mathscr{X}_F$, the network structure is made of undirected relations, the configuration of the Florentine families network is considered to be "less complex" than $\mathscr{X}_A$ or the Incubator

Algebraic Analysis of Social Networks: Models, Methods and Applications using R,
First Edition. J. A. R. Ostoic. Companion website: www.wiley.com/go/ostoic/algebraicanalysis.
© 2021 John Wiley & Sons Ltd. Published 2021 by John Wiley & Sons Ltd.

network made with directed relations. This is because not only there is no direction on the ties in an undirected network-as opposed to the systems with directed edges-but here it is also because the Florentine families network has only two types of relations, which constitutes the simplest case of a multiplex network structure, and is less than $\mathscr{X}_A$.

5.1   Directed Role Structures: Incubator Network A

We start the positional analysis on a multiplex network made with directed relations and actor attributes that is of Incubator network A, $\mathscr{X}_A$. This is a business network made of entrepreneurial firms working in the same physically closed setting, which are related by formal collaboration relations, informal friendship ties, and perceived competition. The third one is a cognitive type of tie, whereas the two first are at least instrumental.

The actor attributes for $\mathscr{X}_A$ are represented by diagonal matrices $\mathbf{A}_A^\alpha$, and they are the adoption of two types of innovations from the World Wide Web; that is, $A, B \in \mathbf{A}_A^\alpha$. This means that from the modeling point of view there are two kinds of attribute related information from the actors in the social system. Appendix A provides details of types of relationships in $\mathscr{X}_A$.

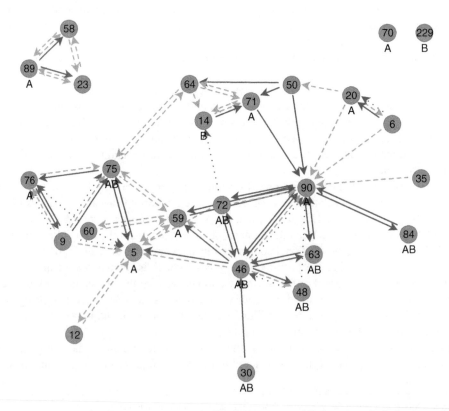

Figure 5.1   Multigraph of Incubator network A, $\mathscr{G}_A^+$. *Solid arcs are collaboration, dashed arcs are friendship, and dotted arcs are perceived competition. Letters below nodes represent actor attributes.*

Incubator network A is depicted as a multigraph in Figure 5.1 where nodes with a circle shape represent the actors with an identifying number. The three types of directed relations are represented by directed edges having different shapes and colors. That is, solid arcs are collaboration relations, dashed arcs are friendship ties, and dotted arcs are perceived competition. The two kinds of actor attributes are given as well, and these are letters below the nodes representing one or two type of innovation adoption.

We can see in the picture that the network has two components, a pair of isolated actors, and some actors are marked with letters that represent the two actor attributes. Although the existence of different components and isolates implies that compound relations are not feasible among all actors, it does not mean, however, that actors not directly related are unable to be correspondent in structural terms as we are going to see with Compositional equivalence.

5.1.1   Social Positions in Network $\mathscr{X}_A$

Finding social positions and role in multiplex networks without losing the multiplicity of the ties is possible with the notion of Compositional equivalence. The first aspect when applying Compositional equivalence in the reduction of empirical multiplex networks is to define alphabet or the collection of generator relations to construct the Role structure. In the case of Incubator network A, $\mathscr{X}_A$ the letters in the alphabet are

$$\Sigma_A = \{ \text{ C, F, K, A, B, D, G, L } \}.$$

where D, G, and L denote C^{-1}, F^{-1}, and K^{-1}, respectively. On the other hand, the actor attributes are represented by diagonal matrices A_A^α, which means that $A = A^{-1}$, and $B = B^{-1}$; that is, the strings that stand for actor attributes are equated prior the relational interlocking as semigroup of relations.

Recall that Compositional equivalence builds on the individual algebras in the system that expresses the social roles from the perspective of the actors. The correspondent roles of the actors in the network that takes the Compositional equivalence definition start from the point of view of the individual actors, and then it culminates with the aggregate level of such particular perspectives.

In order to get the individual perspective of the actors, we still rely on the components found in the Relation-Box, and the first matter to concern when we construct this device is the length of the relational path. Although Breiger and Pattison promote the unrestricted version of the Relation-Box, the substantial arguments made by Winship and Mandel in favor of using the truncated version of $R(W_k)$ with $k \approx 3$ prevails in the establishment of positional systems with Compositional equivalence. One of the main reasons for choosing the truncated version is the relatively high amount of relational types in the alphabet, which in this case includes the transposition of the ties as well. Having unlimited length in the relational paths certainly leads to a potential more complex structure and at this stage of the analysis our aim is to reduce the complexity of different empirical systems.

The simplest version of the $R(W_k)$ is for $k = 1$ and it does not consider compound relations at all; it is just the product of the indexed adjacency matrices for the network relations that are stacked together. In this case the length of the relational dimension equals the number of elements in the alphabet, which for Incubator network $\mathscr{X}_A$ equals to 8 where 6 are for measured social relations and transposes and 2 for the pair of actor attributes.

If we consider relational paths of length 2 or $k = 2$, the amount of possible strings raises to 72, and this means that the length of relational dimension has incremented 9 times. For $k = 3$, the amount of strings is 584, which is 73 times the length of the simplest Relation-Box. Certainly, many of these strings are equal because of the limit in network order, and this means that the relation set of the actors are limited as well. Nevertheless, the size of $\mathbf{R}(\mathbf{W}_k)$ raises exponentially as we can see, and the level of complexity comes along with that.

5.1.2 Modeling $\mathscr{X}_A$ with Compositional Equivalence

To model network $\mathscr{X}_A$ in terms of Compositional equivalence, we construct the Relation-Box with compounds until length 3 or $\mathbf{R}(\mathbf{W}_3)$ with all possible combinations of compounds made with the generators until paths of a defined length k. Due to the Relation-Box structure usually becomes very large with r and $k > 2$, it is not possible to reproduce whole structures of $\mathbf{R}(\mathbf{W}_3)$ in a simple manner. However, we can get a flavor of this configuration by looking at a sample of a Relation.

For instance, Table 5.1 provides a partial representation of the Relation Plane for actor 5 of this Incubator network where the set of primitive relations and all compounds involving C until length 2 are included. Since $\mathscr{X}_A$ has $n = 26$ and the Relation-Box is for $k = 3$, there are additional 25 relation planes in $\mathbf{R}_{\mathscr{X}_A}(\mathbf{W}_k)$ with dimensions 26×584. The array R_5^+ in this Table shows, for instance, that a combination of collaboration ties leads to a loop for this actor and a link towards actor 76. On the other hand, the combination of a collaboration tie with attributes does not affect the structure because if a compound link is produced, then it is contained in the single collaboration tie that is occurring with this actor.

In this sense, from the point of view of actor 5 there is a hierarchy among some of the actors in the system, which is the ego's Person Hierarchy. This configuration is based on the set of tie inclusions among the different types of relations, and it produces a partial order in the network structure in which this particular actor is embedded. Certainly, other actors in the network have a different perspective in the hierarchy of inclusions, and this is because they relate in different ways with the rest of members of the system. If the network is not fully connected such as Incubator network A, this means that the partial order inclusion in $\mathscr{X}_A$ is restricted to the component where the actor belongs to. For isolated actors, there are hardly any inclusions at all with the rest of the network members, and there are only self-inclusions that are a product of the reflexive property of the partial order structure.

As an illustration, Table 5.2 presents some of partial order inclusions corresponding to actors 5 and 64, where only the actors involved in the ordering structure are included. Otherwise the dimensions of the complete partial order table equal the order of the network, with only self-inclusions for the rest of the actors. We can see in these tables that these actors have a different point of view of the network structure where for instance firm 5 is not involved in any inclusion with other actors from the perspective of actor 64 but the converse is true in this case.

Table 5.1 Some links in the Relation Plane for actor 5 in $\mathbf{R}(\mathbf{W}_3)$ for Incubator network A, $\mathcal{L}_A$, R_5^+. This array includes all generators and compound of generators involving collaboration relations in $\mathcal{L}_A$.

	5	6	9	12	14	20	23	30	35	46	48	50	58	59	60	63	64	70	71	72	75	76	84	89	90	229
C	0	0	0	0	0	0	0	0	0	0	0	0	0	0	0	0	0	0	0	0	1	0	0	0	0	0 ...
F	0	0	0	1	0	0	0	0	0	0	0	0	0	1	0	0	0	0	0	0	0	0	0	0	0	0 ...
K	0	0	0	0	0	0	0	0	0	0	0	0	0	0	0	0	0	0	0	0	0	0	0	0	0	0 ...
A	1	0	0	0	0	0	0	0	0	0	0	0	0	0	0	0	0	0	0	0	0	0	0	0	0	0 ...
B	0	0	0	0	0	0	0	0	0	0	0	0	0	0	0	0	0	0	0	0	0	0	0	0	0	0 ...
D	0	0	0	0	0	0	0	0	0	1	0	0	0	1	0	0	0	0	0	0	1	0	0	0	0	0 ...
G	0	0	1	1	0	0	0	0	0	1	0	0	0	1	1	0	0	0	0	1	0	0	0	0	0	0 ...
L	0	0	1	0	0	0	0	0	0	0	0	0	0	0	0	0	0	0	0	0	0	0	0	0	0	0 ...
CC	1	0	0	0	0	0	0	0	0	0	0	0	0	0	0	0	0	0	0	0	0	0	0	0	0	0 ...
CF	0	0	0	0	0	0	0	0	0	0	0	0	0	1	0	0	1	0	0	0	0	0	0	0	0	0 ...
CK	0	0	0	0	0	0	0	0	0	0	0	0	0	0	0	0	0	0	0	0	1	0	0	0	0	0 ...
CA	0	0	0	0	0	0	0	0	0	0	0	0	0	0	0	0	0	0	0	0	0	0	0	0	0	0 ...
CB	0	0	0	0	0	0	0	0	0	0	0	0	0	0	0	0	0	0	0	0	0	0	0	0	0	0 ...
CD	1	0	0	0	0	0	0	0	0	0	0	0	0	0	0	0	0	0	0	0	0	1	0	0	0	0 ...
CG	0	0	1	0	0	0	0	0	0	0	0	0	0	1	0	0	1	0	0	0	0	1	0	0	0	0 ...
CL	0	0	0	0	0	0	0	0	0	0	0	0	0	0	0	0	0	0	0	0	0	0	0	0	0	0 ...
...																										...

Table 5.2 Partial order inclusions from the perspective of actors 5 and 64 in Incubator network $\mathscr{X}_A$. *The matrices here are extracts with the links of the ego with other actors in $\mathscr{X}_A$.*

$\leq$	6	12	20	30	35	46	50	59	63	64	72	84
6	1	0	1	0	1	1	0	0	0	0	1	0
12	0	1	0	0	0	0	0	1	0	0	0	0
20	1	0	1	0	1	1	0	0	0	0	1	0
30	0	0	0	1	0	0	0	0	1	0	0	0
35	1	0	1	0	1	1	0	0	0	0	1	0
46	0	0	0	0	0	1	0	0	0	0	0	0
50	0	0	0	0	0	1	1	0	0	0	0	0
59	0	0	0	0	0	0	0	1	0	0	0	0
63	0	0	0	0	0	0	0	0	1	0	0	0
64	0	0	0	0	0	0	0	1	0	1	0	0
72	0	0	0	0	0	0	0	0	0	0	1	0
84	0	0	0	0	0	1	0	1	1	0	1	1

(a) Actor 5

$\leq$	6	12	20	35	46	48	59	63	72	84
6	1	0	0	0	0	0	0	0	0	0
12	0	1	0	0	0	0	1	0	0	0
20	0	0	1	0	0	0	0	0	0	0
35	1	0	1	1	1	0	0	0	1	0
46	0	0	0	0	1	0	0	0	0	0
48	0	0	0	0	1	1	0	0	0	0
59	0	0	0	0	0	0	1	0	0	0
63	0	0	0	0	1	0	1	1	1	1
72	0	0	0	0	0	0	0	0	1	0
84	0	0	0	0	1	0	1	1	1	1

(b) Actor 64

5.1.3 Cumulated Person Hierarchy $\mathscr{H}_A$

The collection of individual partial order inclusions in the system constitutes the basis for the definition of the Compositional equivalence and for the analysis of roles and positions in the empirical networks. In this sense, the next significant step is to articulate these particular views of the partial order inclusions to the entire network structure where each partial order structure represents particular ego algebras in the system. In the words of Breiger and Pattison (1986), this problem concerns with the aggregation of the partial algebras *across* individuals (p. 224, emphasis in the original).

Because we want to model the network actors by their patterns of relationships, then the resultant structure obtained by aggregation of the partial algebras contributes to the relational structure of the whole network. Such outcome then is based on the different levels of network relations and the combination of these ties among the actors.

The network relational structure in terms of roles is then the setting where different classes of structurally equivalent actors play a similar character in the system, and it is achieved by the union of the Person Hierarchies of the network and by applying a transitive closure on the resultant structure. In this sense, the Cumulated Person Hierarchy of the actors in Incubator network A, $\mathscr{H}_A$, after transitive closure is presented in Table 5.3. This structure for $\mathscr{X}_A$ then results from paths of relations in the system with $k = 3$ at all levels having each a relational contrast, and the attributes of the actors as well.

Since the Cumulated Person Hierarchy represents the relational structure of the network in terms of partially structurally equivalent actors, we are able to determinate classes of actors who are meant to play a similar role in the system based on the actors' different types of tie. The partial order table expresses a hierarchy in the system and it can be represented graphically by a lattice diagram. Lattices are algebraic structures that play an important part in the analysis of partially ordered structures, and we have looked at their properties at the end of Chapter 2.

Table 5.3   Partial order of the Cumulated Person Hierarchy $\mathcal{H}_A$ with Compositional equivalence. This matrix reflects the aggregation of the Person Hierarchies in the network like those in Table 5.2.

≤	5	6	9	12	14	20	23	30	35	46	48	50	58	59	60	63	64	70	71	72	75	76	84	89	90	229
5	1	0	0	0	0	0	0	0	0	0	0	0	0	0	0	0	0	0	0	0	0	0	0	0	0	0
6	1	1	1	1	1	1	0	1	1	1	1	1	0	1	1	1	1	0	1	1	1	1	1	0	1	0
9	1	1	1	1	1	1	0	1	0	1	1	1	0	1	1	1	1	0	1	1	1	1	1	0	1	0
12	1	0	1	1	1	0	0	1	0	1	1	1	0	1	1	1	1	0	1	1	1	1	1	0	1	0
14	1	0	1	1	1	1	0	1	0	1	1	1	0	1	1	1	1	0	1	1	1	1	1	0	1	0
20	1	0	1	1	1	1	0	1	1	1	1	1	0	1	1	1	1	0	1	1	1	1	1	0	1	0
23	0	0	0	0	0	0	1	0	0	0	0	0	0	0	0	0	0	0	0	0	0	0	0	0	1	0
30	1	0	1	1	1	0	0	1	0	1	1	1	1	1	1	1	1	0	1	1	1	1	1	0	1	0
35	1	1	1	1	1	1	0	1	1	0	0	0	0	0	0	0	0	0	0	0	0	0	0	0	0	0
46	0	0	0	0	0	0	0	0	0	1	0	0	0	0	0	0	0	0	1	0	0	1	0	0	0	0
48	0	0	0	0	0	0	0	0	0	1	1	0	0	1	1	1	1	0	1	1	1	1	1	0	1	0
50	1	1	1	1	1	1	0	1	0	1	1	1	1	1	1	1	1	0	1	1	1	1	1	0	1	0
58	0	0	0	0	0	0	0	0	0	0	0	0	1	0	0	0	0	0	0	0	0	0	0	0	0	0
59	0	0	0	0	0	0	0	0	0	0	0	0	1	1	0	0	0	0	0	0	0	0	0	0	0	0
60	1	0	1	1	1	0	0	1	0	1	1	1	0	1	1	1	1	0	1	1	1	1	1	0	1	0
63	1	0	1	1	1	0	0	1	0	1	1	1	0	1	1	1	1	0	1	1	1	1	1	0	1	0
64	1	0	1	1	1	0	0	1	0	1	1	1	0	1	1	1	1	0	1	1	1	1	1	0	1	0
70	0	0	0	0	0	0	0	0	0	0	0	0	1	0	0	0	0	1	0	0	0	0	0	0	0	0
71	1	1	1	1	1	1	0	1	0	1	1	1	0	1	1	1	1	0	1	1	1	1	1	0	1	0
72	1	0	1	1	1	0	0	1	0	1	1	1	0	1	1	1	1	0	1	1	1	1	1	0	1	0
75	1	1	1	1	1	0	0	1	0	1	1	1	0	1	1	1	1	0	1	1	1	1	1	0	1	0
76	1	0	1	1	1	0	0	1	0	1	1	1	0	1	1	1	1	0	1	1	1	1	1	1	1	0
84	1	0	1	1	1	0	0	1	0	1	1	1	0	1	1	1	1	0	1	1	1	1	1	0	1	0
89	0	0	0	0	0	0	0	0	0	0	0	0	0	0	0	0	0	0	0	0	0	0	0	1	0	0
90	0	0	0	0	0	0	0	0	0	0	0	0	0	0	0	0	0	0	0	0	0	0	0	0	1	0
229	0	0	0	0	0	0	0	0	0	0	0	0	0	0	0	0	0	0	0	0	0	0	0	0	0	1

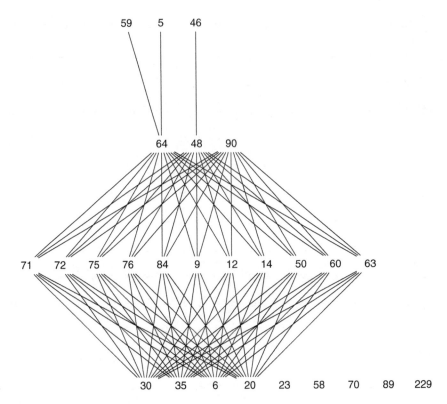

Figure 5.2   Inclusion lattice of actors in $\mathscr{X}_A$ with Compositional equivalence. *This diagram is derived from the Cumulated Person Hierarchy $\mathscr{H}_A$ with k = 3, which constitutes the basis for the partition of the system.*

5.1.4   Positional System $\mathscr{S}_A$

Having defined the lattice structure, we are able to represent the partial order structure of the Cumulated Person Hierarchy as a partial order diagram. Since such hierarchical structure is product of the set of inclusions existing in the system, the meet and join operations serve to reflect such ordering inclusions. Thus in a case where there is an element whose structure is contained in another structure, this implies that this first element is subject to a covering relation by the second element as well.

The inclusion lattice in Figure 5.2 represents the partial order structure of the actors in Incubator network A, whose Cumulated Person Hierarchy $\mathscr{H}_A$ was given in Table 5.3. This type of inclusion lattice is called a *Hasse diagram* and it is also known as *line diagram*, which is a picture of the partial order set of the elements in the relational structure with inclusions as the ordering relation. As a result, the Hasse diagram provides information from the overall structure of the actors in this relational system that is based on their paths of relations, which in this case is under the criteria of the Compositional equivalence definition.

As we can see in the lattice diagram, there are four levels of connected actors in the picture and a set of incomparable elements in this hierarchy; that is, elements $x, y \in (P, \leq)$ are neither

$x \leq y$ nor $y \leq x$. It is important to state, however, that incomparable elements in the inclusion lattice do not necessarily imply that these represent isolated network members, as the actors in the triangle component of $\mathscr{X}_A$ demonstrate. In this case, it means that because the chosen length of the path relations for the construction of $\mathscr{H}$ of the network do not completely satisfies the axioms for the inclusion lattice structure. Actor 89 in $\mathscr{X}_A$ for instance, ends up having just self-relations and therefore is not comparable with the other actors in the Cumulated Person Hierarchy.

The number of levels in the inclusion lattice, including the set of incomparable actors, suggests that the relational structure may be well represented by a similar number of class actors with an equivalent Role Set. In these sense, the next step in the analysis is to perform a blockmodeling of the actors in the network having the matrix of the Cumulated Person Hierarchy as a base, and a hypothesized positional systems with 4 and 5 categories.

As a result, in a case where we permute the Cumulated Person Hierarchy $\mathscr{H}_A$ and make a partition in terms of structurally equivalent actors, then we get the structure that is shown in Table 5.4 with the blocks of relations constitute roles of compositional equivalent actors. Here we can see that there are blocks of ties between and within the different classes of actors and also blocks of empty relations among some of the categories. For instance, if we look at some of the blocks more carefully, we can see that actor 30 prevents one of the blocks from having a perfect fit or fat-fit. On the other hand, the ties from actors **48** and **64** prevent another block from having a zeroblock or lean-fit, but this is if we ignore the diagonal of $\mathscr{H}_A$.

The decision to ignore the diagonal of the partial order matrix in the blockmodeling procedure actually makes a lot of sense, and this is because such ties are just a consequence of the reflexive closure. Consequently, when considering partially ordered sets like this, the errors in the block diagonals do not really affect the structure of the reduced model because what is most relevant for a poset structure is the set of inclusions that are present in the relational system.

In fact if we just consider a model with four positions rather than with five positions and all disconnected actors in Hasse diagram are classed together, then we get a simpler and more manageable structure without affecting the set of inclusions in the poset. In consequence the reduction of the network structure will consider four classes of actors performing different roles in this system.

Rather to proceed with the image matrix of the Cumulated Person Hierarchy $\mathscr{H}_A$, the next step in the blockmodeling process is to apply the positional model to the adjacency matrices of the 8 primitive relations considered in the alphabet for this empirical network. This means that we obtain a separate image matrix for each type of relation, which is based on the union of the ties within the different categories of actors in these structures. In this way, we count with a reduced model for every type of relation where the class membership of the actors is preserved across the relational levels.

In this sense, Figure 5.3 represents the Image Matrices for the set of primitive relations in $\mathscr{X}_A$ that considers different classes of actors, each one playing a determinate role set at every different level in the relational system. These classes correspond to the partition of the permuted Cumulated Person Hierarchy matrix, which means that the categories of actors are correspon-dent in terms of Compositional equivalence that produce a positional system for this network.

We first note that the last two classes in the permuted poset of the Cumulated Person Hierarchy have been merged together in order to be more consistent with the inclusion lattice structure, and thus there are just four classes of actors in the positional system. Moreover, for each relational

Table 5.4 Partitioned poset for the Cumulated Person Hierarchy $\mathcal{H}_A$. There are 2 blocks of relations with inconsistencies, whereas 5 blocks are fat-fit and 16 blocks are lean-fit.

≤	5	46	48	59	64	90	6	20	30	35	9	12	14	50	60	63	71	72	75	76	84	23	58	70	89	229
5	1	0	0	0	0	0	0	0	0	0	0	0	0	0	0	0	0	0	0	0	0	0	0	0	0	0
46	0	1	0	0	0	0	0	0	0	0	0	0	0	0	0	0	0	0	0	0	0	0	0	0	0	0
48	0	1	1	0	0	0	0	0	0	0	0	0	0	0	0	0	0	0	0	0	0	0	0	0	0	0
59	0	0	0	1	0	0	0	0	0	0	0	0	0	0	0	0	0	0	0	0	0	0	0	0	0	0
64	1	1	1	1	1	0	0	0	0	0	0	0	0	0	0	0	0	0	0	0	0	0	0	0	0	0
90	1	0	0	0	0	1	0	0	0	0	0	0	0	0	0	0	0	0	0	0	0	0	0	0	0	0
6	1	1	1	1	1	1	1	1	0	1	1	1	1	1	1	1	1	1	1	1	1	0	0	0	0	0
20	1	1	1	1	1	1	1	1	0	1	1	1	1	1	1	1	1	1	1	1	1	0	0	0	0	0
30	1	1	1	1	1	1	0	0	1	0	1	1	1	1	1	1	1	1	1	1	1	0	0	0	0	0
35	1	1	1	1	1	1	1	1	0	1	1	1	1	1	1	1	1	1	1	1	1	0	0	0	0	0
9	1	1	1	1	1	1	0	0	0	0	1	1	1	1	1	1	1	1	1	1	1	0	0	0	0	0
12	1	1	1	1	1	1	0	0	0	0	1	1	1	1	1	1	1	1	1	1	1	0	0	0	0	0
14	1	1	1	1	1	1	0	0	0	0	1	1	1	1	1	1	1	1	1	1	1	0	0	0	0	0
50	1	1	1	1	1	1	0	0	0	0	1	1	1	1	1	1	1	1	1	1	1	0	0	0	0	0
60	1	1	1	1	1	1	0	0	0	0	1	1	1	1	1	1	1	1	1	1	1	0	0	0	0	0
63	1	1	1	1	1	1	0	0	0	0	1	1	1	1	1	1	1	1	1	1	1	0	0	0	0	0
71	1	1	1	1	1	1	0	0	0	0	1	1	1	1	1	1	1	1	1	1	1	0	0	0	0	0
72	1	1	1	1	1	1	0	0	0	0	1	1	1	1	1	1	1	1	1	1	1	0	0	0	0	0
75	1	1	1	1	1	1	0	0	0	0	1	1	1	1	1	1	1	1	1	1	1	0	0	0	0	0
76	1	1	1	1	1	1	0	0	0	0	1	1	1	1	1	1	1	1	1	1	1	0	0	0	0	0
84	1	1	1	1	1	1	0	0	0	0	1	1	1	1	1	1	1	1	1	1	1	0	0	0	0	0
23	0	0	0	0	0	0	0	0	0	0	0	0	0	0	0	0	0	0	0	0	0	1	1	0	0	0
58	0	0	0	0	0	0	0	0	0	0	0	0	0	0	0	0	0	0	0	0	0	1	1	0	0	0
70	0	0	0	0	0	0	0	0	0	0	0	0	0	0	0	0	0	0	0	0	0	0	0	1	0	0
89	0	0	0	0	0	0	0	0	0	0	0	0	0	0	0	0	0	0	0	0	0	0	0	0	1	0
229	0	0	0	0	0	0	0	0	0	0	0	0	0	0	0	0	0	0	0	0	0	0	0	0	0	1

$$
\begin{array}{cccc}
1 & 0 & 1 & 0 \\
0 & 1 & 1 & 0 \\
1 & 0 & 1 & 0 \\
0 & 0 & 0 & 1
\end{array}
\qquad
\begin{array}{cccc}
1 & 0 & 1 & 0 \\
1 & 1 & 1 & 0 \\
1 & 0 & 1 & 0 \\
0 & 0 & 0 & 1
\end{array}
\qquad
\begin{array}{cccc}
1 & 0 & 1 & 0 \\
0 & 1 & 0 & 0 \\
0 & 0 & 1 & 0 \\
0 & 0 & 0 & 0
\end{array}
\qquad
\begin{array}{cccc}
1 & 0 & 0 & 0 \\
0 & 1 & 0 & 0 \\
0 & 0 & 1 & 0 \\
0 & 0 & 0 & 1
\end{array}
$$

C   F   K   A

$$
\begin{array}{cccc}
1 & 0 & 1 & 0 \\
0 & 1 & 0 & 0 \\
1 & 1 & 1 & 0 \\
0 & 0 & 0 & 1
\end{array}
\qquad
\begin{array}{cccc}
1 & 1 & 1 & 0 \\
0 & 1 & 0 & 0 \\
1 & 1 & 1 & 0 \\
0 & 0 & 0 & 1
\end{array}
\qquad
\begin{array}{cccc}
1 & 0 & 0 & 0 \\
0 & 1 & 0 & 0 \\
1 & 0 & 1 & 0 \\
0 & 0 & 0 & 0
\end{array}
\qquad
\begin{array}{cccc}
1 & 0 & 0 & 0 \\
0 & 1 & 0 & 0 \\
0 & 0 & 1 & 0 \\
0 & 0 & 0 & 1
\end{array}
$$

D   G   L   B

Figure 5.3 Positional system for Incubator network A $\mathscr{S}_A$ with four classes of actors. *Relations* D, G *and* L *are tie transposes, whereas* A *and* B *are for actor attributes.*

type in the network, the image matrix of the tie converse is the transposition of the respective original image, and this means that the structure has been preserved in the reduction process.

Because in this case there is no distinction between the resultant structures from adopting either of the two types of innovations A and B, it means that the composition of any of the other relations with either of these two matrices will yield identical results. As a result, only a single matrix is required to represent the attributes of the actors suggesting, which simply denotes the adoption of a social network service from the Web in generally terms.

The different image matrices constitute an integrated model that represents the positional system of the entire network that is denoted as $\mathscr{S}_A$. Such model is made of compositional equivalent actors in terms of both the multiplexity of ties and paths of relations, and this means that the relational structure that is the product of the combination of these block relations reflects the relational structure of the network in terms of the different role sets of the actors.

5.2 Role Structure Incubator Network A

The positional system $\mathscr{S}_A$ is a form for representation of the network structure in $\mathscr{X}_A$ that is based on the existing patterns of relationships among the actors. In the network positional system, actors are classified into different classes according to a defined equivalence type, and in this way the resulting structure is a simplified model of the original network where its essential structural features are highlighted.

The set of direct and indirect relations of an individual actor is a special type of structural role of the actor. At the aggregate level, the set of associations among primitive and compound relations of the entire relational system characterizes the **Role structure** for the network, which is denoted by $\mathscr{Q}$. In this sense, the network Role structure is expressed as a partially ordered semigroup of the family of role sets in the system, and it constitutes the *role algebra* of the entire network, whereas the Role structure for personal networks corresponds to a *partial role algebra* (Pattison, 1993).

This means that the Role structure of the network is obtained from the semigroup of the role relations occurring in the social system together with the inclusion lattice among the unique relations (White and Reitz, 1983). Since compound ties are also involved in this system, the partially ordered semigroup is therefore a "cumulated" form for Role structure (White, 1963). However, for simplicity we omit this term when we refer to the relational system among the different classes of actors.

The Role structure of the network that is derived in algebraic terms characterizes the *role inter-lock* of the actors (Boorman and White, 1976), and the role interlock represents the algebra of relations whose characteristics can be shared across multiple networks (Pattison, 1980; Breiger and Pattison, 1986). As said, the model to represent the network Role structure comes from the partially ordered semigroup where the multiplication table represents a system of compound roles, and which is called in this context as the *role table* of the network (cf. Lorrain, 1975; Boorman and White, 1976; Wasserman and Faust, 1994). Hence, the two structures represent-ing the role algebra of the network are the right multiplication table and the partial order structure where the associations and the inclusions among the representative string relations are located.

5.2.1   Constructing Role Structures

Before starting with the construction of Role structures for empirical networks such as $\mathscr{X}_A$, we need to address important issues regarding the role tables, both multiplication and par-tial order. From the Small configuration $\mathscr{X}_Z$ in Table 3.8 from Chapter 3 Multiplex Network Configurations, we know that the semigroup of relations can easily result in big and complex structures.

This is even though the network structure have been already reduced considerably in size and with the establishment of the positional systems, and the potential size and complexity of the Role structure is still a major concern both in the exposition of the results and in the analysis of Role structures. For this reason, in the establishment of the Role structure corre-sponding to Incubator network A first we are going to look at the roles representing just the three types of tie measured in $\mathscr{X}_A$ without tie transposes. Then we will consider in the modeling self-relations representing the attributes of the actors or their adoption of Web innovations in this particular case.

Recall that the different roles for Incubator network A correspond to a model considering a four positional system that is depicted in Figure 5.3. Thus the Role structure of this system, which is denoted by $\mathscr{Q}_A$, departs from the positional system $\mathscr{S}_A$, and is represented by image matrices where the rules of the semigroup applies. That is, the partially ordered semigroup of the role relations in $\mathscr{S}_A$ represents $\mathscr{Q}_A$.

Figure 5.4 presents a simple Role structure for $\mathscr{S}_A$ in the form of a matrix multiplication table of the semigroup of relations together with the inclusion lattice of unique relations in this structure. There are three primitive relations R_1, R_2, R_3 and $\mathbf{A}_A^\alpha$ in the semigroup representing social ties and actor attributes in the positional system of network $\mathscr{X}_A$. The partially ordered semigroup representing the Role structure has six unique relations labelled with the words hav-ing the shortest length, and this is because they have a more direct interpretation than those with more length. However, often there is more than one string with identical length connecting precisely the same elements in the system, and for instance we have in this Role structure that CK = KC and FK = KF.

If we take a look at the inclusion diagram, we can see that this lattice structure is not complete, but it is actually a semilattice with the friendship relation as the suprema element. This means that the friendship relationship contains the rest of relations including the attribute relation type

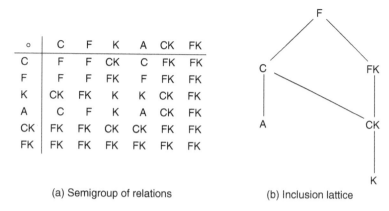

∘	C	F	K	A	CK	FK
C	F	F	CK	C	FK	FK
F	F	F	FK	F	FK	FK
K	CK	FK	K	K	CK	FK
A	C	F	K	A	CK	FK
CK	FK	FK	CK	CK	FK	FK
FK	FK	FK	FK	FK	FK	FK

(a) Semigroup of relations   (b) Inclusion lattice

Figure 5.4 Role structure for $\mathscr{S}_A$ with measured social relations and actor attributes $\mathbf{A}_A^\alpha$ as a partially ordered semigroup.

in of the Role structure of Incubator A. This latter type of tie is also included in the collaboration ties but not in perceived competition. If we consider the paths of relations, we can see that the combination of friendship and competition covers the relational path made of collaboration and competition, and this latter compound relation covers perceived competition, but not the attribute relation.

Although the specific matrices of the unique compound role relations are not included in the picture, we can verify through a Boolean matrix multiplication that none of these two relations is empty, which means that the absorbing behavior from the last element of the matrix semigroup comes from the particular configuration of this string.

5.2.2   Particular Elements in the Role Structure

The analysis of the Role structure continues with the semigroup of relations where we can distinguish some special elements. For instance, the semipath made of F and K in the system is an absorbing element, and this means that whenever a string of tie of the semigroup is concatenated with this semipath, then the semipath remains unchanged; that is $S(FK) = FK$ and $(FK)S = FK$. On the other hand, A constitutes an identity relation where it is rather the string of tie of the semigroup that remains unaffected when is concatenated with this type of tie; i.e. $SA = S$ and $AS = S$. The equations corresponding to the associations among the semigroup elements and the special elements constitute *algebraic constraints* in the system and they have significant consequences in the substantive interpretation of the interlock of the different string of ties.

Another pattern of an algebraic constraint is that of the "Letter laws" in the semigroup of relations (Lorrain, 1975; Boorman and White, 1976), which occur when the compound relations have a path that results from those of their "First" or "Last" string component. For instance, we can see in Figure 5.4 that each composition involving a friendship relation leads either to a friendship itself or to a 2-semipath relation made of friendship and perceived competition. If the compound is represent by FK as it happens in the picture, then the associated pattern corresponds to the *First Letter law*; otherwise it represents the *Last Letter law*.

Similarly, the pattern concerning the 2-semipath made of CK represents a Letter law, which depends on the representative word. In this sense, the picture representation constitutes a Last Letter, whereas it is a First Letter if we consider KC and KF as representatives. Finally, each absorbing element in S represents both the First- and Last Letter since there is no change in the pattern.

The ambiguity in the example results from the commutative character of these strings of tie. However, there is not always an alternative representation or if there is one, it is not commutative, which means that the expressions in the Letter laws are more straightforward. In this sense, later on the substantive interpretation of the Letter laws will be made to the final Role structures of the empirical networks, and this is because they provide insights into the logic of interlock among the different strings of ties occurring in the system.

5.2.3   Role Structure with Relational Contrast

The previous example of the Role structure for $\mathscr{S}_A$ is very useful for illustration purposes; however, there is a pending matter in the analysis and this is the converse of the ties. This is important because these ties have been considered for the positional model of the different Incubator networks and, since the classification of structurally equivalent actors has been made on systems having a relational contrast in it, then the Role structure of the network should also include these types of relations. This means that for consistency the set of relations used to generate the partially ordered semigroup needs to consider all the relations occurring in the positional system of this network, which has been given in Figure 5.3.

As a result, the Role structure of Incubator network A, $\mathscr{Q}_A$, is based on the complete positional system is presented in Table 5.5 as a multiplication or role table in this context. Naturally, the Role structure in this case has more elements than the previous one where we disregarded the tie converses, and there are also more equations to take into account with this new structure. Boyd points out that the tie converses ensure that the semigroup of relations is closed with respect to the inverse (Boyd and Everett, 1999; Boyd, 2000), and this is an important aspect we are trying to achieve in the algebraic analysis of social networks.

For choosing the representative strings in the multiplication table, we are going to apply the criteria mentioned in previously and favor words with the shortest length, then strings where attributes are present, and finally composes without tie transpose when is possible. For example, even though CK = KC = LC = CKA = LAC ... , the commutative character of a 2-path made of collaboration and perceived competition matters for the interpretation of the Role structure, and it will be shown in the representation of this kind of arrangement. Similarly, there is the equation FK = KF = CL but in this case we just consider the first equality in the interpretation because is made of two primitive relations without the converse operation. As a result, we put emphasis simply on the equations involving the strings with the mentioned criteria in the Role structure representation, unless the string of relations has a significant location in the partially ordered structure, in which case other types of equalities are put in parenthesis.

Similarly, the Role structure of $\mathscr{S}_A$, $\mathscr{Q}_A$ is also presented as an inclusion lattice of the unique strings in Figure 5.5. The structure of the inclusion lattice in the diagram of this Role structure has some interesting characteristics that are important to mention. First, we can see that the Hasse diagram is almost a complete lattice in this case, and actually the diagram nearly resembles a distributive lattice, which means that the relational structure of the positional systems for

Table 5.5 Role table of the Role structure $\mathcal{Q}_A$ for Incubator network A with a symbolic format.

o	C	F	K	A	D	G	L	CK	CG	FK	KD	KL	DC	DF	DK	GC	CKD	KDC	DCK	DCL
C	F	F	CK	C	CG	CG	FK	FK	CG	FK	CKD	FK	CG	CG	CKD	CG	CKD	KDC	DCK	DCL
F	F	F	FK	F	CG	CG	FK	FK	CG	FK	CKD	FK	CG	CG	CKD	CG	CKD	CKD	CKD	CKD
K	CK	FK	K	K	KD	KD	KL	CK	CKD	FK	KD	KL	KDC	CKD	KD	KDC	CKD	CKD	CKD	CKD
A	C	F	K	A	D	G	L	CK	CG	FK	KD	KL	DC	DF	DK	GC	CKD	KDC	DCK	DCL
D	DC	DF	DK	D	G	G	DK	DCK	CG	DCL	KD	DK	GC	CG	KD	GC	CKD	KDC	DCK	DCL
G	GC	CG	KD	G	G	G	KD	KDC	CG	CKD	KD	KD	GC	CG	KD	GC	CKD	CKD	CKD	CKD
L	CK	FK	KL	L	KD	KD	L	CK	CKD	FK	KD	KL	DCK	DCL	DK	KDC	CKD	KDC	KDC	CKD
CK	FK	FK	CK	CK	CKD	CKD	FK	FK	CKD	FK	CKD	FK	CKD	CKD	CKD	CKD	CKD	KDC	DCK	DCL
CG	CG	CG	CKD	CG	CG	CG	CKD	CKD	CG	CKD	CKD	CKD	CG	CG	CKD	CG	CKD	CKD	CKD	CKD
FK	FK	FK	FK	FK	CKD	CKD	FK	FK	CKD	FK	CKD	FK	CKD	CKD	CKD	CKD	CKD	CKD	CKD	CKD
KD	KDC	CKD	KD	KD	KD	KD	KD	KDC	CKD	CKD	KD	KD	KDC	CKD	KD	KDC	CKD	CKD	CKD	CKD
KL	CK	FK	KL	KL	KD	KD	KL	CK	CKD	FK	KD	KL	KDC	CKD	KD	KDC	CKD	KDC	KDC	CKD
DC	DF	DF	DCK	DC	CG	CG	DCL	DCL	CG	DCL	CKD	DCL	CG	CG	CKD	CG	CKD	KDC	KDC	CKD
DF	DF	DF	DCL	DF	CG	CG	DCL	DCL	CG	DCL	CKD	DCL	CG	CG	CKD	CG	CKD	CKD	CKD	CKD
DK	DCK	DCL	DK	DK	KD	KD	DK	DCK	CKD	DCL	KD	KD	KDC	CKD	KD	KDC	CKD	CKD	CKD	CKD
GC	CG	CG	KDC	GC	CG	CG	CKD	CKD	CG	CKD	CKD	CKD	CG	CG	CKD	CG	CKD	KDC	KDC	CKD
CKD	CKD	CKD	CKD	CKD	CKD	CKD	CKD	CKD	CKD	CKD	CKD	CKD	CKD	CKD	CKD	CKD	CKD	CKD	CKD	CKD
KDC	KDC	CKD	KDC	KDC	CKD	CKD	CKD	CKD	CKD	CKD	CKD	CKD	CKD	CKD	CKD	CKD	CKD	CKD	CKD	CKD
DCK	DCK	DCL	DCK	DCK	CKD	CKD	DCL	DCL	CKD	DCL	CKD	DCL	CKD	CKD	CKD	CKD	CKD	CKD	CKD	CKD
DCL	DCL	DCL	DCL	DCL	CKD	CKD	DCL	DCL	CKD	DCL	CKD	DCL	CKD	CKD	CKD	CKD	CKD	CKD	CKD	CKD

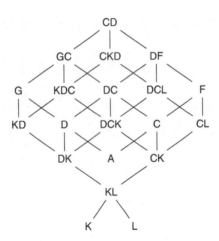

Figure 5.5 Inclusion lattice of the Role structure representing $\mathscr{S}_A$, $L_\pi(\mathscr{Q}_A)$. *Equations in $\mathscr{Q}_A$ are* CK = KC, FK = KF, (CG = CD = FG = FD = GF).

this network is somewhat regular. The top element in the lattice diagram is not just friendship anymore, but a compound made of a collaboration and friendship semipath of length 2, whereas the elements at the bottom are perceived competition and its converse, whose join is a perceived competition semipath of length 2.

If we compare the two lattice diagrams for $\mathscr{S}_A$, which are product of the Role structure whether considering tie converses or not, it is clear that the set of inclusions are preserved in all cases. This suggests that although the relational contrast may play a crucial role in the definition of Compositional equivalence among the actors, the fundamental characteristics of the Role structure without the relational contrast that has been used for the establishment of the positional system appears to remain unchanged in this case.

There are, however, some differences concerning the location of the ties in the diagram structure, and there is also the presence of intermediary ties between the covering relations among perceived competition and the rest of the strings. In this sense, a more consistent way to continue with the analysis of the network is to work with the Role structure of the system that considers the converse of the ties. Later on, through a reduction process of this system, we can then corroborate that the essential structure of the network effectively remains unaffected or not.

We can see in the Role structure from $\mathscr{S}_A$ that the attribute relation representing in this case the adoption of innovations is the identity element in the semigroup of relations, and that in the inclusion lattice diagram this relation is the meeting of the collaboration ties with their converses. This means that, at the aggregate level, not only the attribute relations are included in C and D, but also that a combination of a collaboration tie with the attribute relation leads to collaboration in this system.

Since collaboration C is included in the friendship relation F, implying that whenever there is a formal relation there is also an informal link between the different classes in the network, then a combination of friendship with an attribute relation leads to a friendship relation. Both collaboration and friendship ties with their converses are involved in a semipath compound at the top of the inclusion lattice, which means that this pattern is occurring most in this positional system.

In the case of perceived competition, there is a quite different situation, and one of the reasons is that the attribute relation is not included in this type of tie. This means that not in all cohesive subgroups made of partially structurally equivalent actors having adopted an innovation there is a perceived competition among its members. Besides, whenever there is a path relation made of competition and collaboration among the different classes of actors, there is also a path relation of competition and friendship, but not the converse.

5.3 Undirected Role Structures: Florentine Families Network

Having looked at the establishment of positional systems in terms of Compositional equivalence of a directed multiplex network such as $\mathscr{X}_A$, this time we perform the analysis of a network with *undirected* relations. For this, we construct the positional system of the Florentine families network with Business and Marriage ties together with relevant characteristics acquired from the actors such as the financial Wealth of the families and the number of Priorates they held.

The Florentine families network dataset (Kent, 1978; Breiger and Pattison, 1986; Padgett and Ansell, 1993; Borgatti, Everett and Freeman, 2002) corresponds to a group of people from Florence in the Italian region of Tuscany who had a leading role in the creation of the modern banking system in 15th century Europe. There are two types of social ties in the network that correspond to Business and Marriage relations among 16 Florentine families. The ties are undirected, which does not represent any problem for the Marriage ties but is unfortunate for the Business relations, a circumstance that was remediated by Breiger and Pattison in their analysis by including measures of power such as the Wealth of the families and their number of Priorates.

As a result, the families' financial Wealth and the number of Priorates they hold constitute two sorts of actor attributes in $\mathscr{X}_F$, besides the two mentioned types of social relations. The two types of actor attributes represent in a certain way both the economic and political power of the these families in Florence, and it is worth including this information in the modeling of this network relational structure. The following study of the Role structure for the Florentine families network takes the analysis made in Ostoic (2018), where the purpose was to demonstrate the efficacy of incorporating significant actor attributes in the modeling of networks relational structures.

5.3.1 Positional Analysis of the Florentine Families Network

The classical network made of Business and Marriage relations between the Florentine families is depicted as a multigraph in Figure 5.6 in that century. Here, different shapes in the edges represent the two kinds of relations and the size of the nodes reflects the amount of economic power of each actor. We can see, for instance, that the network has one component and the *Pucci* is the only isolated actor. Besides, the *Strozzi* family is the wealthiest family among these actors, but the *Medici* has more connections that are of different types. In fact, the *Medici* family is the one that held most of strong bonds of the eight patterns combining Business and Marriage ties in this system.

The visualization gives us initial insights into the general social structure where actors are linked, and a force-directed layout algorithm (Fruchterman and Reingold, 1991) has been

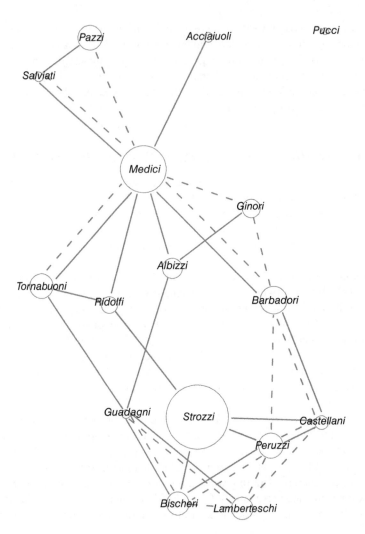

Figure 5.6 Multigraph of the Florentine families banking network in the 15th century, $\mathscr{G}_F^+$. *Solid edges are Marriage relations, dotted edges are Business ties, and node size reflects financial Wealth of the family.*

applied to this graph to avoid crossing edges and also to group together closely related actors. However, we need to implement some computations if we want to look at the network relational structure in a form where the different types of tie are interrelated.

Since the network is undirected, the primitive relations in this case do not contemplate tie transposes. This means that the alphabet with the generator relations for $\mathscr{X}_F$ are

$$\Sigma_F = \{\ \text{B, M, W, P}\ \}.$$

where B, and M represent Business and Marriage relations, whereas W, and P stand of economic Wealth and political power in the number of Priorates.

A crucial part in the modeling of multiple networks is the reduction of the social system where the corresponding relational structure represented by the semigroup results being large and complex, even for small arrangements. For instance, Breiger and Pattison (1986, p. 221) report a semigroup size with an order of 81 for the Florentine families network, and this is only considering the two generator relations without actor attributes. Certainly, it is necessary to work with a more manageable structure in order to obtain better insights in the logic of interlock of this network.

As with the previous network $\mathcal{X}_A$, the reduction of the network implies constructing a relational structure based on a system of roles and positions, and consequently this process leads to the *Role structure* of the network. Thanks to its reduced size, the Role structure is typically a more convenient configuration than the "raw" relational arrangement of the system for a substantial interpretation of the network. A key aspect in the creation of the Role structure is to preserve the multiplicity of the ties, and we know that Compositional equivalence allows us to combine different levels in the relationships.

In this sense, we are going to categorize the actors in the Florentine families network in terms of Compositional equivalence as in the Breiger and Pattison paper. However, in this case we will establish the Role structure of the network with relevant attributes from the actors, and the first step is to look at the structure product of the actors' views of their neighbors' relations in terms of inclusions. Such structure of views and inclusions is represented by the Cumulated Person Hierarchy, which serves as a basis to perform the modeling and the analysis of the Role structure of this particular network with attributes of the families.

5.3.2 Constructing Person Hierarchies, H_F

Applying Compositional equivalence in the reduction of a multiple network structure implies the construction of the Relation-Box, and this is because this array constitutes the basis for the local part of this type of correspondence. Recall that the Relation-Box is defined by the number of actors in the network and the number of string relations that make up the actors' immediate social ties and eventually the combination of these. Then all inclusions from the individual perspectives are combined into a single matrix that stands for the global part of Compositional equivalence.

To illustrate the process of constructing Person Hierarchies we restrict the analysis to the smallest case of the Relation-Box with no compounds that for the Florentine families network has dimensions $16 \times 16 \times 2$. When we look at Figure 5.6, we see that apart from *Pucci*, the actor of the network with the lowest number of connections is the *Acciaiuoli* family who is a pendant actor with a single (reciprocated) tie with the *Medici* family. For the direct contacts in the network without compounds, this means that the personal hierarchy of *Acciaiuoli* just includes their immediate neighbor who is the *Medici* family, and hence the only inclusion in the matrix is a reflexive closure corresponding to this neighbor, while all the other possibilities lack containment.

For a two-chain relationship, the Person Hierarchy of *Acciaiuoli* includes the neighboring of the *Medici* family as well, i.e. the *Albizzi, Barbadori, Ginori, Pazzi, Ridolfi, Salviati, Tornabuoni*, and in this case the *Acciaiuoli* itself. Note that longer paths include not just the rest of the members in the component, but also those actors who take part in the partial order structures

Table 5.6   Cumulated Person Hierarchy $\mathscr{H}_F$ of the Florentine families network $\mathscr{X}_F$ for social ties with $k = 5$.

$\leq$	1	2	3	4	5	6	7	8	9	10	11	12	13	14	15	16
1 *Barbadori*	1	1	1	1	1	1	1	1	1	1	0	0	0	0	0	0
2 *Bischeri*	1	1	1	1	1	1	1	1	1	1	0	0	0	0	0	0
3 *Castellani*	1	1	1	1	1	1	1	1	1	1	0	0	0	0	0	0
4 *Guadagni*	1	1	1	1	1	1	1	1	1	1	0	0	0	0	0	0
5 *Lamberteschi*	1	1	1	1	1	1	1	1	1	1	0	0	0	0	0	0
6 *Medici*	1	1	1	1	1	1	1	1	1	1	0	0	0	0	0	0
7 *Pazzi*	1	1	1	1	1	1	1	1	1	1	0	0	0	0	0	0
8 *Peruzzi*	1	1	1	1	1	1	1	1	1	1	0	0	0	0	0	0
9 *Salviati*	1	1	1	1	1	1	1	1	1	1	0	0	0	0	0	0
10 *Tornabuoni*	1	1	1	1	1	1	1	1	1	1	0	0	0	0	0	0
11 *Acciaiuoli*	1	1	1	1	1	1	1	1	1	1	1	1	1	1	0	0
12 *Albizzi*	1	1	1	1	1	1	1	1	1	1	1	1	1	1	0	0
13 *Ridolfi*	1	1	1	1	1	1	1	1	1	1	1	1	1	1	0	0
14 *Strozzi*	1	1	1	1	1	1	1	1	1	1	1	1	1	1	0	0
15 *Ginori*	1	1	1	1	1	1	1	1	1	1	0	0	0	0	1	0
16 *Pucci*	0	0	0	0	0	0	0	0	0	0	0	0	0	0	0	1

for generators and shorter compounds. Thus for compounds of length 2, we still account for the self-containment for the *Acciaiuoli* family in its Person Hierarchy, etc.

Each actor l in the network has its own Person Hierarchy $\mathbf{H}_l$ that is based on the relational plane of the actor, R_l^+ that contains the primitive relations and the compounds until a certain length. However, these hierarchies are aggregated into the Cumulated Person Hierarchy $\mathscr{H}_F$, which is a single matrix of inclusions among all the network members. For the Florentine families network, the structure of $\mathscr{H}_F$ is represented by the universal matrix, which is certainly by disregarding the isolated actor. The Cumulated Person Hierarchy $\mathscr{H}_F$ makes no differentiation among the actors until it reaches chain of relations of length 4 since it is only from chains of relations with length 5 or more that $\mathscr{H}_F$ produces a distinction among the actors that is a product of their particular inclusions expressed in $\mathbf{H}_l$.

The partial order structure representing $\mathscr{H}_F$ with chains of length 5 is given in Table 5.6, and this set of ordered relations has been reported by Breiger and Pattison (1986, p. 234). The Cumulated Person Hierarchy in this case presents two categories of actors in the network plus the isolated family. One category corresponds to the actors who contain other network members without being contained in them, whereas the other category groups those who are merely contained in other actors without containing them. The partition of this system almost fits the requirements of Structural equivalence, except for the case of the *Ginori* family who is positioned in the same class with the *Acciaiuoli*, *Albizzi*, *Ridolfi* and *Strozzi* even though this actor is not implicated in any inclusions with the rest of the members in this class other than a self-containment.

Therefore, the positional system can have either two classes of collective actors plus the isolated actor, or four classes with pairwise individual positions in the system. Regardless of the option chosen, both reduced arrangements seem to be good representations of the network structure in terms of the patterned social relations, and they serve as the basis for the construction of the Role structure of the Florentine families network. However, a number of attributes from the actors may play a significant part in the establishment of the network positional system, and hence we continue the rest of the analysis of this network by incorporating actor attributes in the establishment of the network Role structure for $\mathscr{X}_F$.

5.3.3   Family Attributes in $\mathscr{X}_F$

The power and influence of the Florentine banking families constitute significant characteristics, and Table 5.7 provides the Wealth and the number of Priorates of Florentine families as reported in Wasserman and Faust (1994, p. 744). Wealth and number of priorates then constitute the two attribute types that either together or individually, are candidates for the modeling of the network positional system and subsequent Role structure. For such type of analysis each attribute is represented as an indexed matrix, and the reduction of the network structure with actor attributes resembles the process applied to the marriage and business relations with Compositional equivalence, except that there are additional generators to the social ties representing the attributes.

We note in Table 5.7 that each category has two columns, one for the absolute values and another that marks the limits of these values according to a cutoff value. In one case we differentiate the "very" wealthy families from the "modestly" rich actors in the network by adopting

Table 5.7   Wealth and number of Priorates in the Florentine families network $\mathscr{X}_F$. *NA stands for "data not available".*

	Wealth (×1000 Lira)	> 40	Priorates number of	$\gtrsim$ 34 (avg.)
Acciaiuoli	10	0	53	1
Albizzi	36	0	65	1
Barbadori	55	1	NA	0
Bischeri	44	1	12	0
Castellani	20	0	22	0
Ginori	32	0	NA	0
Guadagni	8	0	21	0
Lamberteschi	42	1	0	0
Medici	103	1	53	1
Pazzi	48	1	NA	0
Peruzzi	49	1	42	1
Pucci	3	0	0	0
Ridolfi	27	0	38	1
Salviati	10	0	35	1
Strozzi	146	1	74	1
Tornabuoni	48	1	NA	0

a cutoff value of 40000 Lira, which approximates the average of their financial resources. Actually, the mean is 42.56, and the *Lamberteschi* family lies in this limit, but placing this family with the very wealthy class makes more sense in the analysis because outcomes get more transparent.

On the other hand, as regarding the number of Priorates, it seems reasonable to assume that the lack of information implies that these actors did not have a large number of jurisdictions at that time, if at all, and the cutoff lies in the average of the accessible number of priorates that is rounded to 34. As a result, there is a pair of vectors of binary values that make the diagonal of the indexed matrices representing the two actor attributes, and these are additional generators for constructing the network relational structure.

As a result, we proceed with the positional analysis of this banking network by applying Compositional equivalence for grouping the actors, and this time taking actor attributes into the modeling. The difference is that the Relational-Box on which the Person Hierarchies are based now includes the additional generators representing the attribute-based information. An indexed matrix records the data on the diagonal, which means that the different Person Hierarchies in the network include self-containments whenever the actor has the attribute. For example, while the Person Hierarchy of *Acciaiuoli* for immediate ties comprises just the *Medici*, with actor attributes it will include the *Acciaiuoli* family itself when $k = 1$ because this particular actor is politically very powerful with a number of priorates larger than the average. Naturally, the rest of the actors in the network will follow the same logic, and the arrangement of the Cumulated Person Hierarchy will be affected by the different personal views on inclusions, which are restructured due to the presence of actor attributes.

Figure 5.7 shows in a graphic mode the Cumulated Person Hierarchies of the Florentine banking network with Business and Marriage ties together with Wealth, number of Priorates, and also with the two attributes combined. Recall that these pictures, which are also known as "Hasse diagrams", depict the inclusion levels in the hierarchy where the lower bound elements are contained in the upper bound elements whenever there is a link among them. For instance, in each diagram the inclusion ties of the *Medici* family contain the inclusion ties of the *Acciaiuoli* and the *Pazzi* families, whereas in any of the cases there is a containment relation between these last two actors.

It is important to note, however, that although the different levels in the Hasse diagrams try to reflect the ranks in the partial order structures, there can be ambiguities in the placements depending on the diagram structure. For example, the *Guadagni* family is always placed in the intermediary level of the diagrams in Figure 5.7, but in a couple of cases this actor does not contain any other actor in $\mathcal{H}_F$. Likewise, the inclusion ties of *Barbadori* contain the ties of other actors while it is not being contained at all similar to *Medici* and *Peruzzi*, and it may be best depicted at the same level with these actors. Such aspects deal with aesthetics rather than the structural representation of the partially ordered system, however.

All partial orders shown are emerging structures with the smallest value of k. This means that there are "empty cells" among connected actors in $\mathcal{H}_F$ with compounds of such lengths, which allows us to rank classes of actors according to the Compositional equivalence criteria. In the case of the actors' wealth, the structure of $\mathcal{H}_F$ remains unaltered after compounds of length 5, but in the other two cases, the cumulated Person Hierarchies involve a lower number of inclusions with larger k. However, shorter chains of relations imply more truthful individual viewpoints than ordered systems with longer compounds and they are therefore preferred.

These diagrams very clearly show that the attributes of the actors such as their monetary wealth and political power have an impact on the relational structure of this particular network.

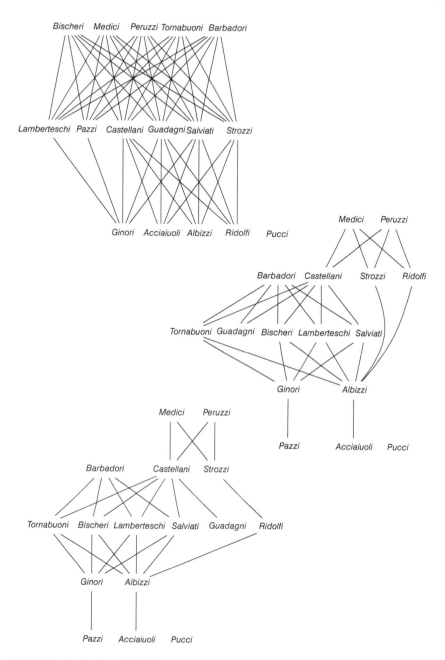

Figure 5.7 Inclusion lattices of $\mathcal{H}_F$ for the Florentine banking network with actor attributes. *Top to bottom: with Wealth, k = 5; with number of Priorates, k = 4; with Wealth & Priorates, k = 4.*

If we look at the diagrams in Figure 5.7, we note that there is a further differentiation in the network in all three cases when considering actor attributes in the modeling. Apart from the isolated actors, whose personal hierarchy corresponds to the null matrix, the Cumulated Person Hierarchy for Wealth clearly involves three levels, whereas there are five levels in the diagrams for the number of Priorates, and for the two attributes together.

As a result, the positional system with Wealth differentiates three categories of actors plus the isolated node where the largest class in the previous classification is now divided into two categories. Thus the personal wealth has a structuring influence in the network, and this makes a lot of sense; the richest actor of the banking network is the *Strozzi* family who is no longer in the same class as the *Acciaiuoli*, *Albizzi*, and *Ridolfi* families, but is placed in another category with other actors having much more social and financial capital.

When we look at the Number of Priorates there is even more differentiation in $\mathcal{H}_F$ than we saw when just considering the Wealth of the actors. Apart from the families who contain most of the network members, i.e. the actors "at the top" (*Medici*, *Peruzzi*, *Barbadori*), and conversely the actors "at the bottom" (*Pazzi*, *Acciaiuoli*, *Guadagni*) who are contained in the rest of the network component, there is ambiguity with the rest of the actors and they can be classed in different ways. We get a similar picture when both attributes are taken together (cf. Fig 5.7c), where the "top" and "bottom" actors in the diagram representing $\mathcal{H}_F$ remain unambiguously placed, whereas the categories of the actors in-between require interpretation. Since the matrices representing both social ties and actor attributes have a commutative character, it means that the order of the different string relations does not affect the structure in the Cumulated Person Hierarchy.

Theory can guide us in the establishment of the categories in the positional system in the two last cases. We also need to determine which of the resulting Role structures product of the positional system provides the best insights into the relational interlock of the multiple network structure. Such aspect constitutes one of the last steps in the modeling of the system and we look at the reduced relational structures of the banking families network.

5.4   Role Structure of the Florentine Families Network

The main challenge in establishing the positional system of the network is to find the sets of collective relations that produce the most meaningful network Role structure. That is, a reduced system that provides an insight into the logic of interlock of the network relations, and this is typically achieved with the Role structure having the smallest possible dimension. The logic of interlock is a kind of rationality that is shaped by different algebraic constraints expressed in the final relational structure where the different types of ties and the relevant actor attributes are interrelated in this case.

Although the class membership with the Wealth attribute with three defined classes of collective actors seems straightforward, there are ambiguities both as regards the amount of Priorates and as regards when the two features are combined. Such uncertainties arise because a number of actors in the network can be classed in different ways according to their respective locations in the partial order structures of $\mathcal{H}_F$, and for the time being we concentrate our analysis on the two cases where political power is involved.

Hence, assuming that the isolated actor of the network makes its own class, we need to categorize the eight actors that are neither at the "top" nor at "bottom" of the hierarchies shown in Figures 5.7b and 5.7c, and in both arrangements the placement of the actors at the different levels aims to reflect the set of containments in the partial order structures with an aesthetical representation in the lattice. That is why *Barbadori* and *Guadagni*, for example, who are unequivocally part of the same class as the top and bottom actors, respectively, are located at intermediary levels in the diagram.

Now we look closer at the in-between actors in the two hierarchies where political power is involved. From Table 5.7, which provides the Wealth and number of Priorates in network $\mathscr{X}_F$, we obtain the assignment of these families with respect to the two attributes. In this sense, the upper and lower vectors in the structure below give for these actors in $\mathscr{X}_F$ the categories for Wealth and number of Priorates, respectively:

Albizzi	*Bischeri*	*Castellani*	*Ginori*	*Lamberteschi*	*Ridolfi*	*Salvia*	*Strozzi*	*Tornabuoni*
0	1	0	0	1	0	0	1	1
1	0	0	0	0	1	1	1	0

Certainly, one option is that all these actors are grouped together onto a single class irrespective of their economic or political power, and in this way we have a positional system with three categories of collective actors for both Priorates, and also for Wealth and Priorates. The arrangements of roles for Business and Marriage are then equal and all the positions are represented by actors who are both very wealthy and powerful in political terms (certainly this by disregarding *Pucci*). This means that the two attribute types are represented in the positional system by identity matrices with no structuring effect in the system of roles. In order to have an effect from the Wealth and the number of Priorates on the Role structure, we need to make a differentiation between classes of actors with respect to these attributes, and this is only possible by having characteristic strings not acting as neutral elements in the construction of the semigroup of relations.

A straightforward way to achieve a structuring effect of diagonal matrices is by separating the actors with "ones" in the intermediate category from the actors with "zeroes" in the vector corresponding to this attribute type. Hence, we end up with a positional system that has four categories of collective actors, and for the number of Priorates, for instance (the second row above), then *Bischeri, Castellani, Ginori, Lamberteschi*, and *Tornabuoni* will make their own class. This means that the attribute string is no longer represented by an identity matrix and the semigroup of the Role structures for Business, Marriage, and number of Priorates will record different compounds of social roles with class attributes. However, the semigroup structure in this case has an order of 13, which is relatively large.

Conversely, if we model the network relational system with both attributes at the same time, we first differentiate the *Strozzi* who is a very powerful family both politically and economically, and second we differentiate *Castellani* and *Ginori* who are actors who are neither very wealthy nor have much political power. By grouping the last two actors into a single class we again avoid having the identity matrix, and the Role structure of the network in this case has just 9 representative strings, which means that we expect a more tractable substantial interpretation of the role interlock than when just considering the Priorates.

The fact that the Role structure gets smaller rather than larger as one would expect with another generator is because the two social roles and both class attributes are equated, and the relational structure of the positional system is then based just on two generators. When we equate roles or attributes, we get a poorly informative Role structure where we need to inter-polate the roles and collective characteristics in the analysis. Besides, assigning *Strozzi* in the central class does not affect the Role structure at all.

A third possibility is to combine the Business and Marriage ties with Wealth in the analysis, in which case the class system of actors takes the levels given in the Hasse diagram of Figure 5.7a. The main advantage of this positional system is certainly that there are no ambiguities in the categorization of actors, which leads to a univocally substantial interpretation of the Role struc-ture. Moreover, the semigroup of role relations is smaller with these generators than with the previous two settings. As a result, we are going to proceed by modeling the Role structure of the banking network with Florentine families' Wealth as the sole actor attribute, and we are aware that a different logic may arise in the Role structure when considering the number of Priorates.

5.4.1   Interlock of Business, Marriage and Wealth Role Relations in $\mathcal{Q}_F$

From a modeling perspective, we look for positional systems that produce a small semigroup structure as they provide more transparent insights into the network relational system than large semigroups of relations. We have already mentioned that the Role structure of the banking net-work with Business, Marriage, and Wealth produces a smaller semigroup than when we consider just Priorates or both attributes together, and the categories in the Cumulated Person Hierarchy are univocally delineated with just the Wealth of the families.

In this sense, Figure 5.8 provides the partially ordered semigroup representing the system with Business, Marriage, and Wealth that are denoted as B, M, and W, respectively. The Role structure, denoted as $\mathcal{Q}_F$, has 8 representative strings, and the different algebraic constraints of the system are expressed by the right multiplication and the partial order tables. One type of constraint corresponds to the composition of role relations in the semigroup, and another one to

○	1	2	3
1	1	1	4
2	1	1	5
3	6	7	3
4	1	1	4
5	1	1	5
6	6	6	8
7	6	6	8
8	6	6	8

≤	M	B	W	MW	BW	WM	WB	WMW
M	1	0	0	0	0	0	0	0
B	1	1	0	0	0	0	0	0
W	1	1	1	1	1	1	1	1
MW	1	0	0	1	0	0	0	0
BW	1	1	0	1	1	0	0	0
WM	1	0	0	0	0	1	0	0
WB	1	1	0	0	0	1	1	0
WMW	1	1	0	1	1	1	1	1

M	B	W	MW	BW
1 1 1	1 1 1	1 0 0	1 1 0	1 1 0
1 1 1	1 1 0	0 1 0	1 1 0	1 1 0
1 1 1	1 0 0	0 0 0	1 1 0	1 0 0

Figure 5.8   Role structure of Florentine families network with financial wealth, $\mathcal{Q}_F$. *Upper panel: par-tially ordered semigroup with Edge table. Lower panel: Generators from the positional system* $\mathcal{S}_F$ *and representative compounds with* $k = 2$ *in* $\mathcal{Q}_F$ *(without isolated class).* B, M, *and* W *represent Business, Marriage relations, and families' Wealth.*

the different inclusions of the representative strings represented by the partially ordered set in the latter structure. Besides, the positional system $\mathscr{S}_F$ of the Role structure without the isolated class is given at the bottom of the figure, and it has the three generators together with a pair of representative compounds until a length of two, which are used for the analysis of the role interlock of the reduced system of $\mathscr{X}_F$.

If we take a look at the role generators in the positional system $\mathscr{S}_F$, we notice that the matrix for Business matches a core-periphery structure where the two central positions are held by the wealthiest families, cf. B and W. Although the correspondence between commercial relations and capital perhaps makes a lot of sense, it is, however, yet a revealing result. On the other hand, the role relations for Marriage correspond to the universal matrix, which means that the marital ties do not follow a particular pattern and they occur both between and within central and peripheral actors. The fact that Marriage acts as absorbing element to Business in the Role structure $\mathscr{Q}_F$ implies that at this aggregated level the composition of these two roles, i.e. BM, MB, MM, and BB, is equated to M, and hence there is no clear pattern in the composition of these social roles alone.

When we examine the relationship between the two social roles and the class attribute that is given at the bottom of Figure 5.8, both BW and MW are equal to the transposes of the commutative composition of such role relations, respectively; i.e. to $(WB)^T$ and $(WM)^T$. This results from the symmetry in the structure of the generators, and in both cases the central positions are maximally connected, but with some activity from the third class, especially with marriage.

Finally, most of the three-chain role relations are equated either to these two-chain Role structures or to the universal relation. In fact the only role relation of length 3 that differs from the above mentioned aggregated compounds is when a wealthy class is related to another wealthy class either by Business or by Marriage, i.e. WBW = WMW, where the two core positions make a clique and the peripheral class does not play any structural role.

Other equations in the Role structure until length three are for Business

$$BW = BWW = (WB)^T = (WWB)^T,$$

for Marriage

$$MW = MWW = BBW = MMW = BMW = MBW =$$

$$(WM)^T = (WWM)^T = (WBB)^T = (WMM)^T = (WBM)^T = (WMB)^T,$$

whereas the remaining three-chain compounds product of right multiplication correspond to the universal relation as M.

Larger compounds are likewise equated to one of the representative strings of the Role structure, and their substantial implications become more restricted as the length of the chain grows. It is, however, the configuration made of generators and compounds of length two and perhaps three that are more relevant for such interpretations of the network relational structure.

5.4.2   Inclusion of Role Relations

Another type of algebraic constraint in the Role structure deals with the inclusions among the role relations, which is among the representative relations product of the set of equations from Chapter 3. The partial order table to the right of the upper panel in Figure 5.8 represents this type of algebraic constraint. The partially ordered structure in this case shows that while Marriage includes all the string roles, Wealth is contained in all the rest of role relations.

In the case of Business ties, this role relation lies between these two extremes, meaning that whenever a commercial relation exists in the network positional system $\mathscr{S}_F$, whether with a wealthy or a less wealthy class of actor, then such role relation exists within a familiar setting; however, at this aggregated level, Marriage relations do not necessarily include Business ties. For the composite string roles, the two-chain role relations correspond to the class attribute and one of the social roles, and these compounds are included in the respective primitive social roles, which in turn contain the representative three-chain role relation.

However, the visual representation of the partial order structure brings nice insights about the configuration of string containments, and Figure 5.9 depicts the partial order table as an inclusion lattice diagram among role strings in $\mathscr{Q}_F$. Even though it was clear already from the positional system of $\mathscr{X}_F$ that for instance Marriage relations occur within a context of Business ties, the inclusion lattice of the Role structure helps us to see the containments among compound relations in $\mathscr{Q}_F$ as well.

From this hierarchy of collective ties we summarize the set of inclusions in the Role structure of Florentine families network for Business, Marriage, and Wealth as:

$$W \leq WBW \leq (\ BW \leq (B, MW) \ \text{ and } \ WB \leq (B, WM)\ ) \leq M.$$

It is worth mentioning that, although the Role structures where the number of Priorates is modelled are not shown here, there is no inclusion among the generators, except for the reflexive closure in the partial order structure. When Priorates and Wealth are modelled together, then the social roles and the class attributes are equated to each other, which means that political authority follows the economic power and vice versa. As a result, by taking the Role structure with the most transparent positional system, which in this case is with the actors Wealth, then we obtain the richest interpretation of the logic in the interlock of the role relations for the Florentine families network.

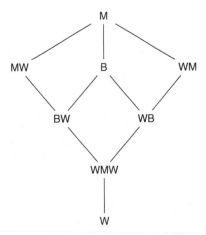

Figure 5.9   Inclusion lattice of the Florentine families network Role structure $\mathscr{Q}_F$ with Wealth. *The symmetry of the diagram is because the network is undirected.*

5.5 Role Structure in Multiplex Networks: Summary

Role structures are reduced relational configurations based on positional systems, which means that there was a strong emphasis on the structural similarity among the involved actors before modeling the relationships in the networks. This was because when considering chains of relations, even with few types of ties and with limited length chains, the number of combinations of unique strings tends to be very large and thus the size of the multiplication table.

Since the modeling of the actors had to be consistent with the algebraic approach used for the analysis of the network Role structure, and it was required to apply an equivalence type that can take into account chain of relations made of different kinds of tie. Compositional equivalence has been an ideal candidate for such task since Compositional equivalence is a partial structural equivalence based on the algebra of relational structures that takes into account the role interlock of the individual actors. These are strong arguments to apply this type of equivalence in the modeling of the actors of both in directed and undirected multiplex networks through their relational structures.

The fact that Role structures build form classes of aggregate actors implied that they represented the links between collective relations of multiple kinds that are occurring in the associated multiplex network structure. The algebraic system of the semigroup allowed constructing the accumulated relational structures for two kinds of empirical networks, one directed and another undirected where partially ordered semigroups revealed different role interlocks of the classes of actors in their positional systems. Partially ordered semigroups express different algebraic constraints of the relational structures where the foundations for a substantial interpretation of the studied multiplex networks lie. A further step in the analysis will be to perform a homomorphic reduction of the semigroup of role relations.

5.6   Learning Role Structure in Multiplex Networks by Doing

5.6.1   Incubator Network A

5.6.1.1   Plotting Multigraph of Figure 5.1

```
# load data for Incubator network A
R> data(incA)


# redefine node / edge / graph characteristics
R> scp <- list(cex = 3, ecol = c("green","orange","red"), lwd = 2, vcol = "#3399FF",
+    vcol0 = "#808080", bwd = .5, pos = 0, fsize = 8, rot = -75, mirrorX = TRUE)


# Fig. 5.1. plot multigraph of 'incA' with force-directed layout
R> multigraph(incA, layout = "force", seed = 22, scope = scp)
```

```
# Relation-Box of Incubator A with transpose relations and actor attributes
# (by default k = 3 where NA in 'tlbs' prevent tie transposition)
R> rb_A <- rbox(incA$net, transp = TRUE, tlbs = c("D", "G", "L", NA, NA))


# data structure in object 'rboxnetA'
R> str(rb_A)

List of 7
 $ w    : num [1:26, 1:26, 1:8] 0 0 0 0 0 0 0 0 0 1 ...
 ..- attr(*, "dimnames")=List of 3
 .. ..$ : NULL
 .. ..$ : NULL
 .. ..$ : chr [1:8] "C" "F" "K" "A" ...
 $ W    : num [1:26, 1:26, 1:584] 0 0 0 0 0 0 0 0 0 1 ...
 ..- attr(*, "dimnames")=List of 3
 .. ..$ : NULL
 .. ..$ : NULL
 .. ..$ : chr [1:584] "C" "F" "K" "A" ...
 $ lbs  : chr [1:26] "5" "6" "9" "12" ...
 $ Note : chr "Transpose relations are included"
 $ Trels: chr [1:3] "D" "G" "L"
 $ k    : num 3
 $ z    : int 584
 - attr(*, "class")= chr "Rel.Box"
```

```
# Cumulated Person Hierarchy of the above relation-box
R> cph_A <- cph(rb_A)
```

```
# plot lattice diagram of Cumulated Person Hierarchy in Fig. 5.2.
# function diagram() requires "Rgraphviz"
R> require(Rgraphviz)
R> diagram(cph_A)
```

```
# Table 5.4 permute poset of the CPH of 'incA'
R> perm(cph_A, clu = c(1,2,3,3,3,2,4,2,2,1,1,3,4,1,3,3,1,4,3,3,3,3,3,4,1,4))
```

5.6.1.2   Positional System $\mathscr{S}_A$

```
# reduce matrices in 𝒳_A to produce its positional system
R> ps_A <- reduc(rb_A$w,
+     clu = c(3,2,1,1,1,2,4,2,2,3,3,1,4,3,1,1,3,4,1,1,1,1,1,4,3,4))
```

```
# Table 5.5 role structure 𝒬_A
R> S_A <- semigroup(ps_A, type = "symbolic")
```

```
# partial order of the strings in 𝒮_A
R> P_A <- partial.order(strings(ps_A))

# Fig. 5.5. Hasse diagram of the poset
R> diagram(P_A)
```

5.6.2   *Florentine Families Network,* $\mathscr{X}_F$

5.6.2.1 $\mathscr{X}_F$ with Actor Attributes

```
# load network and actor attributes data from public repository
# both are Ucinet data format
R> ffnet <- read.dl(file = "http://moreno.ss.uci.edu/padgett.dat")
R> ffatt <- read.dl(file = "http://moreno.ss.uci.edu/padgw.dat")
```

```
# sort attribute labels to match network
R> ffatt <- ffatt[order(rownames(ffatt)), ]
# labels
R> fflbs <- rownames(ffatt)
```

```
# Table 5.7 with actors having wealth above 40
R> dichot(ffatt[,1], c = 40)
```

ACCIAIUOL	ALBIZZI	BARBADORI	BISCHERI	CASTELLAN	GINORI	GUADAGNI	LAMBERTES
0	0	1	1	0	0	0	1
MEDICI	PAZZI	PERUZZI	PUCCI	RIDOLFI	SALVIATI	STROZZI	TORNABUON
1	1	1	0	0	0	1	1

```
# actors having priorates above 34
R> dichot(ffatt[,2], c = 34)
```

ACCIAIUOL	ALBIZZI	BARBADORI	BISCHERI	CASTELLAN	GINORI	GUADAGNI	LAMBERTES
1	1	0	0	0	0	0	0
MEDICI	PAZZI	PERUZZI	PUCCI	RIDOLFI	SALVIATI	STROZZI	TORNABUON
1	0	1	0	1	1	1	0

```
# Wealth relations with a cut-off value of 40
R> ffw <- matrix(0, nrow = 16, ncol = 16, dimnames = list(fflbs, fflbs))
R> diag(ffw) <- dichot(ffatt[,1], c = 40)
```

```
# bind Wealth to network ties
R> ffnetw <- zbind(ffnet, ffw)
```

```
# redefine node / edge / graph characteristics
R> scp <- list(cex = 6, fsize = 8, pos = 0, vcol = 8, ecol = 1, lwd = 2, bwd = .5)
R> scp2 <- list(directed = FALSE, layout = "force", seed = 2, cex = ffatt[,1])

# Fig. 5.6. plot undirected multigraph of 'ffnet' with force-directed layout
R> multigraph(ffnet, scope = c(scp,scp2), ffamily = "serif", fstyle = "bolditalic")
```

```
# Table 5.6 Cumulated Person Hierarchy of ℋF
R> perm(cph(rbox(ffnet, k = 5)), clu = c(11,12,1,2,3,15,4,5,6,7,8,16,13,9,14,10))
```

5.6.2.2   Cumulated Person Hierarchies $\mathscr{H}_F$

```
# Wealth relations with a cut-off value of 40
R> ffw <- matrix(0, nrow = 16, ncol = 16, dimnames = list(fflbs, fflbs))
R> diag(ffw) <- dichot(ffatt[,1], c = 40)
```

```
# bind Wealth to network ties
R> ffnetw <- zbind(ffnet, ffw)
```

```
# construct ℋF from relation-box with 5-paths
R> rb_F <- rbox(ffnetw, k = 5)
R> cph_F <- cph(rb_F)

# Fig. 5.7. above
R> diagram(cph_F)
```

```
# construct diagonal matrix for Priorates
R> ffp <- matrix(0, nrow = 16, ncol = 16, dimnames = list(fflbs, fflbs))

# establish a cut-off value for attributes
R> diag(ffp) <- dichot(ffatt[,2], c = 34)
```

```
# bind diagonal matrix with the network
R> ffnetp <- zbind(ffnet, ffp)
```

```
# construct CPH from the Relation-Box with 4-paths on 'ffnetp'
R> rb_Fp <- rbox(ffnetp, k = 4)
R> cph_Fp <- cph(rb_Fp)

# Fig. 5.7. middle
R> diagram(cph_Fp)
```

```
# bind Wealth & Priorates with the network
R> ffnetwp <- zbind(ffnet, ffw, ffp)
```

```
# construct CPH from the Relation-Box with 4-paths on 'ffnetwp'
R> rb_Fwp <- rbox(ffnetwp, k = 4)
R> cph_Fwp <- cph(rb_Fwp)

# Fig. 5.7. below
R> diagram(cph_Fwp)
```

5.6.3   Role Structure of $\mathcal{X}_F$ with Wealth

5.6.3.1   Positional System $\mathcal{S}_F$

```
# positional system with customized string labels
R> ps_F <- reduc(rb_F$w, clu = diagram.levels(cph_F, perm = TRUE)$clu,
+    slbs = c("M","B","W"))

# remove isolates
R> ps_F <- rm.isol(ps_F)
```

```
, , M                                          , , W

    [,1] [,2] [,3]                                 [,1] [,2] [,3]
[1,]   1    1    1                            [1,]   1    0    0
[2,]   1    1    1                            [2,]   0    1    0
[3,]   1    1    1                            [3,]   0    0    0

, , B

    [,1] [,2] [,3]
[1,]   1    1    1
[2,]   1    1    0
[3,]   1    0    0
```

5.6.3.2 Algebraic Constraints

```
# Fig. 5.8 Edge table of the positional system 𝒮_F
R> edgeT(ps_F)$ET

   1 2 3
1  1 1 4
2  1 1 5
3  6 7 3
4  1 1 4
5  1 1 5
6  6 6 8
7  6 6 8
8  6 6 8


# three-path string relations of 𝒮_F with equations on primitives
R> st_F <- strings(ps_F, equat = TRUE, k = 3)

...

$st
[1] "M"    "B"    "W"    "MW"    "BW"    "WM"    "WB"    "WMW"

$equat
$equat$`M`
 [1] "M"    "MM"    "BB"    "MB"    "BM"    "MMM"   "BBM"   "MBB"   "MMB"   "BBB"   "BMM"   "BMB"
[13] "MBM"  "MWM"   "BWB"   "MWB"   "BWM"

$equat$W
[1] "W"    "WW"    "WWW"

...

attr(,"class")
[1] "Strings"
```

5.6.3.3 Factorization and Congruence Lattice

```
# partial order structure according to the unique string relations
R> P_F <- partial.order(st_F, type = "strings")

    M B W MW BW WM WB WMW
M   1 0 0  0  0  0  0   0
B   1 1 0  0  0  0  0   0
W   1 1 1  1  1  1  1   1
MW  1 0 0  1  0  0  0   0
BW  1 1 0  1  1  0  0   0
WM  1 0 0  0  0  1  0   0
WB  1 1 0  0  0  1  1   0
WMW 1 1 0  1  1  1  1   1
attr(,"class")
[1] "Partial.Order" "strings"
```

```
# Fig. 5.9. lattice diagram of role structure 𝒬_F
R> diagram(P_F)
```

```
# semigroup of relations
R> S_F <- semigroup(ps_F, type = "symbolic")


...
$S
      M   B   W   MW   BW  WM  WB  WMW
M     M   M   MW  MW   MW  M   M   MW
B     M   M   BW  MW   MW  M   M   MW
W     WM  WB  W   WMW  WMW WM  WB  WMW
MW    M   M   MW  MW   MW  M   M   MW
BW    M   M   BW  MW   MW  M   M   MW
WM    WM  WM  WMW WMW  WMW WM  WM  WMW
WB    WM  WM  WMW WMW  WMW WM  WM  WMW
WMW   WM  WM  WMW WMW  WMW WM  WM  WMW

attr(,"class")
[1] "Semigroup" "symbolic"
```

6

Decomposition of Role Structures

6.1 Aggregation and Decomposition

A significant step in the analysis of multiplex network has been the employ of graph homomorphisms, which are particularly useful in the reduction of the network structure by establishing the positional system for the entire configuration. For instance, we applied in Chapter 5 Role Structure in Multiplex Networks a type of correspondence that considers both the different levels in the relations and the combinations of the distinct types of tie. Compositional equivalence applies both for directed and for undirected multiplex networks, which resulted in an arrangement made by classes of partially structurally equivalent actors.

However, the representation of relational structures from multiplex networks typically results in large and complex systems, even if they originate from smaller positional systems. This aspect makes difficult the practical interpretation of these systems, and therefore it is required to account with methods and techniques to reduce (if possible) the relational or role structure into a smaller components. This kind of reduction is a process known as *decomposition*, which leads to an aggregated relational structure that allows making a more straightforward interpretation of the complex system in substantial terms.

A major benefit of counting with aggregated relational structures is that they are easier to compare with each other by looking at a common underlying mechanism that is playing on the configurations of different population networks. Fewer distinctions in the overall structure, however, can come with a cost since the representation of the network relational structure can result being ambiguous and less transparent, particularly when generators are equated with each other. Nonetheless, the potential benefits that the homomorphic reduction of semigroup of relations brings with the additional simplification of the relational and role tables makes the topic of decomposition a crucial part of the analysis of multiplex network structures.

The commutative diagram of Figure 6.1, adapted from Pattison (1981), provides a schema of the analysis process of aggregation and decomposition until and from the decomposition of the role structures. In this diagram, solid arrows are the homomorphisms performed between different types of structures, whereas dashed arrows constitute alternative homomorphisms to reduce the structure that theorems assert exist.

Algebraic Analysis of Social Networks: Models, Methods and Applications using R,
First Edition. J. A. R. Ostoic. Companion website: www.wiley.com/go/ostoic/algebraicanalysis.
© 2021 John Wiley & Sons Ltd. Published 2021 by John Wiley & Sons Ltd.

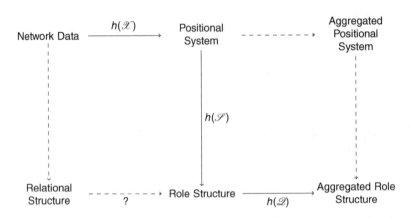

Figure 6.1 Commutative diagram of the network relational structure aggregation process. $h(\mathscr{X})$, $h(\mathscr{S})$, and $h(\mathscr{Q})$ are functions for homomorphic reductions of the algebraic structure where dashed arcs stand for other homomorphisms.

The starting point in the aggregation and decomposition analysis process employs the raw network data $\mathscr{X}$. This network is typically made of different types of relations and can include actor attributes as well. $h(\mathscr{X})$ represents the reduction of the network structure by means of a structurally equivalence type of relation among the network actors. For networks $\mathscr{X}_A$ and $\mathscr{X}_F$, for instance, the correspondence applied has been Compositional equivalence, and this is because it makes it possible to account for systems made of different kinds of relations.

The outcome of the reduction mapping $h(\mathscr{X})$ is a positional system of the network where actors are classified according to their different type of ties. The network Role structure is obtained with mapping $h(\mathscr{S})$, and hence it constitutes the relational structure of the network positional system $\mathscr{S}$. Both the role and the relational structures are represented by the algebraic semigroup object, which carries the structure made of relationships among the network relations.

Finally, $h(\mathscr{Q})$ represents the homomorphic reduction of the semigroup that is made either by a factorization procedure for partially ordered semigroups (Pattison, 1993) or by congruence classes of elements in abstract semigroups (Hartmanis and Stearns, 1966). As a result, the mapping $h(\mathscr{Q})$ produces an aggregation of the network role structure in the form of semigroup factors and maximal homomorphic images. Typically, we are most interested in the maximal "non-trivial" homomorphic reductions of the semigroup system for a substantial interpretation of the network relational structure.

Both Boorman and White (1976) and Pattison (1981) point out that although the blockmodeling of the network is product of a homomorphism, it is an open question whether the role structure is directly product of a homomorphic reduction of the semigroup of relations representing the network relational structure. However, since we count with the semigroups generated by the network blockmodels, we are able to perform the decomposition of these role structures directly through a semigroup homomorphism and obtain an aggregated semigroup that is generated by the positional system of the network. In this sense, the decomposition of the role structures corresponds to the process given in the right part of the commutative diagram with the reduction of the blockmodel semigroups.

6.1.1 Homomorphic Reductions

Even though the positional systems of the networks represent a homomorphic reduction of the raw network data, we define just now this type of homomorphism aimed to simplify algebraic structures. In this sense, the decomposition of the partially ordered semigroups representing the networks relational structures is based on the principles defined next that are aimed to the modeling of the network role relations in order to obtain an aggregation of the role structures.

A *homomorphic reduction of a semigroup* of relations $S(R)$ (Boorman and White, 1976; Breiger and Pattison, 1986) is a partition π of the words in the semigroup structure into equivalent classes $[x_1], [x_2], \ldots , [x_m]$ in such way that for each of the images $h(a) \in [x_i]$, $h(b) \in [x_j]$, and for k indexing any class $[x]$ including i or j:

$$(h(a), \ h(b)) = h(c) \in [x_k].$$

This means that every product between $h(a)$ and $h(b)$ holds for a result $h(c)$ that is assigned to the same class. Obviously, if the images correspond to a semigroup of relations, then the class membership is consistent with the compound operation.

As a result, with a homomorphic reduction of a semigroup of relations we obtain a homomorphic image of the semigroup also called "blockmodel semigroup". This is a quotient semigroup that constitutes a robust approximation to the role structure of the network, and it is important to mention that although other equations have been added to the structure of the homomorphic image, the original equalities in the semigroup are still holding true.

The existence of the homomorphic reduction of the semigroup permits the applying of different theories in the interpretation of the network role structure and the role interlock between the actors. Even though it is also possible to go one step further and construct a semigroup of the aggregated blockmodel that is the homomorphic image of the blockmodel semigroup, in any case it is needed a decomposition method to simplify either the network role structure or the blockmodel semigroup. Different methods exist for this purpose and two of them are introduced in the next section, which can be applied in the decomposition of the resultant role structures.

6.2 Synthesis Rules

A fundamental structure problem in algebra is decomposing an algebraic system into simpler components from which the given structure can be reconstructed by a *synthesis rule*. According to Pattison (1993, cf. also Pattison and Bartlett (1982)) two forms of decomposition or synthesis rules can be applied to the semigroup of relations, and which are introduced next.

6.2.1 Direct Representation

A synthesis rule widely used is of the *direct representation* where the algebra is decomposed into direct components, which are simpler structures that remain with a relatively high degree of independence to each other. The reason for the independence of the direct components is because the elements of the compound structure are in this case members of the Cartesian product structure.

In this sense, for algebras $S_1, S_2, \ldots, S_s$ the *direct product* of $S_1 \times S_2 \times \cdots \times S_s$ is defined as the set of *n*-tuples $(x_1, x_2, \ldots, x_n)$ where $x_i \in S_i$ and for $n, m \geq 0$:

$$(x_1, x_2, \ldots, x_n)(y_1, y_2, \ldots, y_m) = (x_1 y_1, x_2 y_2, \ldots, x_n y_m).$$

The partial order for this kind of product is given for each $i = 1, 2, \ldots, r$ by:

$$(x_1, x_2, \ldots, x_n) \leq (y_1, y_2, \ldots, y_m) \text{ iff } x_i \leq y_i$$

where $x_i, y_i \in S_i$, for $i = 1, 2, \ldots, r$.

If an algebra is a direct product of different relational structures, then the algebra is said to be *directly reducible*. In other words, a directly reducible algebra implies that direct components exist for the given structure. A direct product involves the set of all ordered pairs of elements from the two constituent algebras. The Kronecker product on partially ordered semigroup structures determines a directly reducible partial order of a given semigroup.

6.2.2 Subdirect Representation

In many cases, the direct product as a synthesis rule may not be feasible. For example, a partially ordered semigroup with an odd number of distinct elements are not directly reducible, and for structures having an even number of elements the proportion of directly reducible semigroups are probably small (Pattison, 1993, p. 145). Although a direct representation is desirable and certainly useful, in most of the cases we need to obtain another type of synthesis rule where some overlapping in the product is permitted.

In this sense, another synthesis rule widely used is of the *subdirect representation* of a family of algebras $\{S_i\}$ onto subsets of the algebra. For semigroups S and T, $h : S \to T$ is a surjective homomorphism whenever h is a subdirect product of $S \times T$. Furthermore, given that $\equiv$ is an equivalence relation on a semigroup S, then $\equiv$ is a congruence relation if and only if it is a subdirect product of $S \times S$ Boyd (1991, pp. 87–88).

A subdirect product of partially ordered semigroups is any subsemigroup that is a subset product of the semigroup such that each element in every constituent component semigroup S_i occurs at least once as an element in S. If the algebra is product of a subdirect representation, then it means that the structure is *subdirectly reducible*, and that corresponds to an isotone homomorphism of the semigroup.

If, for example, we consider the case of both $\mathcal{Q}_A$ and $\mathcal{Q}_F$, which are the Role Structures that correspond to $\mathcal{X}_A$ and $\mathcal{X}_F$ for Incubator A and Florentine families networks, the partially ordered semigroups have an even number of elements. This means that they may be directly irreducible; however, it is extremely unlikely that these Role Structures constitute the set of all ordered pairs of elements from other semigroups, which is a condition for a direct representation. Hence, we need to account for a subdirect representation of these semigroups as the form for decomposition of the Role Structures.

It is likely, however, that a given algebraic structure cannot be represented as a subdirect product of "smaller" algebras, which means that such structure is *subdirectly irreducible*. Similarly, any algebra is said to be *directly indecomposable* when it cannot be non-trivially represented as Cartesian products of other algebras. Subdirectly irreducible algebras are usually building blocks for other algebras and they play an important role in the decomposition of Role Structures.

6.3 Lattice of Congruence Relations

If we want to perform the decomposition of the semigroup of relations structures, then we need to refer to the Lattice of Homomorphisms of the Semigroup, which is defined later on in this chapter, and represents the partition of the free semigroup of relations as congruence relations. This type of lattice structure characterizes the existence of direct product representation for the algebra, and inside its configuration is also represented the *Lattice of Congruence Relations* for the free semigroup that correspond to each type of homomorphism in $FS(R)$.

The Lattice of Congruence Relations in a given Role structure $\mathscr{D}$ is a complete type of lattice that is denoted as $L_\pi(\mathscr{D})$, and in this type of lattice both the suprema and the infima elements are partitions of the algebraic structure. In such cases, either everything belongs to the same class or each element makes its own class, and between these extremes there are the non-trivial partitions of the algebraic structure that are ordered by inclusion. Since we denote the partition of the algebraic structure in classes of congruent elements with the partition symbol π and the congruence relation in a semigroup of relations $S(R)$ by the relation π_i or else by the π-relation i, then the homomorphic image of the semigroup is represented as S/π_i.

A family of congruence relations are then represented by the π-relations and—if we consider a semigroup structure S—these are represented by lattice of congruences that is denoted by $L_\pi(S)$. The partially ordered semigroup S is then a direct product of the partially ordered semigroups $S_1, S_2, \ldots, S_s$ if and only if there exist π-relations $\pi_1, \pi_2, \ldots, \pi_r \in L_\pi(S)$ such that the partially ordered semigroup is isomorphic to the corresponding quotient semigroups.

A complete lattice of congruences implies the existence of a suprema and infima elements in the structure such that:

$$\pi_1 \vee \pi_2 \vee \ldots \vee \pi_r = \pi_{\max}$$
$$\pi_1 \wedge \pi_2 \wedge \ldots \wedge \pi_r = \pi_{\min},$$

which in this context are called respectively the *maximal* and *minimal* elements of the lattice structure. Besides, with a direct from for representation we count with a complete Lattice of Congruence Relations having π-relations that are commutative; i.e. $\pi_i \pi_j = \pi_j \pi_i$ for all $i, j = 1, 2, \ldots, r$.

The π-relation lattice $L_\pi(S)$ determines a subdirect representation as well. A partially ordered semigroup S is isomorphic to a subdirect product of the partially ordered semigroups $S_1, S_2, \ldots, S_s$ if there exists π-relations $\pi_1, \pi_2, \ldots, \pi_r \in L_\pi(S)$ such that the partially ordered semigroup is isomorphic to the corresponding quotient semigroups, and also that there exists a minimal element according to the above definition.

This means that a partially ordered semigroup that is subdirectly reducible usually admits a number of different subdirect representations. A determinant aspect in the analysis is to find out how to select the best subdirect representation to characterize the structure being decomposed (Pattison, 1993).

The Lattice of Congruence Relations for a semigroup is represented by $L_\pi(S)$ represents the Lattice of Congruence Relations for a semigroup, and if S is partially ordered, the minimal element in $L_\pi(S)$ or $\pi_{\min}$ is the partial order structure itself, whereas the maximal element of the lattice $\pi_{\max}$ is the universal relation with no structure and considered as a trivial partition of S. If the Lattice of Congruence Relations has a unique element covering the partial order, then $L_\pi(S)$ represents a two-element structure made of the partial order and the one-element semigroup, which means that the partial order structure is subdirectly irreducible; otherwise S is reducible.

6.4 Factorization

A decomposition technique based on subdirect representation for the partially ordered semigroup was proposed by Pattison and Bartlett (1982, cf. Ardu (1995) for the implementation algorithm) as *factorization*. With this procedure, the algebraic structure is factorized into simpler constituent components termed as *Factors*, which are semigroup factors, and where each one constitutes a particular homomorphic image of the algebra. With partially ordered semigroups, Factors constitute isotone homomorphic images that correspond to maximally independent features of the original semigroup of relations, where the quotient semigroups and the reduced partial order tables are part of the *factorizing set* (Pattison, 1993).

The factorization procedure of a given role structure is based on the Lattice of Homomorphisms of the Semigroup, to be defined in the last part of this section, and more specifically on the dual part of this structure found in the Lattice of Congruence relations or simply *Congruence lattice*, $L_\pi(S)$ defined in the previous section. However, since the complete lattice representation of the semigroup is practically unfeasible due to its large size, with this method it is sufficient to establish the minimal elements of the Congruence lattice, which was defined in Chapter 2 and it was denoted as $L_\pi(\mathcal{D})$. If we consider a partially ordered semigroup, then the minimal element of the lattice of congruence relations $L_\pi(\mathcal{D})$ is represented by the partial order structure of the semigroup. This kind of table is then a unique element that stands for a trivial partition of the semigroup configuration where each unique string of relation makes its own class.

A proper reduction of the partially ordered semigroup begins with the structures that cover this minimal element in the Lattice of Congruence Relations, and this is due to the existence of a number of equations in this partially ordered structures. A partially ordered semigroup structure is subdirectly irreducible if and only if its lattice $L_\pi(S)$ has a unique element covering the minimal element, otherwise is subdirectly reducible. In consequence, an important task with the factorization procedure is to identify the particular elements covering the minimal element $\pi_{\min}$.

6.4.1 Atoms and their Meet-Complements

An element π_a that covers $\pi_{\min}$ in the lattice structure such that for all $\pi_i \in L_\pi(S)$:

$$\pi_a > \pi_{\min}, \quad \text{iff} \quad \pi_a \geq \pi_i \text{ and } \pi_i > \pi_{\min} \text{ implies } \pi_a = \pi_i,$$

is called an *Atom*. In this sense, according to this definition of an Atom, there is no other element π_i immediately covering the minimal element unless it is an Atom as well. As a result, if $\pi_a = \pi_i$ then it means that the structure is subdirectly reducible because it has more than one Atom; otherwise the structure is subdirectly irreducible.

Pattison (1993) provides an algorithm to perform the factorization of partially ordered structures in two main steps. First by eliminating the subdirect representations with reducible components, and then by defining a partial ordering on the remaining subdirect representations in terms of efficiency; that is the attention is restricted to the irredundant subdirect forms for representation of the lattice components. The factorization of the partially ordered semigroup S is any minimal irredundant subdirect representation of S.

The identification of Atoms in the Congruence lattice representing the partially ordered semigroup is then a fundamental step with the factorization procedure and in practical terms they represent some small additions to the partial order table where the elements in S satisfy the substitution property after performing the automorphism. However, this means that the type of structure resulting from the extra equations do not differ much from the original formation of relations, and in this sense it is desirable that the basis for the decomposition of partially ordered semigroups has further aggregations to the partial order of the structure.

Atoms are *join-irreducible* elements in the lattice structure, and this means that for $\pi_a \neq \pi_{\min}$ and for all the π-relations in $L_\pi(S)$:

$$\pi_a = \pi_i \vee \pi_j \quad \text{implies that} \quad \pi_a = \pi_i \text{ or } \pi_a = \pi_j;$$

that is, Atoms can not be join product of any of the elements in the lattice.

Meet-irreducible elements are dually defined for π_a as:

$$\pi_a = \pi_i \wedge \pi_j \quad \text{implies that} \quad \pi_a = \pi_i \text{ or } \pi_a = \pi_j.$$

An element π_a^* is the *meet-complement* of a join irreducible element $\pi_a \in L_\pi(S)$ if and only if:

$$\pi_a^* > \pi_{\min} \quad \text{and} \quad \pi_a \wedge \pi_a^* = \pi_{\min},$$

and a meet-complement is *maximal* if there is no other meet-complement element π_i^* such that

$$\pi_i^* > \pi_a^*.$$

The meet-complements are meet-irreducible elements and they share the properties of the partial order structure. This means that each meet-complement is reflexive, transitive, and anti-symmetric.

6.4.2 Lattice of Homomorphisms of the Semigroup

The unique factorization for the Congruence lattice is associated with the set of maximal meet-complements of $L_\pi(S)$, which constitutes a homomorphism of the Congruence lattice onto its factors. If the factorization is not unique; that is the structure does not possess a unique irredundant subdirect representation, then we need to check the kind of structure representing the congruence relations of the semigroup.

If the congruence representation has the form of a distributive lattice, then the decomposition is unique. However, if the lattice structure is modular—a structure given at the end of Chapter 2 in Figure 2.4—and nondistributive then, according to the Kurosh–Ore Theorem, the number of components in any irredundant subdirect representation of the semigroup is independent of such representation and the factorization is not unique (Pattison, 1993, cf. also Birkhoff (1967)).

The collection of meet-complement relations for each Atom in the Congruence lattice corresponds to a certain way of decomposition of the partial order semigroup structure, which is by the *meet-complements of the Atoms*. This kind of configuration gives raise to the *lattice of homomorphisms of the semigroup L(S)*, which is another type of lattice structure that is dual to the Congruence lattice.

In the structure $L(S)$ is the set of elements in the abstract semigroup that is partially ordered by homomorphic mappings resulted from the congruence relations among them. The maximal element in the Lattice of Homomorphisms of the Semigroup is S while the minimal element that is the one-element semigroup resulting from the trivial partition of the semigroup structure by π_{max}. On the other hand, the minimal element in the Congruence lattice is the partial order associated with the semigroup while the maximal element is the universal relation or in this case the "one-element partial order."

The importance of the Lattice of Homomorphisms of the Semigroup $L(S)$ is not only in the decomposition process of a partially ordered semigroup, but it is perhaps even bigger when it comes to compare different relational structures of two or more networks or else overlapping images of the same network.

Finally, a *progressive factorization* process is possible when the successive image matrices generated by the decomposition procedure are subdirectly reducible as well. Although such process can generate a large number of (possibly) smaller subdirect representations, some of these structures product of factorization can result isomorphic to each other even with the extreme case of counting with a single non-trivial subdirect representation such as the one-element partially ordered semigroup. In this sense, the progressive factorization allows to find the smallest non-trivial structures and the algebraic constraints associated with such structures. All the information in these restrictions allows making a straightforward substantial interpretation of the network relational structure and also of the logic in the interlock of the relations.

6.5   Congruences by Substitution Property

Factorization is an efficient way to perform the decomposition of semigroups. However, there are other techniques to reduce semigroup of relations, and one method is that of the substitution property, which is used for the partition of sequential machines as specified in Hartmanis and Stearns (1966).

The main difference between the substitution property and the factorization lies in the type of structure where the equivalence applies. Factorization focuses on the partial order structure where congruences are found through induced inclusions, whereas congruences by substitution property applies directly to the semigroup. This means that this latter method is the one chosen for *abstract* semigroup structures without a partial order.

An example of the decomposition of a semigroup of relations in terms of the congruence property involves machines, and we start by defining a *finite state machine* or *Mealy machine* in algebraic terms as the 5-tuple:

$$M = \langle\, S,\, X,\, Z,\, \delta,\, \lambda\, \rangle$$

where X and Z are the *inputs* and the *outputs*, and S represents the *internal states* of the machine (and where $S_0 \in S$ represents the initial state).

The other two elements, λ and δ, are functions $S \times X \to S$ and $S \times X \to Z$.

A partition π on the sets of M is said to have the *substitution property* if and only if for all $a \in X$

$$s \equiv t(\pi) \qquad \text{implies that} \qquad \delta(s, a) \equiv \delta(t, a)(\pi)$$

where s and t are classes of elements in the π-image matrices of the machine.

Hence, for a partition on S of M each input X maps blocks of the partition into blocks of π. The π-*image* of the state machine is another state machine with the form

$$M_\pi = \langle [B_\pi], X, \delta_\pi \rangle$$

where for $\{B_\pi, B'_\pi\} \in \pi$

$$\delta_\pi(B_\pi, x) = B'_\pi \qquad \Leftrightarrow \qquad \delta(B_\pi, a) \subseteq B'_\pi.$$

6.6   Aggregation of Role Structures in $\mathscr{Q}_A$

Now we apply the factorization procedure in the decomposition of relational structures, and we start with the Role Structure that corresponds to the Incubator network A, $\mathscr{Q}_A$. The Role Structure $\mathscr{Q}_A$ is given in Chapter 5 Role Structure in Multiplex Networks, and it corresponds to a directed network with three types of social relations with relational contrast and two kinds of actor attributes. Hence, this relational structure has more components than Role Structure of Florentine Families Network, for example.

The Role Structure $\mathscr{Q}_A$ is represented by the partially ordered semigroup where actor attributes are recorded in a relational form, and the inclusion lattice diagram constitutes the representation form used for the partial order structures. Although we can appreciate the different levels of inclusion among the strings of relations, the specific inclusions in the semigroup structure are not specified, and this kind of information is the starting point in the factorization procedure since it provides the connections, or rather the lack of them, which make the other structures inside the Congruence lattice.

Moreover, most of the semigroups for the Role Structures have been given with the right multiplication table alone. We still need to consider the complete representation of the semigroup structures, and this is because the semigroup of relations also represents the role structures and, similar to the partial order, they are subject to simplification through a decomposition process. Thus the decomposition of the Incubator networks is based on the partially ordered semigroups of the role structures representing the empirical systems of the study.

Two main structures make this type of partially ordered set. First, the semigroup of relations that reflect the products of the combinations among the distinct strings of role relations, and then the partial order table that provides the set of inclusions between pairs of role relations in the system. In this sense, we are going to look at the factorization procedure in detail for Incubator network A, and then we take other systems of directed and undirected multiplex networks.

In this sense, the partial order table for $\mathscr{Q}_A$, which represents a hierarchy of role relations of the Incubator network A, is given in Table 6.1, and this structure constitutes the minimal element in the lattice of Congruence relations for the role structure of $\mathscr{X}_A$, which is represented

Table 6.1 Partial order table for the role structure of Incubator network A, $PO(\mathscr{Q}_A)$. *This set of inclusions is the basis for the algebraic constraint in* $\mathscr{Q}_A$ *with the hierarchy of role relations.*

≤	1	2	3	4	5	6	7	8	9	10	11	12	13	14	15	16	17	18	19	20
1 = C	1	1	0	0	0	0	0	0	1	0	0	0	1	1	0	1	0	0	0	0
2 = F	0	1	0	0	0	0	0	0	1	0	0	0	0	1	0	0	0	0	0	0
3 = K	1	1	1	0	1	1	0	1	1	1	1	1	1	1	1	1	1	1	1	1
4 = A	1	1	0	1	1	1	1	0	0	1	0	0	0	1	1	0	1	0	0	0
5 = D	0	0	0	0	1	1	0	0	1	0	0	0	1	1	0	1	0	0	0	0
6 = G	0	0	0	0	0	1	0	0	1	0	0	0	0	0	0	1	0	0	0	0
7 = L	1	1	0	0	1	1	1	1	1	1	1	1	1	1	1	1	1	1	1	1
8 = CK	1	1	0	0	0	0	0	1	1	1	0	0	1	1	0	1	1	1	1	1
9 = CG	0	0	0	0	0	0	0	0	1	0	0	0	0	0	0	0	0	0	0	0
10 = FK	0	1	0	0	0	0	0	0	1	1	0	0	0	1	0	0	1	0	0	1
11 = KD	0	0	0	0	0	1	0	0	1	0	1	0	0	0	0	1	1	1	0	0
12 = KL	1	1	0	0	1	1	0	1	1	1	1	1	1	1	1	1	1	1	1	1
13 = DC	0	0	0	0	0	0	0	0	1	0	0	0	1	1	0	1	0	0	0	0
14 = DF	0	0	0	0	0	0	0	0	1	0	0	0	0	1	0	0	0	0	0	0
15 = DK	0	0	0	0	1	1	0	0	1	0	1	0	1	1	1	1	1	1	1	1
16 = GC	0	0	0	0	0	0	0	0	1	0	0	0	0	0	0	1	0	0	0	0
17 = CKD	0	0	0	0	0	0	0	0	1	0	0	0	0	0	0	0	1	0	0	0
18 = KDC	0	0	0	0	0	0	0	0	1	0	0	0	0	0	0	1	1	1	0	0
19 = DCK	0	0	0	0	0	0	0	0	1	0	0	0	1	1	0	1	1	1	1	1
20 = DCL	0	0	0	0	0	0	0	0	1	0	0	0	0	1	0	0	1	0	0	1

as $L_\pi(\mathscr{Q}_A)$, and from which the rest of the inclusions are based. On the other hand, the maximal element of this lattice structure corresponds to the universal matrix where the overall set of string relations in $\mathscr{Q}_A$ in this case makes a single class of elements where no string of tie covers another relation.

Each of the absences in the inclusions within the partial order structure constitute an induced or generating inclusion to the partial order, which gives rise to a particular π-relation in the Lattice of Congruence Relations. However, many of the induced inclusions may be equal and they will produce identical π-relations, which means that the Congruence lattice structure for the system needs to consider just the different types of congruence relations.

6.6.1 Atoms with Meet-Complements in Role Structure $\mathscr{Q}_A$

The factorization procedure of a partially ordered semigroup starts by making additions to the partial order table, which is the minimal element of $L(S)$. Such additions are generating inclusions that will lead to—most likely—different π-relations in which the partition and the aggregation of the multiplication table is based on. The aim is to be able to apply the substitution property in the semigroup, and this is typically done with additional generating inclusions

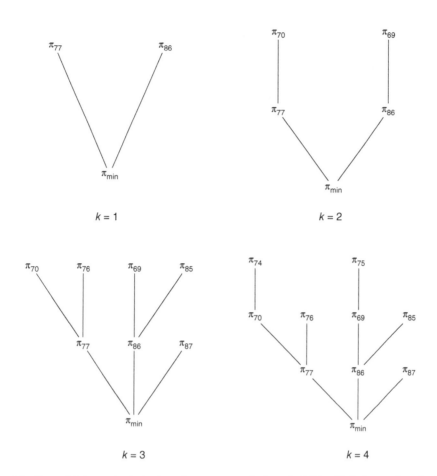

Figure 6.2   Potential Atoms in the lattice of congruence relations, $L_\pi(\mathcal{Q}_A)$ with k-shortest length of additions to the partial order. *The Atoms of $L_\pi(\mathcal{Q}_A)$ are π_{77}, π_{86}, and π_{87}.*

to the partial order. Hence, each "zero" in the partial order structure has the potential to produce a particular aggregated structure but, unless the partially ordered semigroup is small, some generating inclusions will lead to identical results.

In the case of the relational structure that corresponds to Incubator network A, $\mathcal{X}_A$, the factorization procedure produce in total 257 induced inclusions to the partial order of $\mathcal{Q}_A$ from which 104 are unique. These unique generating inclusions then constitute partition relations that can be partially ordered by inclusion making up the elements of the Congruence lattice for this Role Structure. Hence, an important aspect in the factorization process is that the different π-relations belong to the Lattice of Homomorphisms of the Semigroup, which is a complete lattice composed by join-irreducible and meet-irreducible elements. Recall that the former are the Atoms of the lattice and identifying these elements is crucial for a successful factorization of the partially ordered semigroup.

Figure 6.2 represents ordered structures of the potential (and actual) Atoms in the Congruence lattice of $\mathcal{Q}_A$, which are the shortest generating inclusions to the partially ordered structure that correspond to this network. For instance, π-relations 77 and 86 correspond to the shortest generating inclusions, which are 11, 14 and 10, 16; each one with six additions to the partial order. For $k = 2$; that is, for the second shortest generating inclusions, there is also a pair of elements that covers the previous two π-relations. This means that these elements are not Atoms because they are not join-irreducible. However, for $k = 3$, the π-relation 87 that corresponds to the generating inclusion 1, 17 is a join-irreducible element and hence an Atom, and this is despite the fact that it has more additions than the joint-reducible π-relations 70 and 69.

For $k = 4$, there are no more join-irreducible elements appearing apart from the latest inclusion that corresponds to π_{87}. Hence, we could conclude that the set of join-irreducible elements or Atoms in the Congruence lattice has three elements. However, even if there still the possibility of counting with more Atoms in the lattice structure, we are able to look at the complete Congruence lattice of $L_\pi(\mathcal{D})$ made from the induced inclusions to the partial order structure $\mathcal{D}$. We evidence in this case that the lattice structure has just three Atoms by looking at the Congruence lattice given in Figure 6.4.

6.6.2 Congruence Lattice $L_\pi(\mathcal{D}_A)$

The Congruence lattice of the Role Structure $\mathcal{D}_A$ is product of the factorization procedure applied to the relational system that corresponds to $\mathcal{D}_A$. From this process, we get the list of induced inclusion to the partial order structure given a set of generator relations, and in this case there are in total 257 induced inclusions to the partial order of $\mathcal{D}_A$ (which are not shown here). From the overall potential induced inclusions to the partial order of this role structure, there are in total 104 that are unique, and which are the elements in the Congruence lattice of $\mathcal{D}_A$.

Recall that the elements that are immediate successors of the minimal element constitute the Atoms that in this case is of $L_\pi(\mathcal{D}_A)$. These elements then represent the smallest additions to the partial order within the Lattice of Homomorphisms of the Semigroup, $L(S)$. A significant information that the Atoms provide lies in their "complements" in the lattice structure, and this is because among these complements there are meet-irreducible elements that perform the greatest non-trivial aggregations in the semigroup structure. As a result, the decomposition of the partially ordered semigroup is typically in terms of meet-complements of the Atoms, and these elements establish the most dramatic decomposition of the partially ordered semigroup.

The meet-complements to the Atoms are the congruence relations that comply with the requirements given in Equation 6.3, and the fact that $L_\pi(\mathcal{D}_A)$ has more than one Atom implies that the factorization procedure produces at least three representations of the simplified structure. This implies as well that the resultant structure product of the decomposition of the semigroup will carry different *logics* that are superposed on each other.

The collection of meet-complements for each Atom makes a particular structure inside the Congruence lattice $L_\pi(\mathcal{D})$, and the resultant structure is a complete lattice diagram where the minimal element is the Atom itself, and the maximal element is the maximal meet-complement of the Atom. Figure 6.3 represents the collections of meet- complements of the three Atoms in $L_\pi(\mathcal{D}_A)$, which are generating inclusions (11, 14), (10, 16), and (1, 17). Two Atoms have a 48 π-relations that are their meet-complement, whereas the third Atom, which is (1, 17), has in total 56 meet-complement relations in $L_\pi(\mathcal{D}_A)$. These lattice diagrams provide, however, the representative generating inclusions to the partial order, and we can see that two Atoms do not produce a unique maximal meet-complement. In such cases, it is possible to create a

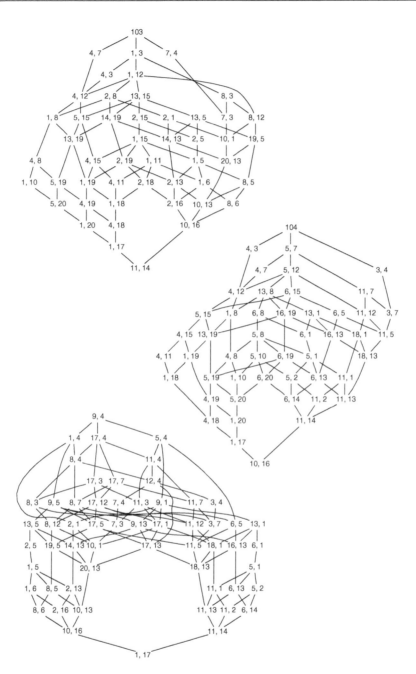

Figure 6.3 Partial order of the meet-complements of the Atoms in $L_\pi(\mathscr{Q}_A)$. *Upper:* π_{77}, *middle:* π_{86}, *bottom:* π_{87}.

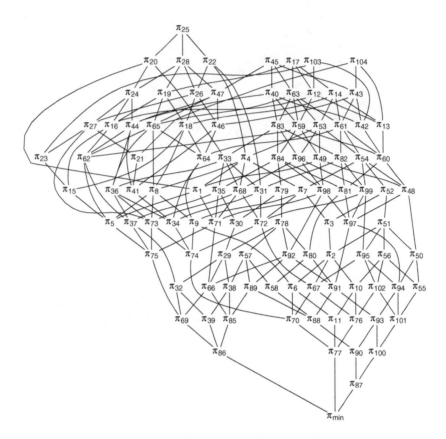

Figure 6.4   Congruence lattice for the Role structure of $\mathscr{S}_A$, $L_\pi(\mathscr{Q}_A)$. Elements π_{103} and π_{104} correspond to the added inclusions in the meet complements of π_{77} and π_{86}.

complete lattice structure with the union of the maximal meet-complements of the Atom, and as a result, elements 103 and 104 represent two additional meet-complements to the Atoms that are maximal.

If we look at the lattice of meet-complements, we can see that two of the three lattice diagrams resemble each other in structure. This matches up with the fact they have a similar size in the collection of their meet-complements; however, all three generated structures have common elements: 18 congruence relations are shared between π_{77} and π_{86}, and 19 congruence relations are common between these Atoms two with π_{87}. Such aspect visibly indicates a type of structure that is shared and which has the partial order of the Role Structure as their source.

Figure 6.4 provides a picture of the Congruence lattice for the Role Structure of the positional system $\mathscr{S}_A$. This lattice structure is a Hasse diagram, which is made of π-relations that represent all the unique induced inclusion for the Role Structure that correspond to $\mathscr{S}_A$. Recall that π-relations stand for partition relations, which implies that each induced inclusion to the partial order "induces" a partition. This means that π-relations are congruence relations, and the Congruence lattice for the Role Structure of $\mathscr{S}_A$ corresponds to the Lattice of Homomorphisms of the Semigroup in $\mathscr{Q}_A$.

The minimal element in the Congruence lattice of $\mathscr{Q}_A$ is the partial order of the Role Structure representing $\mathscr{S}_A$, whereas the maximal element would be the universal relation, which is a trivial partition since all elements belong to the same class. All the other elements between these congruences are partition relations representing the unique induced inclusion to the partial order ordered by inclusion.

The labelling of the distinct π-relations in the Congruence lattice is arbitrary, and it does not represent any type of ordering. What is more important is the partial order structure that is reflected in the entire lattice diagram, and which is made from different partition relations product of the induced inclusions to the partial order of the Role Structure $\mathscr{Q}_A$.

6.7   Role Interlock of Incubator Network A

6.7.1   Factorizing Set

One of the main goals in the identification of the meet-complements of the Atoms is to represent the factors of the semigroup as maximal independent simple homomorphic images. In this sense, from observations made to the partial order of the meet-complements of the Atoms we verify that each one of the diagrams for the partial order of the meet-complements is a complete lattice with the partial order as the $\pi_{\min}$ relation, and the maximal complement of the Atom as $\pi_{\max}$. That is, the maximal meet-complement of π_{87} is π_{25}, whereas the maximal elements for π_{77} and π_{87} are π_{103} and π_{104}, respectively.

After the two additions made to a couple of Atoms, every lattice representation of the Atom has a single maximal element. This means that there are different ways to decompose the Role Structure representing the Incubator network A, one for each Atom, and the unique maximal meet-complements represented by the collection $\{\pi_{25}, \pi_{103}, \pi_{104}\}$ will define a unique factorization for $\mathscr{Q}_A$. It could be the case, however, that an Atom has had more than one maximal meet-complement and in that case the simplified representation would not have been unique.

Once we have detected the maximal meet-complements of each Atom, we need to pay special attention to the type pf configuration present in these structures. Recall that a congruence relation in $L_\pi(S)$ is product of additions made to the partial order and it is a relation that is both right and left compatible to S. In this sense, any congruence relation induces a homomorphism to the semigroup structure where some strings of relations made up congruence classes. This implies that the type of mapping induced for the elements in the Congruence lattice that are covering the partial order is an isotone homomorphism of the semigroup.

Figure 6.5 provides the matrices of maximal meet-complements for each Atom in $L_\pi(\mathscr{Q}_A)$ ordered by increasing string length as in Table 6.1. To the right of each matrix is the same structure after applying an automorphism according to the definition made in (SE) for Structural equivalence. This means that the blocks of relations for the partition with this equivalence criterion are made of either fat-fit or lean-fit blocks; i.e. either a block with perfect fit full of ones, or else a zeroblock.

The meet-complement for Atom 1 π_{77} is given at the top of Figure 6.5, then in the middle is the meet-complement for Atom 2 π_{86}, and finally at the bottom is the matrix of meet-complement for Atom 3 π_{87}. When the automorphism is applied then it is possible to see the different classes of elements in the meet-complements matrices, which allow us making a reduction of these relations where the three Atoms correspond to three Factors.

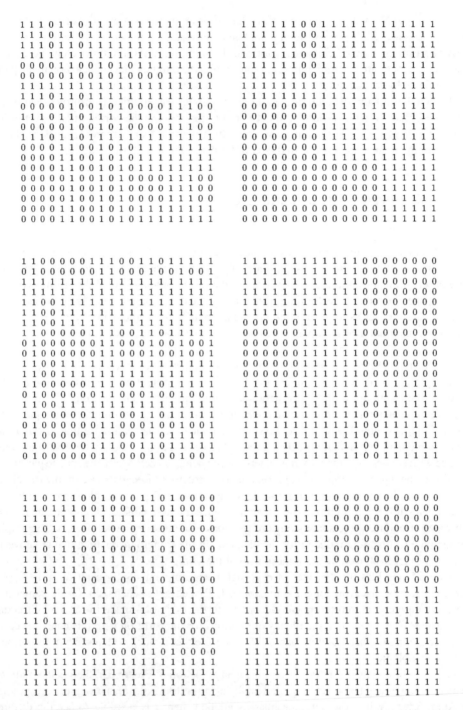

Figure 6.5 Matrices of meet-complements of Atoms in $L_\pi(\mathcal{Q}_A)$ with automorphisms. *Top, middle, and bottom:* π_{103}, π_{104}, *and* π_{25}, *which are meet-complements for Atoms* π_{77}, π_{86}, *and* π_{87}, *respectively.*

Since the congruence relations are both right and left compatible to the elements in the semi-group, a perfect structural correspondence is guarantee with these configurations, which implies that the elements of the Role Structure are partitioned into related classes of perfect structurally equivalent elements. Hence, the permuted matrices of the maximal meet-complements of the Atoms not only gives raise to a certain partition of the Role Structure, but also to the partial order structure of the aggregated role system. This kind of information is important when it comes to look at the hierarchy of relations in the isotone homomorphic images of the system that is based on Structural equivalence of the strings of relations.

In this sense, the matrices in Figure 6.5 product of automorphism evidence that two of the three matrices of meet-complements produce a partition of the Role Structure into 4 classes of elements, whereas the last structure splits the partially ordered role structure into 2 classes of string of relations. Since these matrices are actually partial order "tables", then the structure of the inclusion lattice will reflect the perfect structural correspondence as Figure 6.6 where each level correspond to a "block" with their respective equations, which in this case stand for role relations.

Because of this partition of structurally equivalent role relations, there are additional equations to the Role Structure that are product of the meet-complements of the Atoms for $\mathcal{Q}_A$. The additional equations in the three Factors are the following:

Factor 1, $\mathcal{Q}_A / \pi_{103}$

$$C = F = K = CK = FK = KL$$

$$A = L$$

$$D = DC = DF = DK = DCK = DCL$$

$$G = CG = KD = GC = CKD = KDC$$

Factor 2, $\mathcal{Q}_A / \pi_{104}$

$$C = CK = DC = GC = KDC = DCK$$

$$F = CG = FK = DF = CKD = DCL$$

$$K = A$$

$$D = G = L = KD = KL = DK$$

Factor 3, $\mathcal{Q}_A / \pi_{25}$

$$C = F = A = D = G = CG = DC = DF = GC$$

$$K = L = CK = FK = KD = KL = DK = CKD = KDC = DCK = DCL$$

Disregarding the special case of actor attributes, the three possibilities for primitives in the Role Structure of $\mathcal{S}_A$, that is C, F, and K, are first that all generator role relations are equated, then that generators are not equated, and finally that some generators are equated. Certainly, the usefulness of such information will depend on the relational content of each primitive relation. Since the semigroup structure is partially ordered in this case, the different partitions of the role relations are reflected in the hierarchy of relations as well, which adds a significant information for the interpretation of the role interlock of the system.

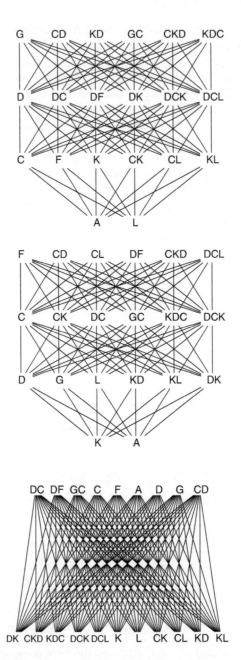

Figure 6.6 Inclusion relations reflecting the meet-complements of Atoms in the Congruence lattice $L_\pi(\mathcal{Q}_A)$. *Top, middle, and bottom: π_{103}, π_{104}, and π_{25} or meet-complements for Atoms π_{77}, π_{86}, and π_{87}.*

Table 6.2 Factor representations for the aggregated Role Structures of the positional system $\mathscr{S}_A$.

	Homomorphic image	Partial order

Factor 1, $\mathscr{Q}_A/\pi_{103}$

	C	A	D	G				
C	C	C	G	G	1	0	1	1
A	C	A	D	G	1	1	1	1
D	D	D	G	G	0	0	1	1
G	G	G	G	G	0	0	0	1

Factor 2 , $\mathscr{Q}_A/\pi_{104}$

	C	F	K	D				
C	F	F	C	F	1	1	0	0
F	F	F	F	F	0	1	0	0
K	C	F	K	D	1	1	1	1
D	C	F	D	D	1	1	0	1

Factor 3, $\mathscr{Q}_A/\pi_{25}$

	C	K		
C	C	K	1	0
K	K	K	1	1

Aggregated versions of the Role Structure for the Incubator network A are possible with all the additional equations made on the strings of relations in $\mathscr{Q}_A$ with three different "logics of interlock." Two of these versions of the Role Structure have an order of 4 and the last representation has an order of 2. Clearly, the third Factor provides the last possible nontrivial subdirectly irreducible representation of the Role Structure, whereas the other two representations could still be factorized if they are not subdirectly irreducible structures. Nevertheless, although there might exist some ambiguity in these representation forms, the order of the semigroup structure—which was of 20 before—has decreased considerable and hence its complexity as well. This aspect is crucial when it comes to the theoretical interpretation of the relational system and the modes of interlock that are occurring in the networks under study.

Table 6.2 presents the aggregated Role Structures for Incubator network A that correspond to the three different factors of $\mathscr{Q}_A$. These forms for representations include both the maximal independent homomorphic images of the Role Structures and the associated reduced partial order table. The aggregate Role Structures in this sense permit one to appreciate the role interlock of the different ties that are occurring in the positional system of the network that in this case considers the attributes of the actors in a relational form as well.

We can see in that each of the aggregated Role Structures has an absorbing element, which is the friendship relation in Factor 2 and the converse of this type of tie in Factor 1, whereas relation K is the absorbing element in the third Factor. On the other hand, the table of actor attributes acts as an identity element in Factor 1 and also in Factor 2, since this type of relation

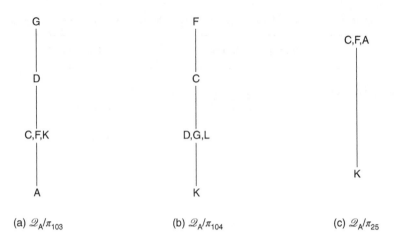

(a) $\mathcal{Q}_A/\pi_{103}$

Figure 6.7   Hierarchy of relations in the homomorphic reduction of the three factors in Role Structure $\mathcal{Q}_A$.

in this case is equated with relation K, which is the identity in Factor 2. Finally, relation C for collaboration acts as identity in the third Factor.

The partial order in the different Factors provides significant information as well. In fact, the partial orders in Factors 1 and 2 are isomorphic structures, and there is a correspondence among the absorbing and neutral elements in their respective homomorphic images. This is a clear sign that confirms a common logic in the interlock of the role relations in the system and that is product of the factorization procedure. The different factors in this sense are hypotheses about the interlock of the role relations that are occurring in some parts of the network. These forms for role interlock are overlapping rather than competing and in this way they complement each other.

A more revealing interpretation of these aggregated Role Structures comes from structural theories. For instance, the Role Structure $\mathcal{Q}_A/\pi_{25}$ expresses the Strength of Weak Ties hypothesis, which has been presented in Chapter 3 Multiplex Network Configurations. Recall that according to the interpretation of this theory in terms of relational composition, a relation is considered as "strong" if functions as an identity in the multiplication table, whereas the relation is considered as a "weak" if this tie functions as a zero element in the algebraic structure.

For instance, relation K is considered as a weak type of tie in the homomorphic image of the third Factor $\mathcal{Q}_A/\pi_{25}$ while the measured relations including the actor attribute are regarded as strong types of tie because they are equated together. This means that whatever combination of ties that includes relation K or its converse leads to this weak type of tie, whereas ties of C, F, or A that stands for actor attributes results just in these latter relational types. A more detailed understanding of the interlock in these relations is, however, obtained from the other two Factors resulted from $\mathcal{Q}_A$, and this is because there are more differentiations in the different types of tie in this Role Structure.

6.7.2   Hierarchy of Relations in $\mathcal{Q}_A$

The aggregated Role Structures resulting from the factorization procedure are product of an isotone homomorphism on the semigroup of the role relations. In this sense, the homomorphic images are not just abstract structures, but there is a hierarchy in the relations as well, which comes from the set of inclusions expressed in the partial order table for the Role Structure.

Figure 6.7 represents as Hasse diagrams the partial order relations in the factor representation of the Role Structure in $\mathscr{S}_A$. These forms of representation then characterize the set of inclusions in the role relations of the network that corresponds to each Factor, and they evidence common features among the three aggregated Role Structures. Factors 1, 2, and 3 correspond respectively to $\mathscr{Q}_A/\pi_{103}$, $\mathscr{Q}_A/\pi_{104}$, and $\mathscr{Q}_A/\pi_{25}$

For instance, the isomorphism between the partial order structures in Factors 1 and 2 has palpable evidence with the diagram representation, and this is because the distinct relations at each level in one structure correspond to the transpose of the relation in the other. Certainly the homomorphism between the two inclusions sets take into account the minimal element in both structures that correspond to the identity relation in the semigroup structure.

The fact that the partial order structures of the two first Factors are isomorphic to each other implies that a single interpretation is well suited for both outcomes. In this sense, if we take the kinds of relations measured in the empirical networks, which are collaboration, friendship and competition, together with the adoption of innovations, then there is a partial order on the different strings types of tie, which can be summarized as follows:

$$C \leq F$$

$$K \leq C$$

$$K \leq F$$

$$A \leq C$$

$$A \leq F$$

$$K \leq A \quad \text{or} \quad A \leq K.$$

And by induction we have the subsequent hierarchy of relations, which expresses algebraic constraints occurring in the system:

$$(K \leq A \quad \text{or} \quad A \leq K) \leq C \leq F.$$

This means that, at the aggregate level in the role relations of Incubator network A, whenever a collaboration relation is made, then such tie is made among classes of actors who have also an informal relation, and this is because the collaboration role system is contained in the friendship system of role relations. Similarly, if there is a perceived competition among classes of actors then there are also both formal and informal ties between these actors.

Regarding the attributes of the actors, and hence the adoption of innovations, this type of relation correspond to a diagonal matrix rather than an adjacency matrix. The hierarchy of relations involving this kind of self-tie must consider this characteristic when it comes to the interpretation of the containment results.

Nonetheless, although the isomorphism between the diagrams of Factor 1 and 2 allowed us to deduce important algebraic constraints involving the measured relation types, we need to take into account that the decomposition of the role relations has been made using the converse of the ties. This implies that the different hierarchy of relations are represented by their respective Factors as in Figure 6.7 where the types of tie and their converses cannot necessarily be represented as a unique partial (or linear) order on the ties when they are represented as a single structure.

6.8 Progressive Homomorphic Reduction of Factors in $\mathcal{Q}_A$

Certainly, the partial order of the relationships constitutes an important part of the algebraic constraints inherent to the system, and which reflect the containments among the role relations occurring in $\mathcal{S}_A$. Yet the homomorphic images of the Factors provide more detailed description about how the different role relations are interrelated, and from which additional algebraic constraints crystallized in the equations among the strings expresses the logic in the role interlock of the relational system.

In the case of the third Factor $\mathcal{Q}_A/\pi_{25}$, this structure expresses the Strength of Weak Ties hypothesis where relation K or perceived competition is a weak type of tie, whereas collaboration (and hence friendship and the adoption of innovations) are strong ties; i.e. C = F = A. This is in accordance with the partial order of this Factor that is given in Table 6.2 and which represents a center-periphery structure where the center or the strong type of relation is recipient of the of all ties while it does not connected towards the periphery.

However, for Factors $\mathcal{Q}_A/\pi_{103}$ and $\mathcal{Q}_A/\pi_{104}$, there still the possibility to reduce the semigroups of relations representing the homomorphic images with a *progressive factorization* or a *progressive homomorphic reduction*. If we take a look at the semigroups in these Factors, we can tell at a glance that the two structures can be further reduced to a half of their orders. Relations C and A in Factor 1 are closely associated, and relations D and G make together another category in this role structure.

The same situation also occurs in the second Factor with two classes of strings that are the pairs (C, F) and (K, D) respectively. In fact the partially ordered structures product of these factor representations can be factorized as well, and the interpretation of the role interlock in this network is made by accounting for the non-trivial images that are *maximal* as the structure representing the Factors.

The factorization of the homomorphic images from Factor 1 and Factor 2 bring identical partial order tables where there is a single inclusion from the second to the third element. The lattices of congruence relations is depicted to the right of Figure 6.8a, and this lattice structure is common to the two Factors. The lattice has both a minima and a maxima element, and also a single Atom that corresponds to the first partition relation. Naturally, the partial order structure corresponds to π_{min}, and the rest of the elements of the lattice including the maxima are additions to the corresponding partial order structure.

More specifically, the π-relations that are additions in the partial order table, which have been given already in Table 6.2, as a pair (row, column):

	Factor 1	Factor 2
$\pi_1 =$	(4, 3)	(2, 1)
$\pi_2 =$	(3, 1)	(4, 3)
$\pi_3 =$	(1, 2)	(1, 4)
$\pi_{max} =$	(3, 2)	(1, 3)

Recall that congruences lattices with a single Atom are subdirectly irreducible. This means that in both Factors the Atom itself constitutes a maximal homomorphic image of the semigroup structure. The set of π-relations implies an additional equation in the quotient semigroup that is D = G for $\mathcal{Q}_A/\pi_{103}$, and C = F for $\mathcal{Q}_A/\pi_{104}$, respectively.

To the right of the lattice of Figure 6.8 are the corresponding homomorphic images with their respective equations. This means that we could use in the case of Factor 2 relation F or

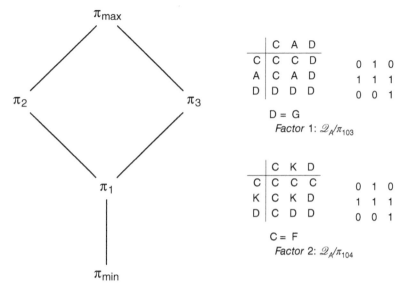

(a) L_π ($\mathcal{Q}_A/\pi_{103}$ and π_{104}) with Images and partial order

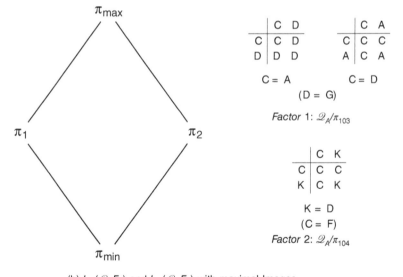

(b) L_π ($\mathcal{Q}_A F_1$) and L_π ($\mathcal{Q}_A F_2$) with maximal Images.

Figure 6.8 Progressive factorization of factors in the Role Structure $\mathcal{Q}_A$ with maximal non-trivial images. *Top and bottom: Two steps in the homomorphic reductions of Factors $\mathcal{Q}_A/\pi_{103}$ and $\mathcal{Q}_A/\pi_{104}$.*

friendship as the representative type of tie in the structure instead of C or collaboration. The reason to choose this type of tie in the factor representation is that F has an absorbing character in this semigroup structure of $\mathcal{Q}_A/\pi_{104}$ and also because collaboration is contained in friendship. However, for the interpretation of the role interlock can also be applied to the collaboration ties where theory should guide the election of representative strings.

The homomorphic images in both Factors $\mathcal{Q}_A/\pi_{103}$ and $\mathcal{Q}_A/\pi_{104}$ are still subdirectly decomposable structures, and the factorization procedure of these homomorphic images produce once more Congruence lattices with identical constitution. Figure 6.8b gives the resulting lattice of π-relations in both Factors, and the additional equations are also given in the figure where the equations inherited from the previous step in the factorization are in parentheses.

We can see in the picture that this latter lattice structure has two Atoms, and these complement each other. This means that the two congruence relations define a unique factorization for the two Factors, whose maximal independent irreducible structures are given afterwards. Here $\mathcal{Q}_A/\pi_{103}$ has two images with a two-element core-periphery structure, whereas both π-relations in $\mathcal{Q}_A/\pi_{104}$ produces a reduced system with two role relations as well.

As with the previous step in the factorization process, the π-relations in the lattice structure that are additions to the partial order tables in Figure 6.8 are:

	Factor 1	Factor 2
$\pi_1 =$	(3, 1)	(3, 2)
$\pi_2 =$	(1, 2)	(1, 3)
$\pi_{max} =$	(3, 2)	(1, 2)

Through the factorization process depicted in Figure 6.8 then we arrive to different types of association involving two representative relational types in $\mathcal{Q}_A$, and these three structures represent maximal non-trivial Image matrices with the role interlock of this social network. In the case of Factor 1, $\mathcal{Q}_A/\pi_{103}$, there are two additional equations on the structure; first the differentiation between the collaboration relation C and its converse D disappears, while in the second representation the actor attributes A is equated with C. The structures of these images are isomorphic to each other and they just support the homomorphic representation given for $\mathcal{Q}_A/\pi_{103}$ where a new hierarchy of relations is produced

$$A \le (C = F = K).$$

In the case of Factor 1, the final structure that corresponds to the decomposition of Factor 2 differentiates C and F on one hand, from K and A on the other hand. Since F = C, then the structure results in a type of logic interlock that apparently contradicts the homomorphic image of $\mathcal{Q}_A/F_3$ (cf. Table 6.2). This is because C now represents a weak type of tie, whereas K is regarded as a strong tie. Certainly, such an outcome reflects the difference in the concatenation of these types of tie in these Factors where we have the equations C ∘ K = C in $\mathcal{Q}_A/\pi_{104}$, whereas in $\mathcal{Q}_A/\pi_{25}$ this composition is C ∘ K = K.

Part of the explanation of these outcomes lies in the equations that each Factor bears with it. The additional equation in Factor 2 meant that C = F (plus K = D) and, although perceived competition is contained in the friendship relation, K appears to be more dominant than F in $\mathcal{Q}_A/\pi_{104}$. This is due to the absorbing character of this last type of relation in the initial factor representation, which is the closest characterization of the Role Structure of the network.

The Role structure of Incubator network A has important aspects to take into account; first, an isomorphism between two of the Factors, and the third Factor that is subdirectly irreducible. Then the resultant image structures are immune to actor attributes in this particular case, and this is because the identity matrix stands for this type of relations, which means that it has a neutral character in the semigroup representing the network role table.

One must consider the construction of the network positional system as the starting point for making the analysis of the role structure in substantial terms. The categorization of actors is a crucial part in the process, and this is because semigroups are very sensitive to role relations with a different structure. Composition equivalence is a useful approach for dealing with networks made of multiplex relations, but this can be done with other approaches including statistical ones. In the next two sections, we are going to perform a similar analysis with other real-world multiplex networks.

We have just seen the decomposition procedure exhaustively of the Role structure corresponding to Incubator network A $\mathcal{Q}_A$. Now we briefly look at some aspects of the Role structures of the other two Incubator networks, B and C, which are systems with directed ties having an equal relational content. This is not only to have a better idea of different decomposition processes but also because at the end, are these reduced structures that are object for comparison in Chapter 11. Later on, we will look at the Role structure of the Florentine families network, which is an undirected structure as oppose to the Incubators.

6.9   Role Structure for Incubator Network B

6.9.1   Factorization of $\mathcal{Q}_B$

Another example of a progressive factorization outcome is found in the Role Structure that corresponds to **Incubator network B** or $\mathcal{Q}_B$. The multigraph of the network itself $\mathcal{X}_B$ is depicted in Chapter 11 (cf. Figure 11.1) where we can see that this network is multiplex and directed. Incubator network B $\mathcal{X}_B$ has the same relational content as Incubator network A, and the nature of the actors is similar to $\mathcal{X}_A$ as well.

The details of the positional analysis of $\mathcal{X}_B$ is given in Appendix B, but it is worth mentioning here that the positional system $\mathcal{S}_B$ produces a Role Structure of order 41, which is an even number. Because of its size, the semigroup of role relations is not much informative and it is not shown here. The Cayley graph of this complex Role Structure, however, can result being more enlightening than the multiplication table and it is depicted with a force-directed layout on the cover of this book.

Cayley graphs depicted using the force-directed layout algorithm "tend" to place both the identity and the empty elements of the semigroup structure at the center of the picture. This is because the identity element is connected to each one of the unique primitives, which in turn generate sets of compounds. The zero element of the semigroup, on the other hand, connects in the same manner to the rest of string relations and therefore the attractive and repulsive forces that characterizes the force-directed layout algorithm places this node element at a pretty much equal distance to the rest of elements. This is the case of the Role structure $\mathcal{Q}_B$ and aggregated images.

What is particular with the factorization of the Role Structure of Incubator network B is that, even though the Lattice of Homomorphisms of the Semigroup produce more than one Atom—and hence a similar number of meet-complements of the Atoms—all non-trivial homomorphic images, which are 38 in total, are isomorphic to each other. As a result, there is a *single* logic interlock in the aggregated Role Structure corresponding this network and this is in contrast with the previous case with the role structure $\mathcal{Q}_A$ where there are three different logics.

Figure 6.9 gives both the inclusion lattice and the Cayley graph of the smallest homomorphic image where structure exists. The order of this structure is 11, which means that the number of elements in $\mathcal{Q}_B$ has been reduced almost four times. The partial ordering of string relations produces a complete inclusion lattice structure where generators C and F are equated and constitute the maximal element in the lattice, while the minimal element is the empty string, which is the composition of K and B.

Below the inclusion lattice is the Cayley graph, with a force-directed layout showing the links between string relations. As mentioned before, the Cayley color graph is a graphical representation of the network relational and role structure where both vertices and edges stand for the ties. Like the multigraphs representing the empirical Incubator networks, each type of social relation is depicted with a characteristic shape and color, and the color of the edge that corresponds to C "interpolates" the colors used for relations C and F. On the other hand, the two actors attributes use variations of the blue color used in the corporate brand of the two adopted Wed innovations for the edges in the graph.

Because there is an equation among generators, it is important to mention that vertex C in the Cayley graph actually represents both C and F, and also vertices represent the two types of actor attributes, A and B. It results striking the symmetry in both structures that correspond to this Incubator network compared with the previous inclusion lattices of the representative role relations for both $\mathcal{Q}_A$ and $\mathcal{Q}_F$.

Moreover, if we take a look at the Cayley graph of Figure 6.9 and the Cayley graph on the book cover, there is a significant simplification in the complexity of the role system represented by $\mathcal{Q}_B$ while keeping its essential structure in the aggregated configuration. This is precisely what decomposition aims to do. In the case of performing a further factorization of this structure, all relations will equate with each other, meaning that there is no differentiation among them.

For the sake of completeness, the multiplication table of the aggregated Role Structure is given in Appendix B. It is easier to see in this table than with the Cayley graph that, for example, all generators and compounds until length 2 are reflexive, and the difficulty in the graph is because the loops stand for different kind of relations other than just the reflexivity property of the strings.

6.9.2   Congruence by Substitution Property in $\mathcal{Q}_B$

In the case of the role structure of Incubator network B $\mathcal{Q}_B$, the non-trivial maximal homomorphic image of $\mathcal{S}_B$ with the substitution property technique separates A from the rest of the string relations. This means that the actor attribute that is represented by B is equated to the social and cognitive role ties C, F, and K, respectively.

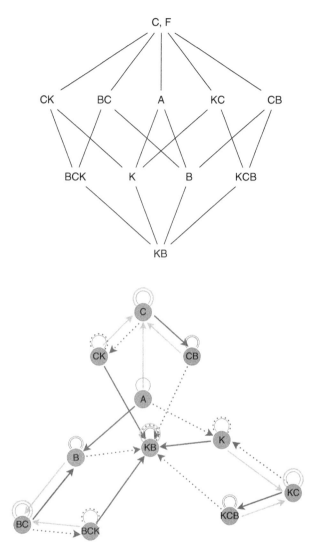

Figure 6.9 The minimal non-trivial aggregated Role Structure of $\mathcal{Q}_B$. *Top: Inclusion lattice with representative string role relations. Bottom: Cayley graph of the reduced role table* $\mathcal{Q}_B$ *with* C = F.

The reduced structure has therefore two elements, the role table evidences the core-periphery structure where the representative string C is at the core, and the actor attribute represented by A is at the periphery. In terms of the Strength of Weak Ties theory, this means that the social role relations constitutes the strong type of tie and the actor attributes is the weak type of relations. The fact that there is a single non-homomorphic image of the congruence classes in $\mathcal{S}_B$, it reinforces the possibility that this kind of structure is actually happening in this network.

6.10   Role Interlock of Incubator Network C

In the case of Incubator network C, the construction of the positional system $\mathscr{S}_C$ is given in B, and Ostoic (2020) has more details for the analysis of this particular Role structure. The positional system $\mathscr{S}_C$ has three categories of actors who are compositional equivalent; that is the reduction of the network starts with the creation of the Relation-Box using tie transposes and compounds with length until $k = 3$.

A couple of significant equations are for the actor attributes A = B on one hand, and for collaboration relations and its transpose or C = D on the other hand. The rest of the equations in system $\mathscr{S}_C$ involve longer compounds and therefore they are not informative.

6.10.1   Decomposition of $\mathscr{Q}_C$

We focus in this section on the decomposition of Role structure $\mathscr{Q}_C$ through a factorization procedure with induced inclusions to the partial order. In this case, the decomposition of $\mathscr{Q}_C$ produces a factoring set with two elements, and these Factors are based on 87 unique π-relations excluding the partial order.

Figure 6.10a provides two parts of the Congruence lattice $L_\pi(\mathscr{Q}_C)$ showing the Atoms among the 5 π-relations at the bottom, and also their meet-complements among the 14 π-relations at the top of the diagram. These two fragments are important in the analysis since they allow locating the smallest non-trivial configurations of the Role structure semigroup. We notice as well that $L_\pi(\mathscr{Q}_C)$ is a complete lattice with both $\pi_{\min}$ and $\pi_{\max}$ order relations.

Clearly, the Atoms are partition relations π_{76} and π_{68}, and the maximal meet-complements of the Atoms are partition relations π_{87} and π_{86}. In the case of the maximal element of the Congruence lattice that is π_{34}, it is not a meet-complement of any Atoms and therefore it does not provide structural basis for making a decomposition of the Role structure even though it may produce a smaller homomorphic Image. This means that the factorization of the semigroup through meet-complements produces at this stage two logics of interlock in this relational structure with identical partial order structure.

The lattice diagrams of the partial order structures for the meet-complements of the Atoms are given in Figure 6.10b with equations of generators and 2-path compounds. These partially ordered semigroup structures with an order of seven then constitutes another factoring set from a progressive factorization.

We can see that both structures are isomorphic to each other, but the labelling differs in most of the cases; in the first Factor representation F is contained in C, while in the other Factor representation these two types of relations are equated. On the other hand, from the equations of the second Factor given in the lattice diagram, these equate distinct generators of the semigroup structure as given in Fig. 6.10. The two homomorphic Images are, however, isomorphic in all essential respects, even though there are differences both in equations and in inclusions.

There is yet a possibility for a further reduction of the two Factors through a progressive factorization, which in this case brings four maximal homomorphic Images after two rounds in each of the structures. The maximal non-trivial Images with equations until length $k = 2$ is also given in Appendix B where we can see that in most of the cases the formal relation C is equated to the informal type of tie F. Besides, both the perceived competition relation K, and the adoption of innovations or the actor attributes represented by A differentiates from the formal collaboration and informal friendship ties in the Role structure of Incubator network C.

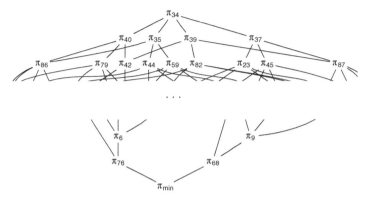

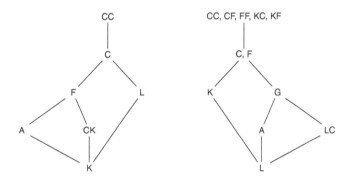

(a) Two fragments of $L_\pi(\mathcal{Q}_C)$ with Atoms and their meet-complements.

(b) Lattice diagrams of the partial order structure for π_{86} and π_{87}.

Figure 6.10 Ordering structures in the decomposition of $\mathcal{Q}_C$. Atoms π_{68} and π_{76} have π_{86} and π_{87} as meet-complements. Equations in Factor to the left are CF = C, FF = F, KC = L, and KF = K.

This means that the logic of role interlock is in this case not univocal and there is some overlapping with the distinctions mentioned above. The interpretation of the logic of interlock for $\mathcal{Q}_C$ is applied with these aggregated structures, and this is since, after the two additional rounds, the progressive factorization results in the one-element semigroup that is considered a trivial structure.

6.11   Role Interlock of $\mathcal{Q}_F$ for Florentine Families Network

Another Role structure we have constructed in Chapter 5 corresponds to the Florentine Families network, $\mathcal{Q}_F$, which is a social system that has both multiplex relations and actor attributes. The Florentine families network $\mathscr{X}_F$ has *undirected* relations with a different relational content, which makes this network different from the Incubator networks we have seen so far. The most significant difference is that there is no need to apply relational contrast to $\mathscr{Q}_F$ as we did with directed structures. Since this network has undirected ties, the corresponding partially ordered structure of the Cumulated Person Hierarchy always is symmetric, and this aspect clarifies its interpretation.

The first step is reducing the network structure has been the establishment of a positional system, and this was done by applying the Compositional equivalence criterion where the different types of tie and actor attributes have been incorporated in the performance reduction. We also have from Chapter 5 Role Structure in Multiplex Networks the process in constructing the Role structure $\mathscr{Q}_F$ and where Figure 5.8 represents it as a partially ordered semigroup with an Edge table. The decomposition of this Role structure is made through a factorization procedure where we first obtain the induced inclusions to the partial order, and then the partition relations.

Figure 6.11 provides the Congruence lattice with the partition relations where π_{min} corresponds to the partial order structure, and the induced inclusions corresponding to each of the 12 π-relations are

$$(1,2),\ (4,2),\ (6,2),\ (1,3),\ (4,3),\ (6,3),\ (8,3),\ (1,4),\ (1,5),\ (1,6),\ (1,7),\ (1,8)$$

As expected, the Congruence lattice product of the factorization has symmetric character since the network is undirected. The Congruence lattice in this case has two Atoms that are π_2 and π_3, and the meet-complements of these Atoms are π_{10} and π_8, respectively. These meet-complements are the immediate upper bounds of the Atoms as well, and the other partition relations in the Congruence lattice are not meet-complements because they cover the Atoms through π_1.

As a result, π-relations 8 and 10 define the aggregated Role Structures of the Florentine Families network with two superimposed logics, which are given at the center and to the right of Figure 6.11. The two aggregated Role Structures are given as partially ordered semigroups with a symbolic format where the partial ordering of the Factors are represented as lattice diagram with the hierarchy of role relations and the respective equations.

It is straightforward to see in both logics that the set of inclusions among primitive relations remains unchanged, and hence Business ties are still under the context of Marriage relations. It is just the placement of compounds WB(= WM) and BW that differs in the hierarchy of role relations. Yet, since W represents financial Wealth that is an actor attribute, these two compound relations mostly have the same meaning in substantial terms.

6.11.1 Congruence Classes in Role Structure $\mathscr{Q}_F$

There at least two options for a further decomposition of these two Factors. Since the Role structures are also partially ordered, a progressive factorization of these is feasible. However, there is also the possibility to decompose the *abstract* Role Structure of the Florentine Families network. That is, the network relational structure without the information given by the partial order.

In the second case, the generating elements of the congruence classes in the Role Structure are the representative strings inside the semigroup of relations, which means that the maximum number of ways the system is decomposed cannot be higher than the order of the semigroup. In many cases, however, there are fewer decomposition options and, as with the factorization of

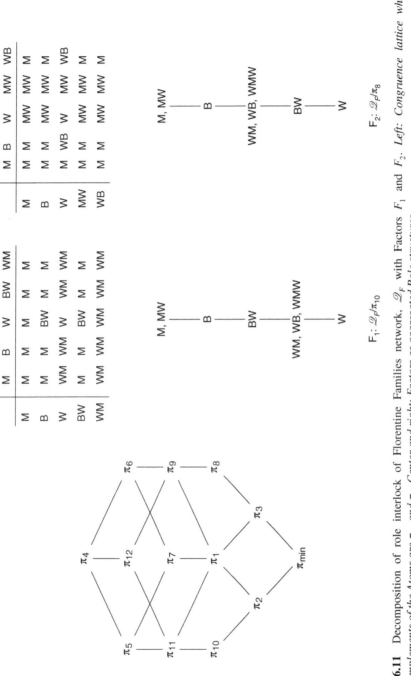

Figure 6.11 Decomposition of role interlock of Florentine Families network, $\mathcal{D}_F$ with Factors F_1 and F_2. *Left: Congruence lattice where meet-complements of the Atoms are π_{10} and π_8. Center and right: Factors as aggregated Role structures.*

	M	W	WM
M	M	M	M
W	WM	W	WM
WM	WM	WM	WM

M = B = BM

Factor Image for $\mathcal{Q}_F F_1$

	M	W	MW
M	M	MW	MW
W	M	W	MW
MW	M	MW	MW

M = B = WB

Factor Image for $\mathcal{Q}_F F_2$

	M	W
M	M	M
W	M	W

M = B = BW = WB

Image for $\mathcal{Q}_F F_1$ and $\mathcal{Q}_F F_2$

	M	B
M	M	M
B	M	M

M = W = BW = WM

Congruence Image for $\mathcal{Q}_F F_1$

	M	B	W
M	M	M	W
B	M	M	W
W	M	M	W

M = WB   and   W = MW

Congruence Image for $\mathcal{Q}_F F_2$

Figure 6.12   Congruence classes in the decomposition of Role structures for the Florentine Families network, $\mathcal{Q}_F$ with equations. *Factor Images above appear both in the progressive factorization and in congruence. The homomorphic Image at the center is common for the two aggregated Role structures from Figure* 6.11.

the partially ordered semigroup, some of the outcomes are equal, which means that the amount of interlock logics is also decreased.

For the sake of completeness, we are going to perform both a progressive factorization of the two Factors as partially ordered structures in $\mathcal{Q}_F$, and a further decomposition of the abstract role tables in Factors through congruence classes. We will see a connection in these two decomposition strategies when they end up with an aggregated homomorphic reduced Image matrix that is common for both Factors of $\mathcal{Q}_F$, yet with some overlapping.

Figure 6.12 presents the aggregated role structures from the two Factors corresponding to $\mathcal{Q}_F$. Factor and congruence Images correspond to aggregated role tables product of factorization and through congruence relations, respectively. First are homomorphic reduction of the Factors given in Figure 6.11 after a progressive factorization, and where in both cases the social role relations are equated. The two Images are unique in the further factorizations, and they also appear in the decomposition of the abstract role structures through congruence relations.

In the case of the aggregation process through congruence relations, the abstract semigroups brought overlapping structures including the two Factor Images, and this differs from the factorization results. The Image matrices in Figure 6.12 are non-trivial structures overlapped to the Factor Images, and the homomorphic Image at the center of the picture is common for the two aggregated Role structures from Figure 6.11. This shared structure appears in the decomposition by congruence of both Factors, and represents the Strength of Weak Ties theory; however, there are overlapping logics since some congruence classes has both $M \circ W = W$ and $W \circ M = W$, which implies that this result is not univocal.

6.12   Reduction Diagram

A special visualization device that allows us to review the overall decomposition process of a given relational or role structure is of the *reduction diagram* (Pattison, 1993). This form of depiction resembles the Congruence lattice L_π except that L_π has the partial order as the minimal element of the lattice structure that is been reduced "upwards". In the reduction diagram, however, is the semigroup the structure the maximal element in the lattice diagram to be reduced "downwards". If that the configuration to reduce is a semigroup object, the minimal element in the reduction diagram is to the trivial homomorphic reduction of the one-element semigroup, which typically is not included in reduction diagrams.

Figure 6.13 provides two examples of a reduction diagram representing the decomposition process of Role structures of Incubator networks $\mathscr{X}_A$ and $\mathscr{X}_B$ (cf. Tantau, 2013, for the pictures production). In the two cases, the maximal element of the reduction diagram is a partially ordered semigroup, which is the algebraic structure to reduce. However, the reduction diagram serves to illustrate the decomposition of an abstract semigroup or, in principle, to other types of algebraic structures as well.

Each level in the reduction diagram represent in these cases a different step in the progressive factorization of the algebraic structure. Dashed arcs indicate omitted steps in the progressive homomorphic reduction between the elements. Omissions can be sometimes required in order to arrive into a more informative image structure, or because a homomorphic reduction of the structure with additional equations has not been achieved.

In the case of the Role Structure of Incubator network A $\mathscr{Q}_A$, the factorization process is illustrated at the top of Figure 6.13, where the labelling in the π-relations corresponds to Table 6.2 and to Figure 6.8. The symbol π_a^* in the reduction diagram means that the resultant structure arrived from the different maximal meet-complements of the Atoms in the respective Congruence lattice.

For the Role Structure of Incubator network B $\mathscr{Q}_B$, which is at the bottom of Figure 6.13, the whole process is given in more general terms. Initially, the factorization produces six decomposed role structures from which three correspond to the 1-element semigroup that is isomorphic to the universal relation U.

The other three factors are subject to decomposition through a progressive factorization process that brought 38 partition relations in total that resulted in them being isomorphic with each other. It is this precisely this maximal homomorphic image that is represented as a Cayley graph in Figure 6.9 and which is considered as the "most reduced" non-trivial representation of the Role structure. Even if the decomposition of this configuration either with factorization or with

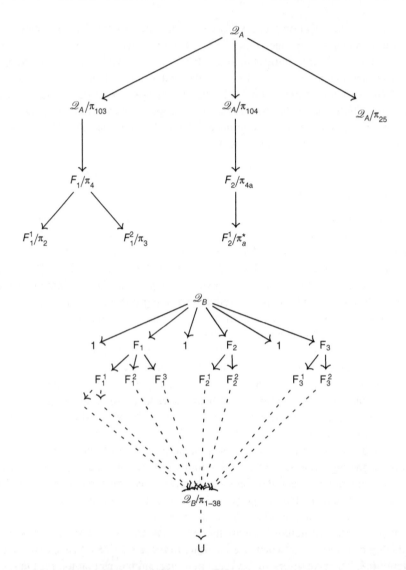

Figure 6.13 Reduction diagrams with the decomposition process of Role structures. *Above: $\mathcal{D}_A$ that after a progressive factorization, there are three non-trivial aggregated Factors. Below: $\mathcal{D}_B$ where the reduced structure ends up with a single aggregated representation.*

congruence relations bring the one-element semigroup, there still a dashed arc to the infima since we do not know for sure whether there are other reduced structures covered by $\mathcal{D}_B/\pi_{1-38}$.

Lastly, is it worth mentioning that it is also possible to depict the decomposition process of the Florentine families Role structure with a reduction diagram even if part of the decomposition of $\mathcal{D}_F$ was made by applying the substitution property to the abstract algebraic structure. In fact, finding congruence classes can be a strategy made in combination with the factorization procedure.

6.13   Decomposition of Role Structures: Summary

The decomposition of the role structures certainly simplified the configuration resulted from the interrelation between role relations in the empirical systems. In this way, it was easier to realize the logic of the interlock among these structures and pave the way for making a substantial interpretation of the multiplex network structures.

In the case of the Role structure for Incubator network A, $\mathcal{Q}_A$, the factorization procedure brought three different logics of interlock where each logic corresponds to a structure that arises from the meet-complement of an Atom. Atoms are irreducible elements in the lattice of congruence relations with different partitions of the semigroup structure. The presence of more than one Atom means that the Role structure is subdirectly reducible.

For the Role structure of Incubator network B, $\mathcal{Q}_B$, there were even more number of Atoms (and hence meet-complements of these), but half of them reduced the Role structure to the trivial 1-element semigroup that correspond to the universal relation. Nonetheless, the progressive factorization of other homomorphic images of the Role structure produced several Factors, which in turn resulted in a single logic of interlock making the substantial interpretation of this multiplex network more transparent.

The third Role structure decomposed has been of the Florentine families network, $\mathcal{Q}_F$ having undirected role relations. The decomposition in this case was made through congruences by the substitution property of the abstract semigroup, and not through the factorization of the partially ordered structure as with the Role structure $\mathcal{Q}_A$ of the respective Incubator network.

6.14   Learning Decomposition of Role Structures by Doing

6.14.1   *Factorization of Role Structure* $\mathcal{Q}_A$

```
# Table 5.5 representing role structure 𝒬ₐ  (Chapter 5)
R> S_A
# Table 6.1 representing Lₐ(𝒬ₐ)
R> P_A
```

```
# factorization by induced inclusions to PO(𝒬ₐ), the partial order of 𝒬ₐ.
R> ii_A <- fact(S_A, P_A)

# partition (π) relations of the factorization with the poset
R> pr_A <- pi.rels(ii_A, po.incl = TRUE)
```

```
# potential Atoms with k-shortest induced inclusion to the partial order, k ⩽ 4
R> for(l in seq_len(4)) {
# factorization with the potential Atoms with customized k
+     pa_A <- fact(S_A, P_A, patm = TRUE, k = l)$patm
# select potential Atoms in the partial order table
+     P_paA <- partial.order(pr_A, type = "pi.rels", po.incl = TRUE, sel = pa_A)
# labels of the lattice with the potential Atoms
+     lbsa <- vector()
+     for(i in pa_A) lbsa <- append(lbsa, parse(text = (paste0("pi[",i,"]"))))
# the minimal element
+     lbsa <- append(lbsa, expression(pi[min]))
# Figs. 6.2 order structure of the potential Atoms, customized labels
+     diagram(P_paA, lbs = lbsa, fsize = 20)
+   }
```

```
# now factorization of 𝒬ₐ with meet complements of Atoms
R> ii_A <- fact(S_A, P_A, atmc = TRUE)

# record meet complements of Atoms in the induced inclusions
R> mc_A <- ii_A$atmc

$'11, 14'
 [1]   1    5   12   13   15   16   23   29   30   32   35   37   38   39   42   48   49
[18]  50   55   57   58   59   60   62   66   69   71   75   78   79   80   81   84   85
[35]  86   87   88   89   90   91   92   93   94   97   98  100  101  103

$'10, 16'
 [1]   2    3    6    7    9   10   11   13   21   31   34   41   42   43   46   48   50
[18]  51   52   54   55   56   58   60   61   64   67   70   72   74   76   77   80   81
[35]  82   87   88   90   91   93   94   95   97   99  100  101  102  104

$'1, 17'
 [1]   1   2   3   4   5   6   7   8   9  10  11  15  16  18  19  20  21  22  23  24  25  26  27
[24]  28  29  30  31  32  33  34  35  36  37  38  39  41  44  46  47  62  64  65  66  67  68  69
[47]  70  71  72  73  74  75  76  77  85  86
```

```
# plot
R> for(k in seq_along(mc_A)) {
# partial ordering structure of meet-complements in π-relations
+     P_mcA <- partial.order(pr_A, type = "pi.rels", po.incl = TRUE, sel = mc_A[[k]])
# labels
+     lbsmc <- vector()
+     for(i in mc_A[[k]])
+     lbsmc <- append(lbsmc, parse(text = (paste0("pi[",i,"]"))))
# the minimal element
+     lbsmc <- append(lbsmc, expression(pi[min]))

# Figs. 6.3  meet-complements of the Atoms (customized labels)
+     diagram(P_mcA, lbs = lbsmc, fsize = 12)
+     }
```

6.14.1.1 Congruence Lattice

```
# partial order of π-relations in 'pr_A'
R> P_prA <- partial.order(pr_A, type = "pi.rels", po.incl = TRUE)

# create the congruence lattice labels
R> lbsppr <- vector()
# each element in 'P_prA' is a π-relation
R> for(i in seq_len(nrow(P_prA)))
R> lbsppr <- append(lbsppr, parse(text = (paste0("pi[",i,"]"))))
# last π-relation is the partial order; i.e. the minimal element
R> lbsppr[length(lbsppr)] <- expression(pi[min])

# Fig. 6.4  congruence lattice of role structure (customized labels)
R> diagram(P_prA, lbs = lbsppr)
```

6.14.1.2 Decomposition

```
# decomposition of 𝒬_A with the meet-complements in π-relations
R> SsA <- decomp(S_A, pr_A, type = "mca", reduc = TRUE)

# clustering of strings with the three factors
R> SsA$clu

[[1]]
  C  F  K  A  D  G  L  CK  CD  CL  KD  KL  DC  DF  DK  GC  CKD  KDC  DCK  DCL
  1  1  1  2  3  4  2  1   4   1   4   1   3   3   3   4   4    4    3    3

[[2]]
  C  F  K  A  D  G  L  CK  CD  CL  KD  KL  DC  DF  DK  GC  CKD  KDC  DCK  DCL
  1  2  3  3  4  4  4  1   2   2   4   4   1   2   4   1   2    1    1    2

[[3]]
  C  F  K  A  D  G  L  CK  CD  CL  KD  KL  DC  DF  DK  GC  CKD  KDC  DCK  DCL
  1  1  2  1  1  1  2  2   1   2   2   2   1   1   2   1   2    2    2    2
```

```
# Image matrices and partial order tables in 6.3

R> SsA$IM                                       R> SsA$P


 [[1]]                                            [[1]]
   C A D G                                          C A D G
 C C C G G                                        C 1 0 1 1
 A C A D G                                        A 1 1 1 1
 D D D G G                                        D 0 0 1 1
 G G G G G                                        G 0 0 0 1

 [[2]]                                            [[2]]
   C F K D                                          C F K D
 C F F C F                                        C 1 1 0 0
 F F F F F                                        F 0 1 0 0
 K C F K D                                        K 1 1 1 1
 D C F D D                                        D 1 1 0 1

 [[3]]                                            [[3]]
   C K                                              C K
 C C K                                            C 1 0
 K K K                                            K 1 1
```

6.14.1.3   Figure 6.6: Hierarchy of Relations in the Homomorphic Reduction of Factors

```
# Factor 1: 𝒟_A π_103
R> diagram(SsA$PO[[1]], lwd = 2, lbs = c("C,F,K","A","D","G"), fsize = 20)

# Factor 2: 𝒟_A π_104
R> diagram(SsA$PO[[2]], lwd = 2, lbs = c("C","F","K","D,G,L"), fsize = 20)

# Factor 3: 𝒟_A π_25
R> diagram(SsA$PO[[3]], lwd = 2, lbs = c("C,F,A","K"), fsize = 20)
```

6.14.1.4   Progressive Decomposition of Factors in $\mathscr{D}_A$

```
# factorization of Factor 1 𝒟_A π_103   (same process for Factor 2 𝒟_A π_104)
R> SAf1 <- fact(as.semigroup(SsA$IM[[1]]), SsA$PO[[1]], atmc = TRUE, patm = TRUE)

# partition relations with partial order
R> SAf1_pr <- pi.rels(SAf1, po.incl = TRUE)
```

```
# plot diagram Factor 1: 𝒟_Λ π_103
R> diagram(partial.order(SAf1_pr, type = "pi.rels", po.incl = TRUE))
```

```
# decomposition of 1st factor image matrix
R> SAf11 <- decomp(as.semigroup(SsA$IM[[1]]), SAf1_pr, type = "mca", reduc = TRUE)
```

```
# progressive factorization of 1st factor: induced inclusions and π-relations
R> SAf1p <- fact(SAf11$IM, SAf11$P, atmc = TRUE, patm = TRUE)

# partition relations with partial order
R> SAf1p_pr <- pi.rels(SAf1p, po.incl = TRUE)

# decomposition of factorized 1st factor with π-relations
R> SAf111 <- decomp(as.semigroup(SAf11$IM), SAf1p_pr, type = "mca", reduc = TRUE)
```

6.14.2   Decomposition of Florentine Families Role Structure $\mathcal{Q}_F$

6.14.2.1   Factorization of $\mathcal{Q}_F$

```
# from chapter 5 we have the partially ordered semigroup

R> S_F                                   R> P_F

...
     M   B   W   MW  BW  WM  WB  WMW           M  B  W  MW BW WM WB WMW
M    M   M   MW  MW  MW  M   M   MW       M    1  0  0  0  0  0  0  0
B    M   M   BW  MW  MW  M   M   MW       B    1  1  0  0  0  0  0  0
W    WM  WB  W   WMW WMW WM  WB  WMW      W    1  1  1  1  1  1  1  1
MW   M   M   MW  MW  MW  M   M   MW       MW   1  0  0  1  0  0  0  0
BW   M   M   BW  MW  MW  M   M   MW       BW   1  1  0  1  1  0  0  0
WM   WM  WM  WMW WMW WMW WM  WM  WMW      WM   1  0  0  0  0  1  0  0
WB   WM  WM  WMW WMW WMW WM  WM  WMW      WB   1  1  0  0  0  1  1  0
WMW  WM  WM  WMW WMW WMW WM  WM  WMW      WMW  1  1  0  1  1  1  1  1

attr(,"class")                           attr(,"class")
[1] "Semigroup" "symbolic"               [1] "Partial.Order" "strings"
```

```
# get induced inclusions to the partial order for factorization
R> ii_F <- fact(S = S_F, P = P_F, atmc = TRUE, patm = TRUE)

# produce partition or π-relations from induced inclusions
R> pr_F <- pi.rels(ii_F, po.incl = TRUE)

# plot congruence lattice in Fig. 6.10 with partial ordering of the partition
R> diagram(partial.order(pr_F, type = "pi.rels", po.incl = TRUE))
```

```
# decomposition by meet-complement of the Atoms with reduction
R> fct_F <- decomp(S = S_F, pr = pr_F, type = "mca", reduc = TRUE)

# Factors as abstract semigroups in Fig. 6.10
R> fct_F$IM
```

```
$IM[[1]]                                         $IM[[2]]
      M   B   W  BW  WM                                 M   B   W  MW  WB
M     M   M   M   M   M                           M     M   M   M  MW  MW   M
B     M   M  BW   M   M                           B     M   M  MW  MW   M
W    WM  WM   W  WM  WM                            W     M  WB   W  MW  WB
BW    M   M  BW   M   M                           MW    M   M  MW  MW   M
WM   WM  WM  WM  WM  WM                            WB    M   M  MW  MW   M
```

6.14.2.2   Congruence Factors Abstract Semigroup

```
# decomposition of Factor 1 by congruence classes
R> cg1_F <- cngr(S = as.semigroup(fct_F$IM[[1]]), uniq = TRUE)
R> decomp(S = as.semigroup(fct_F$IM[[1]]), pr = cg1_F, reduc = TRUE)


...
$IM
$IM[[1]]                                         $IM[[3]]
      M   W  WM                                         M
M     M   M   M                                   M     M
W    WM   W  WM
WM   WM  WM  WM                                   $IM[[4]]
                                                        W   M
$IM[[2]]                                          W     W   M
      M   B                                       M     M   M
M     M   M
B     M   M                                       ...
```

```
# decomposition of Factor 2 by congruence classes
R> cg2_F <- cngr(S = as.semigroup(fct_F$IM[[2]]), uniq = TRUE)
R> decomp(S = as.semigroup(fct_F$IM[[2]]), pr = cg2_F, reduc = TRUE)


...
$IM
$IM[[1]]                                        $IM[[3]]
    M   W  MW                                       M  W  B
M   M  MW  MW                                   M  M  W  M
W   M   W  MW                                   W  M  W  M
MW  M  MW  MW                                   B  M  W  M

$IM[[2]]                                        $IM[[4]]
    M  W                                            W  M
M   M  W                                        W  W  M
W   M  W                                        M  M  M
                                                ...
```

6.14.3   Decomposition of Role Structure $\mathcal{Q}_B$

See Appendix E.

7

Signed Networks

7.1 Structural Analysis of Signed Networks

A special case of a multiplex network structure is that of a **signed network** where the multiplicity in the ties lies in different *signs* or *valences* that different types of relations can have. Typically, signs are of an affective character, but valences can arise from instrumental ties as well, and even from relations having a cognitive character.

Significant positional analyses in signed structures of the Sampson Monastery classic dataset (Sampson, 1969) have been made by Breiger et al. (1975), White et al. (1976), and Fienberg et al. (1985), whereas de Nooy et al. (2005) detected structural balance and clusterability on this signed network as well. Algebraic analyses of signed social networks have been performed for a while, and these have been through different semiring structures (Harary et al., 1965; Harary, 1994; Batagelj, 1994; Doreian et al., 2004). However, in most of the cases the analysis was made with artificial network data in order to obtain prototypical structures that reflect a theoretical model where empirical signed networks rarely fit in. We devote this chapter mainly for the algebraic study of signed networks through two types of semirings where the theoretical framework is based on structural balance.

The analysis in algebraic terms takes the empirical system Incubator network A, $\mathscr{X}_A$, and it is made in the context of social influence processes where each bundle pattern represents a specific type of interaction between the actors that has the potential of mediate. Since the phenomenon to analyze involves a social influence process, social ties are meant to contribute the actor decision to adopt an innovation in this particular case. In other words, a bundle pattern represents a specific type of interaction between the actors that has the potential of mediate a social influence process, which contributes in the decision to adapt an innovation (cf. Ostoic, 2017, for an extended version of the analysis in algebraic terms that includes the Sampson dataset).

Each of the seven bundle class patterns from Chapter 3 represents a structural tendency in the network based in the amount and the direction of the ties except for the null dyad where no relationship occurs. We are aware, however, that signed networks can occur in different contexts with the condition that the multiplicity of ties among network members has at least opposite signs.

Algebraic Analysis of Social Networks: Models, Methods and Applications using R,
First Edition. J. A. R. Ostoic. Companion website: www.wiley.com/go/ostoic/algebraicanalysis.

7.2   Social Influence Process

In the network analysis literature, a *social influence process* is also known as "contagion" or "persuasion" and it occurs when a social actor adapts his behavior, attitude, or belief to other individuals in the social system. With contagion an actor uses another actor as his frame of reference and takes their opinions and attitudes into account either directly or indirectly. When a pair of actors are related in a direct way in the system, then this implies that a contagion process takes place through a *direct communication* between them. This means that the two actors share attributes of an innovation when they socialize with each other, and as a result the probability of adoption increases by their frequency of interaction. Naturally, the actor who is taken as frame of reference should already be an adopter of the novelty.

In network studies of social contagion, direct communication corresponds to a body of theories dealing with *cohesion* in the social system where the main task is to find cohesive subgroups of actors in the network, which can be of order two or also groups with more than two actors. On the other hand, indirect ties between the actors through a common neighbor represent a potential source of social influence process in the system as well. What is important in this case is that the actors are related in a similar way in the network, which implies that they are regarded as equal in a structural manner.

Actors who are structurally alike in a network of relations occupy the same position and they are meant to play a similar role in the relational system. In this case, the underlying mechanism playing in the adoption of the innovation is *comparison* rather than direct communication, where the main task is to make operational the notion of network positions according to a chosen measure of *equivalence* among the actors.

Both comparison and direct communication are then two distinctive underlying mechanisms in a social influence process that can be studied in complex systems like multiplex networks, and at the same time be related to actors' changing attributes by the network members. In this sense, we can take a look at the correlation between cohesion influence and the adoption of innovations, for example, in the context of multiple relations, and as well with comparison influence later on.

Finally, even though complex structures have gained increasing attention within social network analysis, Lazega and Pattison (1999) and Lazega (2001) have already made an empirical analysis of multiplexity involving competition and cooperation in a social setting, and in these works the two types for collective action are shaped by dyadic and structural forms of exchange including reciprocity. A comprehensive treatment of the dynamical processes in multiplex networks that includes diffusion and social influence processes is found in Barrat et al. (2008), whereas the exposure through social media in large-scale networks is reported by Bakshy et al. (2015) with a cross-cutting content at each stage in the diffusion process, and measurements for ideological homophily that are applicable to interpersonal ties. Besides, Salehi et al. (2014) make a survey of diffusion processes in multi-layered networks where influence in static structures is based on diverse threshold models.

7.2.1   Cohesion Influence

One argument in the social influence process is that the communication mechanism is a useful way to understand a phenomenon like the adoption of innovations. Many authors have supported the importance of communication in the social influence process since the early days of structural analysis. For instance, Lazarsfeld and his colleagues (1948) studied political preferences in presidential campaigns and pointed out that personal influence is more powerful than media, and evidences from field studies and laboratory posit that communication is a locomotive force in social structure and a source of pressure toward uniformity (Festinger, 1950) and influence to reach agreements (Back, 1951). A systematic theoretical foundation for social influence through communication was provided by (Homans, 1961) in his series of studies of small group dynamics, and in a social network context, Friedkin (1984) points out that structural cohesion models are based on communication pressure toward uniformity when there is a positively valued interaction between the actors.

In social network analysis, cohesion is a concept that is defined as the strength of the tie density among a subset of actors, where cohesive subgroups are those actors in the system of relations among whom there are relatively strong, direct, frequent, or positive relations. The key structural tendency for cohesion is *mutuality*, which means that the ties existing between two actors and which make the relational bond must have at least a mutual character.

As a result, if we take the dyadic patterns of the bundle census results, we corroborate that only collaboration and friendship ties as such represent a direct communication between the actors. Perceived competition on the other hand is a cognitive type of tie that indicates an alleged contest among two actors where no social interaction is necessarily taking place, which implies that the mechanism for this relation type does not correspond to direct communication.

In order to characterize cohesion, we take the graph representation of Incubator network A $\mathscr{X}_A$ depicted in Figure 5.1 in Chapter 5 Role Structure in Multiplex Networks; this time, however, we are going to consider just the triangle component at the top left of the multigraph and only the friendship relations. Then we can see that there are three actors who are mutually regarded as friends, and this means that this configuration in the network represents a subgraph that is maximal connected given that the connectivity property holds for all the elements there. A maximal connected subgraph is known as a *clique* of actors in social network analysis, and it represents the strongest case of a cohesive subgroup. In this sense, the triangle of actors in $\mathscr{G}^+\mathscr{X}_A$ constitutes a clique of friendship in Incubator network A.

A clique is one of the earliest conceptualizations of group structure, and Luce and Perry already in year 1949 provided a method based on matrix multiplication to identify cliques in a given network. We note that a reciprocated dyad, i.e. a configuration made up with two nodes and two mutual ties between them, is also a maximal connected subgraph; however, a clique needs to have at least three nodes according to the definition. Since mutuality is the distinctive property of cliques and on which cohesion is based on, the associated matrices in the computation to find cliques must be symmetric. The operation of matrix multiplication permits the tracing of chains of relations in the network and thus channels of communication where novel information may flow. In this way, the mechanism of social influence is extended from a direct communication to persuasion in cohesive subgroups with indirect communication via chains of relations.

However, the method of matrix multiplication in finding cliques as original formulated has significant drawbacks for the analysis. First, each type of relations is treated separately, meaning that the multiplexity of ties is disregarded, and the other concern is that when cliques actually arrive, the chains of relations can have different length for each type of tie. This implies that a substantial interpretation of the social influence results among the actors will correspond to different stages of communication in the specific types of tie, and in some cases the gap between the levels can be quite large and thus unpractical for the analysis. In this sense, rather than focus on individual relationships separately, we need to look at the simultaneous combination of multiple ties in the searching of cohesive subgroups in complex systems such as multiplex networks.

7.2.2   Comparison and Influence

The analysis this time takes into account an "indirect" persuasion in the social system rather than cohesion influence through a "direct" communication among the actors. Thus the type of social influence considered now is expected to occur through a comparison mechanism. Recall that with comparison actors use other actors they feel to be similar as their frame of reference, and this means that the more similar the actors are related in the system, the higher is the probability to expect an analogous behavior from them; in this case with respect to the adoption of the same kinds of innovations. One of the reasons to expect an analogous behavior is because actors that are structurally equivalent in a network of relations are facing similar constraints and opportunities within the system in the near future.

A fundamental distinction between communication and comparison lies in the type of structural effect that each mechanism is based on; communication namely requires that each bond in the network has a mutual character, but in the case of comparison mutuality is not a requisite at all. On the contrary; if the system is built in a hierarchical way, then mutuality will be absent almost certainly while comparison can easily take place.

The comparison mechanism is naturally related to the perceived competition among actors in empirical social systems. By combining this type of tie with formal collaboration, it is then possible to see, for example, how the adoption of innovation is driven by the interplay of collaboration and competition between the entrepreneurial firms in Incubator network A. In this way, we could complement the exposure results from cohesion influence that can take different types of tie in $\mathscr{X}_A$.

The effects related to the comparison mechanism are the structural tendencies associated with signed graphs, and an appropriate analysis for such tendencies requires configurations that combine the possible valences at the dyadic level. Thus while the building blocks for the study of cohesive influence via formal and informal ties in multiple networks are the different patterns at the dyadic level, the analysis of indirect persuasion through comparison and collaboration is based on higher level structures like triads and so on.

In this sense, even though a social relation starts with the dyad, a structure that has a true character of society has the triad as its smallest structure (Simmel, 1950). Simmel also affirms that while the critical characteristics of the group change completely by adding a third person to the dyads, "the further expansion to four or more by no means correspondingly modifies the group any further," and a clear example of this is the genesis of the family (1950, p. 138). As a result, while there is a great structural difference between a dyad and a triad, the difference between a triad and a social circle is less dramatic.

It is through relations among three elements that important phenomena like transitivity is deduced; however, transitivity is a generalization of another structural tendency known as structural balance, whose model is quite regularly conformed by empirical social networks such as Incubator network A. Both transitivity and balance can have a considerable role in the collaborative structure among firms and the adoption of innovations, and for this reason these structural tendencies play a key part in the study.

An effective analysis of structural balance can be made through algebraic techniques and thus we will get the flavor of the power of these kinds of models in order to study the adoption of innovations in the context of cooperation and competition ties. Therefore, the next section is devoted to the presentation of the structural balance theory and its relationship to algebraic structures.

7.3 Structural Balance

Heider as the theory of Cognitive Dissonance (Heider, 1946, 2013) proposed a significant model of structural balance when he studied diverse situations that arise when *positive* and *negative* ties are present in small configurations, particularly in triads or triples of actors and the relations between them. Although in his original formulation Heider considered both shared and individual attitudes to a certain object, since Newcomb (1961) the principles have been applied to structures with three actors, where usually social psychologists use *affective* relations with terminal values such as "like" and "dislike" among people in their analyses (e.g. Sampson, 1969; Newcomb, 1961).

Besides the mentioned affective relations, there are other types of ties in human social networks like collaboration or communication between the social actors. In this case, the ties have an *instrumental* character, where the main goal is to access to important resources by means of social ties. The distinction between instrumental and affective ties is associated with the two ideal basic types of relations proposed by Tönnies and Loomis (1940, 1st ed. 1887) in which *Gesellschaft* is a sociality maintained by instrumental relations, whereas *Gemeinschaft* is inherent connectedness of a community of individuals.

In this sense, it is also possible to apply the principles of the structural balance theory to these other kinds of relations or even to a combination of affective and instrumental ties. For instance, if we take a look again at Incubator network A, $\mathscr{X}_A$ (cf. 5.1 in Chapter 5), it is clear that while a collaboration relation between two firms is an instrumental sort of tie, the friendship relations constitute affective ties. Now in the case of the third type of relations, which is perceived competition, this is a cognitive type of tie that has an instrumental value for the actors.

In order to illustrate the principle of the structural balance theory, we start with the smallest configuration made of a set of three actors with undirected ties. Recall that this is according to Simmel where the true character of society begins, and not with the dyad. Hence, if we consider that the actors in the triad are either p-adjacent or n-adjacent, then four possible situations can arise in the system combining such two valences. The structures associated with these situations are represented as signed graphs in Figure 7.1 with triads having solid edges representing positive ties and dashed edges representing negative relations.

For instance, if we take measured relations in $\mathscr{X}_A$, then the positive and the negative ties correspond to cooperation relations and perceived competition, respectively; that is, p → C and n → K. We can see in the triad configuration of Figure 7.1 that in the triad from the left all actors cooperate with each other, whereas the triad on the extreme right every actor regards

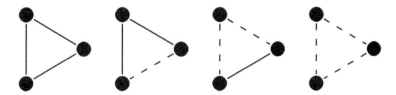

Figure 7.1   Four possible situations in a triad with undirected positive (solid) and negative (dashed) ties. *The second configuration from the left is imbalanced, the fourth to the right is clusterable, and the other two structures are balanced.*

their neighbors as competitors. Between these poles, there are the other possibilities where both cooperation and competition occur among the actors.

Cartwright and Harary's Structure Theorem (1956, also Harary et al. (1965)) states that a signed graph is *balanced* if and only if its nodes can be divided into two mutually exclusive groups of actors—although a group can be empty—where all positive edges are within each group and all negative edges are between the groups. This means that when negative ties exist in a network of relations, then a balanced structure is obtained with the polarization of the population system.

In this sense, if we take a look to the four situations with triads, an assumption taken with the two kinds of tie just considered is that when all actors collaborate with each other then it will produce a balanced situation in the system. However, when everyone regards the other actors as competitors then the situation is less clear. Heider asserts that a balanced triad with two negative ties is obtained with *either* a positive or a negative third relation, but there is a strong tendency towards a closure with a positive relation (1956, p. 206; cf. also Davis (1967)). In this sense, a complete triad made of competing actors where there is no cooperation among them implies that the group is imbalanced.

Now the second configuration from the left in the picture with the triadic situations tells us the following story: "two actors collaborate with a third one while they are competing with each other" or, equally, "an actor collaborates with two competing neighbors." According to the balance theory this type of structure represents a stressed situation and therefore the group is imbalanced. Lastly, the triad with a single collaboration tie says that "two collaborating actors have a common competitor," which implies that there is no tension and the structure is considered as balanced. We will see that the fact a complete triad in the network has one or two negative ties have determinant implications for the whole system in terms of structural balance.

According to Heider's theory of cognitive balance, structures that are imbalanced have an inherent tension and are prone to change, whereas balanced configurations have a more stable structure. This implies that network structures have a tendency to be balanced over time, and we will expect more triads like the first and third one from Figure 7.1 and less of the other two configurations of this picture. However, Doreian and Krackhardt (2001) in studying triads of directed ties found out that there are some triples for which the classical the structural balance hypothesis is not supported or it is even contradicted. Hence, balance mechanisms are not the only mechanisms and need not to be the dominant mechanisms in the evolution of signed networks (2001, p. 57). One mechanism that might drive the evolution of signed structures is a "competition mechanism," where two n-adjacent actors in the triadic configuration are rivals for the third actor.

Despite the existence of multiple mechanisms in the dynamics of signed networks, any alteration in the network structure may be interrelated with the changing attributes of the actors. After all, balanced configurations without inherent tension foster communication among its members and hence they are a potential source for social influence, and for this reason it is perfectly justifiable to see in terms of structural balance the correspondence between the implied type of structure of the network and the changing attributes of the actors.

The formal framework of balance theory is based on relations that have opposite valences from a substantive point of view, and both affective and instrumental types of ties are consistent with the principles of balance theory. For instance, Baldassarri and Diani (2007) point out that organizational interdependence can be characterized in terms of balance theory where instrumental ties take the form of *transactions*. On the other hand, the actors do not need to be individual persons, and the balance schema is useful for understanding tie formation at the interorganizational level (Kilduff and Tsai, 2003). In this sense, the application of the structural balance theory in the analysis of a network like $\mathscr{X}_A$ seems to be reasonable, since the fact that the firms are embedded in balanced or imbalanced network structures may have crucial implications in the innovation adoption process.

7.3.1   Balance and Relational Composition

The theory of structural balance has been conceived by using both positive and negative ties among social entities in a system, and the triadic representations made of these kinds of tie given previously are essentially different types of closures occurring in social networks. Thus the valence of a given relation in the triad is a product of the other two signs of the ties in the structure, where the tie closure suggests a substantive outcome that is highly plausible to take place in the resultant balanced structures, and which is derived from a transitivity effect.

In this sense, it is possible to characterize the correspondence of signed relations in terms of relational composition as chains or paths, and this option is very convenient when it comes to deal with larger structures than the triad. For instance, if we consider just positive and negative types of tie within the classical balance theory, then there are four possibilities to obtain balanced structures. If we take a collaboration relation in form of a potential working partner together with a perceived competition tie as valences p and n respectively, then the different combinations of these two valences in terms of relational composition are characterized in the following manner:

$$p \circ p = p \qquad \text{\textit{"partners of my partners are my partners"}}$$
$$p \circ n = n \qquad \text{\textit{"partners of my competitors are my competitors"}}$$
$$n \circ p = n \qquad \text{\textit{"competitors of my partners are my competitors"}}$$
$$n \circ n = p \qquad \text{\textit{"competitors of my competitors are my partners"}}$$

Essentially, such characterizations as with Rapoport (1963) typically use "friendship" and "enmity" relations, which are both cognitive types of tie, and we apply in this case partners and competitors. A "partner" results from a formal collaboration tie between two actors, while "competitors" are those perceived as such even if there is another type of relation with them like a formal or informal contract. Since the case of Incubator network A $\mathscr{X}_A$, for example, refers to a closed physical setting where the actors have their day work, the last assertion actually makes sense with this relational content.

These are equations that arise from the triadic situations given in Figure 7.1 where–looking from left to right–the first triad characterizes the closure of positive ties. On the other hand, the composition of two negative ties represents the last configuration, whereas the other two equations relation p and n characterize a balanced structure that corresponds to the third triad in the picture.

According to Davis (1967) the last composition is an example of "coalition formation," i.e. the tendency for those with a common enemy to become friends, which depends on the number of subsets in the system; such statement holds for two subsets of actors but not with more than two subsets (p. 186). Davis in fact noted that the population in social network structures can be split into more than two subsets of actors and yet be considered as structurally balanced. In such case, the important aspect to take into account is that the within relations in each cluster are positive and the between relations among clusters are negative. Davis called such groups as *plus-sets*, and a graph containing two or more plus-sets is said to be *clusterable*, which is a special case of a balanced structure.

From this alternative perspective, three actors connected by negative ties represent a structurally balanced configuration when each individual is regarded as a plus-set. Thus in terms of relational composition

$$n \circ n = n,$$

together with the statement applying the tie used in the example

   "the competitors of my competitors are my competitors"

hold since they imply a clusterable structure.
The lack of consensus with the last equation suggests that there is an ambiguity situation with respect to this type of structure, and the sign of the tie closure depends on the definition of what is considered to be a structurally balanced configuration. Such ambiguity within the structural balance schema can be tackled by considering as well *ambivalent* relations in the analysis. Hence, an *extended balance* theory includes also ambivalence in the set of valences of the social system. In this case, the determination of the structurally balance in the social system corresponds to an extension of the *classical balance* in which there are positive and negative ties only.

As a result, according to this extended version of balance, the last equation can be expressed in terms of

$$n \circ n = a,$$

which is translated into

   "the competitors of my competitors are [in fact] ambivalent,"

where "ambivalent" means that this actors are both competitors and partners at the same time. In this sense, the incorporation of the ambivalent relation in the extended balance schema permits the limitations of the classical balance theory to be overcome when it comes to represent the evaluative relationships occurring in real social structures in a more realistic manner. Certainly, if we consider the lack of closure in the social system, then we need to include in the analysis the indifference attitude as another valence as well.

It is important to note that in the extended version of the balance theory the ambivalent relation acts as an absorption element in the relation composition, and this means that the product of a with either p or n will always result in ambivalent relation. In the case of the other two valence types, they still satisfy the principle of "antithetical duality" in which if the opposite tie is applied twice to a sign, then the original sign is obtained (Cartwright and Harary, 1956, p. 28).

All the rules of association for balanced structures can be expressed in the form of a square matrix table where the elements in the valence set correspond to its rows and columns, and the products of these signs in terms of relational composition are represented by the entry values in the array. In this sense, Table 7.1 provides two hypothetical structures in Balance theory that are the classical and the extended versions. We can see that the advantages of this form of representation are pretty evident. The matrix allows summarizing all combinations of the valences in one structure, where is possible to perform some operations in order to generalize

Table 7.1 Classical and extended configurations of the Structural Balance theory. p, n, *and* a *stand for positive, negative, and ambivalent ties, respectively.*

o	p	n
p	p	n
n	n	p

o	p	n	a
p	p	n	a
n	n	a	a
a	a	a	a

(a) Classical   (b) Extended

the different structural balance theories to configurations, which are larger than the triads we have seen so far.

Within this vein of though, Cartwright and Harary (1956) observed that paths in a signed graph are positive if they have an even number of negative edges, and are negative otherwise (in this case zero is considered to be an even number). In this sense, these authors extended the work of Heider in the sense that it is possible to look in an iterative manner at the closed chains or at the closed paths or semipaths existing in the relational composition table and see whether the circuit, cycle, or semicycle in the network is balanced or not. Flament (1963) asserts that a complete graph is balanced if and only if all its triangles are balanced, which means that the whole system can be evaluated in terms of balance by looking at just determinant smaller structures.

Another form for extension to the structural balance theory was made by Doreian and Mrvar (1996, 2009) when they proposed a method for establishing the partition structure of a signed network. This analytical mindset is based on the network raw data rather than transformations of it, and that is close to an "exact" balance structure as is possible. With this extension of the structural balance theory, positive and negative ties are permitted to appear everywhere in the system and not only within subsets and between subsets of actors respectively.

Furthermore, a model can be specified beforehand indicating the absence of the ties among the subsets. This allows identifying complex systems that are more consistent with the signed structure of Incubator A. Doreian and Mrvar (2009) are more specific in pointing out the inconsistencies of the traditional structural balance theory since there are processes such as mediation, differential popularity, and internal subgroup hostility that might be playing in the dynamics of signed networks.

Nevertheless, the relational composition tables allow the incorporation of the ambivalent relation in the analysis and not just consider positive and negative types of tie in the model. In fact, such form for representation make a specific type of an algebraic structure having its own characteristics where the transformation of the data is possible in order to see whether the signed network is balanced or clusterable. For this reason, in the next section we introduce important algebraic structures that are suitable for an efficient analysis of signed networks in terms of the structural balance theory and for the study of multiple networks in general.

7.4   Semirings for Structural Balance

As it has been shown in Chapter 2 Algebraic Structures, any type of algebraic specification contains a set of elements endowed with operations on it. Both the elements set and the operations then constitute the signature of a given specification, which is defined by a number of rules or axioms that the elements of the set and the operations must conform. The signature and the axioms make up together the "sort" of algebra in which several types of algebraic structures exists, and where two algebras belonging to the same sort are said to be "similar" algebras.

There is a variety of algebras having different degrees of complexity, but we start by defining an algebraic object with relative few rules and flexible enough to represent diverse types of relational structure useful for the analysis of multiple networks. Besides, this algebraic system is fundamental in building other types of algebraic structures for the analysis of social phenomena like comparison influence.

Semirings are useful for the analysis of social multiple networks and more specifically this type of algebraic structure allow us to see whether a signed graph is balanced or clusterable (Harary et al., 1965; Davis, 1967; Batagelj, 1994; Doreian et al., 2004). Recall that a signed graph is a form for representation of a special type of multiple networks where relations have different signs or valences in it. Signed graphs have been typically used when considering relations that are either positive or negative, and less often with patterns having both signs at the same time. However, in the context of multiplex networks it is plausible that there exists a mixture of ties pattern made of distinct valences.

For instance, if we take the entrainment of two different types of tie between a pair of nodes as in Figure 3.4a, we can see that these ties have different levels that are occurring at the same time. If the ties have opposite signs, then it means that the nodes are linked by an ambivalent relation that is neither positive nor negative but both of them. In fact, positive, negative, and ambivalent relations are the three signs that stand for the valence set in the representation of the balance theory made in Table 7.1. In addition, there is certainty the possibility that there is no relation among a pair of actors, and the valence type neither is positive, negative, nor ambivalent, but absent or indifferent.

As a result, if we consider the analysis of a signed network structure there are four possible outcomes to take into account, which are p, n, a, and o that represent positive, negative, ambivalent, and the absent relation respectively. In this sense, the underlying set for the balance is determined by the function:

$$Q \rightarrow \{ \, p, n, a, o \, \},$$

which represents the values defined in $\mathcal{V}$ for a_{ij}^{σ} in $\mathbf{A}^{\sigma}$ already introduced in Chapter 1 Structural Analysis with Algebra, and also to the extended version of the structural balance schema.

7.4.1 Valence Rules for Balance Semirings

There are different rules of composition within the semiring structure that determine the valence of the paths or chains in the signed networks according to the type of structural balance chosen. In this sense, by taking the observations of Cartwright and Harary (1956), and the definitions made by Batagelj (1994), and Doreian et al. (2004), it is possible to construct a semiring that corresponds to balance theory in signed networks with the following definitions:

   o: there is no path
   n: all paths are negative
   p: all paths are positive
   a: there is at least one positive and at least one negative path

These characterizations are made in terms of sequences of directed ties in the system and the resultant structure is called *balance semiring*. Hence, ambiguity in the relationship is obtained as long the n and p valences are combined in the sequence of ties, and the existence of ambivalent relations implies that the structure is not considered as balanced in the classical sense of the definition.

It is also possible to represent the clustering version of the structural balance theory with the semiring structure. We need this to redefine the characterizations made for the balance semiring and add another valence-type 'q' that is a particular type of an ambivalent sequence in $\mathscr{G}^\sigma$. As a result, the definitions of the valence types for a *cluster semiring* in directed signed graph are:

o: there is no path
n: there is at least one path with exactly one negative edge, and no path with only positive edges
p: there is at least one path with only positive edges, and no path with exactly one negative edge
a: there is at least one path with at least one ambivalent edge
q: each path has at least two negative edges

Note that Batagelj et al. (1992) define the ambivalent sequence as "[there is] at least one path with only positive edges, and at least one path with exactly one negative edge" if we consider the sentence with paths and edges. However, since a is regarded as an absorbing element in terms of composition, then it is sufficient to have an ambivalent relation in the sequence in order to characterize the type of cycle as ambivalent as well.

Tables in 7.2 provide the rules for the semiring operations in terms of the structural balance theory; that is balance and cluster semiring structures. There is a pair of multiplication tables,

Table 7.2   Semiring for extended Structural Balance. *Absent relation is o, and q is an intermediate valence for cluster semiring.*

·	o	n	p	a
o	o	o	o	o
n	o	p	n	a
p	o	n	p	a
a	o	a	a	a

+	o	n	p	a
o	o	n	p	a
n	n	n	a	a
p	p	a	p	a
a	a	a	a	a

(a) Balance semiring

·	o	n	p	a	q
o	o	o	o	o	o
n	o	q	n	n	q
p	o	n	p	a	q
a	o	n	a	a	q
q	o	q	q	q	q

+	o	n	p	a	q
o	o	n	p	a	q
n	n	n	a	a	n
p	p	a	p	a	p
a	a	a	a	a	a
q	q	n	p	a	q

(b) Cluster semiring

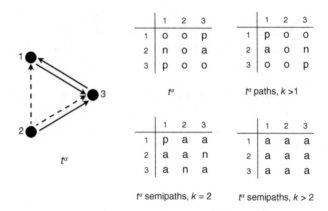

	1	2	3
1	o	o	p
2	n	o	a
3	p	o	o

t^α

	1	2	3
1	p	o	o
2	a	o	n
3	o	o	p

t^α paths, $k > 1$

	1	2	3
1	p	a	a
2	a	a	n
3	a	n	a

t^α semipaths, $k = 2$

	1	2	3
1	a	a	a
2	a	a	a
3	a	a	a

t^α semipaths, $k > 2$

Figure 7.2   Balance semiring structures for a signed triadic configuration t^α having four types of valences.

also called *Cayley tables*, for the two semiring operations, which are then defined for the clas-sical balance and also for clusterability. We observe in the tables that for both the balance and the cluster semirings o represents the identity element and a is the zero element under addition, whereas under multiplication o rather acts as absorbing element and p is the identity element.

An example of the application of the balance semiring rules within a signed triadic configura-tion denoted as t^α is given in Figure 7.2 with the presence of positive, negative, and ambivalent relations. For instance, the multiplication and addition operations in the signed triadic configu-ration work as follows:

For vertices 2 and 3, although the relation $(2, 3)$ is ambivalent, there are three kinds of 2-path relationships, which are (n, p), (o, a), and (a, o). The multiplication operation on these paths results in valences n, o, o, and with the addition operation on these signs the final outcome results being n. Actually, longer paths in $(2, 3)$ will result in n as well, and this is also true for the rest of the vertices in this configuration. In the case of semipaths, the relationship between these vertices has an ambivalent sign from 3-chains and longer.

On the other hand, there are any incoming paths to vertex 2, except for (o, o) that results in valence o after the balance semiring multiplication and addition operations. Since vertex 2 in t^α is not an isolated element, then when considering semipaths there is both a negative and an ambivalent relation as incoming and outgoing chain involving this vertex.

A similar analysis applies to the rest of the vertices in t^α, and if we consider semipaths rather than paths then it manifests the absorbing character of the ambivalent sign to the rest of the structure (cf. Figure 7.2). As a result, for this signed triadic structure all paths larger than one are invariant, whereas semipaths of lengths 3 or more lack structure.

Finally, there is also the possibility to apply the cluster semiring operations, and in this case paths longer than 3 and semipaths longer than 4 lack structure since all relationships have an ambivalent character. Because of the "devastating" effect of a, the practical application of the two kinds of semirings in the analysis of signed social networks is limited. However, it is still possible to detect a "weak" clusterable balanced structure, which is made mainly with a mix of ambivalent and absent paths and chains among the members of the signed system.

7.4.2   Closure Operations in Semirings

Because one of the purposes with the algebraic analysis is to work with closed systems, it is necessary to perform additional closure operations to the semiring elements. One type operation is symmetric closure, and the other unary operation considered on the semiring structure is transitive closure. Both types of closure have been introduced at the beginning of Chapter 2. However, in the case of signed networks it is required some care with respect to the closure operations since there is an ambiguity situation when there is a relational pattern with a multiplexity character.

In the case of the *symmetric closure* in signed graphs, the absence of a relation implies an identity and consequently this element takes the valence value from the tie transposition. On the other extreme, valence p prevails over the rest of the valences when there is an asymmetry of ties; then it comes q over the remaining types of valences in the cluster semiring, and finally a overcomes valence n. In this sense, when the symmetric closure operation is applied on the valence matrix, then it means that we consider semipaths instead for paths in all the definitions given above regarding the semiring construction, and in all cases the edges are undirected.

For *transitive closure*, this operation is similar to the definition made at the beginning of Chapter 2 Algebraic Structures, except that this type of operation is applied just on the positive ties in the semiring structure. More precisely, the transitive closure rule applies on the p relations in X^σ both in the balance and in the cluster semiring structures that are product of multiplication and addition operations. One reason for this application is because the two operations in the semiring structures are transformation on the entire configuration of the signed network and p prevails over o and the other valence types as well.

With definition of balance and cluster semiring operations in place, in the next section we apply the operations involved in these algebraic systems to the study of an empirical signed network structure. The analysis is going to be made in the context of social influence processes on the signed structure that corresponds to Incubator network A, $\mathcal{X}_A$. This signed structure is made of collaboration C and perceived competition K ties or $\mathscr{G}^\sigma(R_1, R_3)\mathcal{X}_A$, which for brevity is denoted as $\mathcal{X}_A^\sigma$.

The outcome of the two semiring structures provides two "settings" in which the adoption of innovations among the actors is contextualized in substantial terms, and each setting for the analysis is then a configuration made of contrasting relation types where social actors are embedded in.

7.5   Balance and Comparison Influence

It is time now to apply the semiring structures for structural balance to the analysis of the comparison mechanism in Incubator network A, $\mathcal{X}_A$. In this case the type of configuration of interest for comparison is made by collaboration ties and perceived competition ties among the entrepreneurial firms in the measured incubator centers where the first type of tie is considered as a "positive" relation, whereas the second one as a "negative" kind of tie. Again, no moral judgement bears on these terms but rather the valences denote a change of sign in the relationship.

We can express the correspondence between the observed relations and the valence types of the signed graph in more formal terms as $\upsilon_1 \to R_1$, $\upsilon_2 \to R_3$. In this case, the third valence type is the ambivalent relation, which is the entrainment of collaboration and competition ties and it is expressed as $v_3 \rightarrow R_1 \cup R_3$. Finally, the absence of these kinds of relation in the network constitutes the fourth valence type and hence $v_4 \rightarrow \emptyset$.

Putting into the language of relational bundles, the assignment of the four valence types from $\mathscr{X}_A$ and which are part of $\mathbf{A}^\sigma$ are characterized taking the definitions made in Chapter 3 as follows:

$$a_{ij}^\sigma \rightarrow \mathsf{p} \quad \Leftrightarrow \quad v_1 = \{ B_{ij}^A, B_{ij}^R, B_{ij}^X \}$$

$$a_{ij}^\sigma \rightarrow \mathsf{n} \quad \Leftrightarrow \quad v_2 = \{ B_{ij}^A, B_{ij}^R, B_{ij}^X \}$$

$$a_{ij}^\sigma \rightarrow \mathsf{a} \quad \Leftrightarrow \quad v_3 = \{ B_{ij}^E, B_{ij}^M, B_{ij}^F \}$$

$$a_{ij}^\sigma \rightarrow \mathsf{o} \quad \Leftrightarrow \quad v_4 = \{ B_{ij}^N \}$$

In this sense, the asymmetric and reciprocal dyads represent for an ordered pair of actors either a positive or a negative relation depending on the relational content of the tie, whereas the null dyad corresponds to the absence of a relation for both parts. Regarding the patterns with a multiple character, the tie entrainment bundle represents an ambivalent relation because ties at different levels produce the directed bond. In the case of the tie exchange, this pattern is just that, namely an exchange of a positive tie with a negative relation with no ambivalence. On the other hand, the mixed bundle is an exchange of either a positive or a negative tie with an ambivalent relation, and finally a full bundle is an exchange of two ambivalent relations. Hence, these last two bundles are regarded as ambivalent in nature.

Naturally when the bundle class pattern has a mutual character, then the assignment of valence values is made to both the ordered pair and its converse. For the non-mutual patterns such as the asymmetric dyad and the tie entrainment bundle, only the ordered pair is assigned with an ambivalent tie and the converse of the tie has the o valence. The mixed bundle is an special case that combines an ambivalent valence tie with the respective converse, which is either p or n. Finally, the bundle class in this context is a mutual ambivalence valence type.

The substantial implications of the relational bundles are closely related to the relational content of the ties. For instance, if we take directed collaboration ties in the system, then the influence is manifest, since

$$(i,j) \in R_1$$

means that there is a direct influence from actor i towards actor j. However, in the case of the perceived competition relation the social influence is in the other way around, and this is because the actor *perceived* as a competitor has a potential influence on the actor who *perceives* the competition. In this case,

$$(i,j) \in R_3$$

rather means that actor j exercises influence on actor i.

Nonetheless, such situation is consistent with the valence assignment in terms of relational bundles. As just said, an entrainment of these two types of tie means that implies that the influence is in both directions, whereas a tie exchange means that the influence—in this special case—is made in one direction only. While the first pattern still has an ambivalent character, the tie exchange pattern does not have ambivalence regardless the course of the social influence. This means that the characterization of the valence types through bundle classes is independent of the relational content of the ties.

7.5.1   Weak Balanced Structures

According to the Structure Theorem of Cartwright and Harary, what makes a signed network balanced or clusterable is that the relations within the group members are positive, and the relations between groups are negative. However, in the case of incomplete graphs, which correspond to most of empirical networks, the absence of a tie or else the o valence does not prevent a relational system to be structurally balanced; that is in accordance with the role of such type of sign as an identity relation.

On the other hand, the ambivalent relation acts as an absorbing element in the relational composition, and this implies that the closed paths and chains will tend to have an increase ambivalent character as the length of the implied sequences grows up. As a result, any structure having ambivalent relations can have plus-sets of actors connected through only negative ties and there is some ambiguity either within the subsets of actors or in the relations between these groups.

If we take a look at the valence matrices representing both the collaboration and competition ties existing in network $\mathscr{X}_A^{\sigma}$ (and as well in networks $\mathscr{X}_B$ and $\mathscr{X}_C$ in Chapter 11 Comparing Relational Structures), we can see that there exist ambivalent relations when ties C and K occur at the same time. This means that the systems considered for the analysis of comparison influence are neither balanced nor clusterable in strictly terms.

However, it is possible to find subsets of correspondent actors located in an "approximately" balanced or clusterable network or a *weak clusterable* balance structure. That is, these are systems that allow having some inconsistencies to the ideal structurally balanced structures, but still hold important characteristics in the configuration of the signed networks. The term "inconsistency" is used in Doreian et al. (2004) that differentiates from "error" or the faults found in imbalanced networks.

In a weak balanced structure, the different subsets of actors are called *factions* that, although they are not fully related structurally balanced structures, yet they have the capability to reveal important structural features of the signed networks. In this sense, we are going to pursue the analysis of the comparison mechanism by trying to identify factions of correspondent actors in $\mathscr{X}_A^{\sigma}$ through algebraic semirings in a loosely version of the structure for the balance theory.

7.6   Looking for Structural Balance

Because the semiring structures require some transformation of the network data, then the study of structural balance based on this type of algebraic structure implies checking for closed paths or chains in the signed graph. This means that we need to account for connected components in the system under study that in this case have collaboration and competition ties in $\mathscr{X}_A^{\sigma}$. Hence, we start the empirical analysis of comparison influence by looking at the structure of contrasting relations in Incubator network A, $\mathscr{X}_A^{\sigma}$, and we bear in mind that the types of relations involved in the analysis are R_1 and R_3 (or C and K) in $\mathscr{X}_A$.

Figure 7.3 presents the signed graphs of relational systems of the Incubator network A where collaboration and competition ties are reported. A significant consequence of disregarding R_2 (or F) in $\mathscr{X}_A$ is that the pair of actors (6, 20) is separated from the main component of the network, whereas actors 12 and 35 are no longer part of the system. We can see in Figure 7.3 that this pair of actors forms a dyadic component having a tie entrainment bundle. On the other hand,

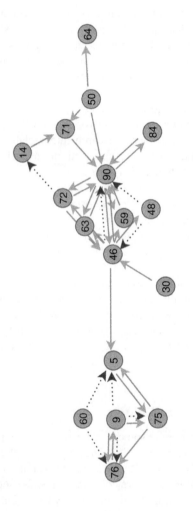

Figure 7.3 Signed graph of the main component in $\mathscr{L}_A^{\sigma}$. Solid arcs are for C (collaboration) and dotted arcs for K (perceived competition).

the triangle component of this network becomes an open semipath with two collaboration ties emanating from actor 89. The rest of the actors and ties remain in the big component, which is obviously sparser since the informal friendship ties are no longer taken into account for the analysis.

Since the system is disconnected, and because one component has only positive relations while the other one has just an ambivalent relation, then a great part of the study of structural balance ends up with just the larger component of this network.

This fact can be seen if we perform the cluster semiring operations once on the valence matrix representing $\mathscr{X}_A^{\cdot\sigma}$, and which is given in Table 7.3. For the sake of clarity, the absent relations are not printed in the table, which is also presented in a permuted form in order to differentiate the three components of the system. However, when all the defined valences are included, then such configuration represents the semiring structure that is product of 2-semicycles, which stands for closed sequences of length 2 in the network.

The information provided by the diagonal of the semiring structure reflects the closed sequence of ties in the system at the given length. In this sense, it is clear that the system is not balanced or clusterable because the within ties in the plus-sets have to be just positive. However, if we consider the component made of three actors, we can see that this configuration is structurally balanced, whereas the dyadic component is completely ambivalent with such closed sequences.

In the first case, the component constitutes a maximal connected subgraph with respect a relation p as a consequence of transitive closure, whereas ambivalence gets a mutual character in the dyadic component product of the symmetric closure operation. As a result, one of the components of the system is structurally balanced with either the balance or the cluster semiring.

Now if we consider the main component, the cycles in this structure present three types of valences; although most of the closed paths are positive, there also a couple of ambivalent closed sequences, and one cycle with valence q. This last case corresponds to actor 60, who is related in the system only through perceived competition, and this means that each path from this node has two negative directed edges, and that is the exactly definition for this type of valence with the cluster semiring structure.

Indeed, Table 7.3 has been given for illustration purposes only, and the algebraic analysis of a signed network through semiring structures treats each component of the network separately since speaking in strictly terms they constitute a differentiated system. Hence, the remaining part of the analysis of the comparison mechanism in this particular network is going to focus just in the large section of $\mathscr{X}_A^{\cdot\sigma}$.

7.6.1 Balance Semiring in Signed Network $\mathscr{X}_A^{\cdot\sigma}$

When the analysis takes the balance semiring in terms of semicycles; that is with all the described semiring operations, then the structure of this component is fully ambivalent with closed semipaths of length 4. However, if we consider paths rather than semipaths, then it is possible to see some factions of correspondent actors in the component after cycles of similar length.

Table 7.3 Valence ties in the cluster semiring of $\mathscr{L}_A^{\sigma}$ with 2-semicycles. *Absent relations* 0 *are omitted in the table.*

	5	9	14	30	46	48	50	59	60	63	64	71	72	75	76	84	90	23	58	89	6	20
5	p	a	p	p	p	p	p	p	n	p	p	p	p	a	a	p	p					
9	a	a	p	p	c	p	p	p	c	p	p	p	c	a	a	p	a					
14	p	p	p	p	c	p	p	p		p	p	c	p	p		p	a					
30	p	p	p	p	p	p	p	p		p	p	p	p	p		p	p					
46	p	c	p	p	p	a	p	a		a	p	p	p	p		c	a					
48	p		p	p	p	p	p	p	c	p	p	p	p	p	p	p	p					
50	p	p	p	p	p	a	p	p		p	p	p	p	p		p	p					
59	p	p	p	p	p	p	p	p		p	p	p	p	p		p	p					
60	n	n			n	a									n							
63	p	p	p	p	a	p	p	p		p	p	p	p	p		p	p					
64	p	p	p	p	p	p	p	p		p	p	p	p	p		p	p					
71	p	p	p	p	p	c	p	p	c	p	p	p	a	c		p	p					
72	p	c	p	p	p	a	p	a		p	p	a	p	p		p	p					
75	a	a	p	p	p	p	p	p	n	p	p	p	p	p		p	p					
76	a	a				a						a	a		p							
84	p	p	p	p	p	c	p	p		p	p	p	p	p			p	p	p	p		
90	p	p	a	p	a	a	p	p		p	p	p	p	p		p		p	p	p		
23																p	p	p	p	p		
58																p	p	p	p	p		
89																p	p	p	p	p		
6																					a	
20																						a

Both semiring structures are given in Table 7.4 where the elements of the balance semiring for cycles have been permuted in order to differentiate correspondences that are present in the system component, and where k stands for the length of the closed sequences of ties. Hence, both the cycles and semicycles have in this case a length of 4.

In this sense, the different correspondences among actors revealed by the balance semiring structure allow us to make a partition of the signed network where equivalent actors constitute a faction inside the system. Thus, based on the results of Table 7.4b, which present a meaningful structure by considering cycles of relations rather than closed semipaths or chains, Figure 7.4 presents the signed graph with the different factions, which are based on the outcomes from the balance semiring with closed paths. This graph also includes the type of innovation that each actor has adopted—as in the case of the system of the strong bonds—and this is because such structure constitutes the appropriate setting where the adoption and influence process can be contextualized in terms of structural balance.

It is important to mention that one of the main reasons to consider cycles of relations for the analysis of innovation adoption through comparison is because what characterizes this mechanism is the asymmetric character of the relationships in the social influence process. This is in contrasts with direct communication where mutuality is the key condition for an influence process. In this way it is possible to complement the results of innovation adoption and cohesion influence from the previous chapter with the outcomes provided by the balance semiring structure.

Figure 7.4 represents a partition of $\mathcal{X}_A^\sigma$ in terms of approximately structural balance, and this diagram shows that the main component of the network has two clearly separated subgraphs, each constituting different factions. It is also evident from the picture that the asymmetric tie between actors 46 and 5 acts as a bridge that connects the two subgraphs. However, the lack of mutuality in this pattern prevents the formation of a cycle between these parts, and therefore two different factions are constituted where a social influence process can occur only in one way, which departs from the larger faction or "green" to the second largest one or "red."

The lack of mutuality is the cause in the formation of the individual factions associated with the large component as well. As a result, two peculiar single actors constitute factions on their own. For instance, actor 64 is pendant that receives a single collaboration tie without contributing to the component, whereas actor 60 only perceives competition ties with two firms inside faction red but a cycle cannot occur with this actor because there are neither incoming collaboration ties nor perceived competition to this firm.

If we consider purely the comparison mechanism in the analysis of innovation adoption in $\mathcal{X}_A^\sigma$, a more interesting picture is given by actors 30 and 50 that together make a faction even if they are not directly related. Indeed, actor 30 is equivalent in structural terms to actor 50, which means that this latter actor is prone to adapt the attributes of the former actor and adopt both types of innovations because they face up to similar structural conditions in this network.

If we take just the perceived competition ties occurring in the network, an influence process through the innovation B can take place only within the larger faction green, and more concretely via actors 46 and 14 who are the frame of reference of actors 48 and 72, respectively. Otherwise a social influence process for this innovation type can happen via direct and indirect cooperation ties through the other neighbor adopters. Note that it is in this case that the bridge between actors 46 and 5 constitutes an important channel where innovation B diffuses into the firms located in faction red via collaboration. Then, if friendship relations were not considered

Table 7.4   Balance semiring structures, main component of signed network $\mathscr{X}_A^{\circ}$ with $k = 4$.

	5	9	14	30	46	48	50	59	60	63	64	71	72	75	76	84	90
5	a	a	a	a	a	a	a	a	a	a	a	a	a	a	a	a	a
9	a	a	a	a	a	a	a	a	a	a	a	a	a	a	a	a	a
14	a	a	a	a	a	a	a	a	a	a	a	a	a	a	a	a	a
30	a	a	a	a	a	a	a	a	a	a	a	a	a	a	a	a	a
46	a	a	a	a	a	a	a	a	a	a	a	a	a	a	a	a	a
48	a	a	a	a	a	a	a	a	a	a	a	a	a	a	a	a	a
50	a	a	a	a	a	a	a	a	a	a	a	a	a	a	a	a	a
59	a	a	a	a	a	a	a	a	a	a	a	a	a	a	a	a	a
60	a	a	a	a	a	a	a	a	a	a	a	a	a	a	a	a	a
63	a	a	a	a	a	a	a	a	a	a	a	a	a	a	a	a	a
64	a	a	a	a	a	a	a	a	a	a	a	a	a	a	a	a	a
71	a	a	a	a	a	a	a	a	a	a	a	a	a	a	a	a	a
72	a	a	a	a	a	a	a	a	a	a	a	a	a	a	a	a	a
75	a	a	a	a	a	a	a	a	a	a	a	a	a	a	a	a	a
76	a	a	a	a	a	a	a	a	a	a	a	a	a	a	a	a	a
84	a	a	a	a	a	a	a	a	a	a	a	a	a	a	a	a	a
90	a	a	a	a	a	a	a	a	a	a	a	a	a	a	a	a	a

(a) Semicycles

	5	9	75	76	14	46	48	59	63	71	72	84	90	30	50	60	64
5	a	a	a	a	o	o	o	o	o	o	o	o	o	o	o	o	o
9	a	a	a	a	o	o	o	o	o	o	o	o	o	o	o	o	o
75	a	a	a	a	o	o	o	o	o	o	o	o	o	o	o	o	o
76	a	a	a	a	o	o	o	o	o	o	o	o	o	o	o	o	o
14	a	a	a	a	a	a	a	a	a	a	a	a	a	o	o	o	o
46	a	a	a	a	a	a	a	a	a	a	a	a	a	o	o	o	o
48	a	a	a	a	a	a	a	a	a	a	a	a	a	o	o	o	o
59	a	a	a	a	a	a	a	a	a	a	a	a	a	o	o	o	o
63	a	a	a	a	a	a	a	a	a	a	a	a	a	o	o	o	o
71	a	a	a	a	a	a	a	a	a	a	a	a	a	o	o	o	o
72	a	a	a	a	a	a	a	a	a	a	a	a	a	o	o	o	o
84	a	a	a	a	a	a	a	a	a	a	a	a	a	o	o	o	o
90	a	a	a	a	a	a	a	a	a	a	a	a	a	o	o	o	o
30	a	a	a	a	a	a	a	a	a	a	a	a	a	o	o	o	o
50	a	a	a	a	a	a	a	a	a	a	a	a	a	o	o	o	o
60	a	a	a	a	o	o	o	o	o	o	o	o	o	o	o	o	o
64	o	o	o	o	o	o	o	o	o	o	o	o	o	o	o	o	o

(b) Cycles

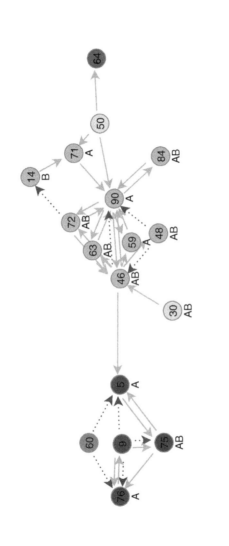

Figure 7.4 Factions in the main component of $\mathcal{D}_A^\sigma$. Graph representation without inconsistencies with factions {5, 9, 75, 76}, {14, 46, 48, 59, 63, 71, 72, 84, 90}, {30, 50}, {60}, and {64} based on C and K relations.

in the analysis, this link should be a case of the strength of a weak tie according to the theory formulated by Granovetter (1973), and which is described with more detail in Chapter 3 Multiplex Network Configurations.

Regarding the adoption of innovation A, this innovation type can be acquired by non-adopters through comparison inside faction red, from actor 60 to faction red, and also in the dyadic component (although this is not shown in this graph). Actor 9 perceives three competitors inside faction red, all of them adopters of innovation A; the other two perceived competition ties related to this faction comes from actor 60 who is also a non-adopter firm, and in the dyad component there is a perceived competition from actor 6.

However, it is hard to justify whether the combination of perceived competition with collaboration ties increases the chance of contagion as with actors 6 and 9, or rather the apparent uncertainty of actor 60 (who has no collaboration ties) motivates this firm to imitate the behavior of actors inside faction red. There may be other factors playing in the decision process of the actors to adopt an innovation, and for example the competition mechanism proposed by Doreian and Krackhardt (2001) could be useful to explain possible closure ties in this subgraph if this actor has reciprocated ties. Hence, from the structural point of view this particular circumstance remains an open question.

7.6.2 Cluster Semiring in Signed Network $\mathscr{X}_A^\sigma$

Even though we do not report the results here, it is worth mentioning that the cluster semirings of $\mathscr{X}_A^\sigma$ brings some structural features as well. When applying all semiring operations, a stable structure is found in semicycles of length 4 (as with balance semirings). Here the network is partitioned into two factions, one made of node 60 alone, and the other faction is with the rest of the nodes. This partition is according to the spirit of a clusterable structure since the groups are linked by a negative type of valence, even though the within relations in one of the groups has an ambivalent character.

Finally, the cluster semiring with 4-cycles gives a structure where, besides the factions that are product of closed semipaths with the cluster semiring, nodes 48 and 64 are also distinguished in the main component. As Figure 7.4 evidences, this last actor is a pendant with no outgoing tie to the component, and that is the reason why no cycles can occur with this actor. On the other hand, node 48 distinguishes from the fact that this actor has no outgoing collaboration ties with the rest of the firms, which means that the possible closed paths from this node necessarily have a negative sign.

The study of the comparison mechanism in the adoption of innovations with semiring structures in $\mathscr{X}_A^\sigma$ serve to complement the results from an influence process with direct communication in the relational systems; that is through direct contact among the actors. However, the relational content in this case is substantially different; while in cohesion influence all ties are positively evaluated, with the comparison mechanism we are counting with ties having divergent signs. Therefore the analysis for comparison requires a different approach than for direct communication, and we benefit both from algebraic structures and theories of structural balance in order to establish the context where social influence and the behavior of the actors such as the adoption of innovations taken place.

The algebraic methods applied to the structural balance theory permitted to find groups and factions inside the network components where competition was contrasted with collaboration, and which constituted the settings for the analysis of the comparison mechanism in multiple networks. The premise is that balanced settings are expected to be more stable over time than imbalanced structures. Then such principle allowed us to support the predictions made about under which conditions the relational structure has stood proxy for social influence and innovation adoption in the empirical networks with contrasting relations.

Although we have succeed to make an analysis of systems having different tie valences, the relationship between the associated structure and the actors attributes was still made through a correlation of the results, both with cohesion influence and comparison. This meant that we have treated these two aspects as separate entities and we have overlooked the fact that the adoption of innovations can be an integrated part of the relational system. In this sense, the rest of the study is going to take such a radical different approach in the analysis of the adoption of innovations with the available data of the business incubator centers and more generally with other types of attribute-based information in systems having relations at different levels like the Incubator networks.

7.7   Signed Networks: Summary

Signed structures represent relations having different signs or valences, which means that they constitute special cases of multiplex network structures. In this Chapter, we performed different analyses of signed structures within social influence processes. Two forms for social influence are cohesion and comparison, which take direct communication and structurally balanced settings as their generating mechanisms.

We employed an algebraic framework for the analysis of comparison with the use of the semiring structure. A semiring is an algebraic object with two operations, multiplication and addition, to represent signed networks in terms of balance or in terms of clustering.

As illustration, we applied semirings to one Incubator network in order to look at influence channels. This was in the context of innovation adoption among entrepreneurial firms with the combination of collaboration and competition ties. Some actors in the network resulted being structurally correspondent in terms of balance semiring making a faction or a weakly balanced cluster even if these actors did not share a tie.

7.8   Learning Signed Networks by Doing

7.8.1   Signed Structures in Figure 7.1

```
# define node / edge / graph characteristics
R> scp <- list(directed = FALSE, cex = 14, vcol = 1, lwd = 10, ecol = 1, rot = -30,
+    showLbs = FALSE)

# plot different undirected signed triads with a scope
R> multigraph(c("1, 2", "1, 3", "2, 3"), scope = scp)
R> multigraph(list(c("1, 2", "1, 3"), "2, 3"), scope = scp)
R> multigraph(list("2, 3", c("1, 2", "1, 3")), scope = scp)
R> multigraph(list(NULL, c("1, 2", "1, 3", "2, 3")), scope = scp)
```

7.8.2   Balance Semiring Structures in a Signed Triad

```
# create a signed structure made of 3 vertices and sort vertices labels
R> sgn <- transf(list(p = c("1, 3", "3, 1", "2, 3"), n = c("2, 1", "2, 3")),
+    type = "toarray", sort = TRUE)

# re-define node / edge / graph characteristics
R> scps <- list(signed = TRUE, cex = 12, vcol = 1, lwd = 10, ecol = 1, bwd = .75,
+    swp = TRUE, showLbs = TRUE, fsize = 45, rot = -30)

# plot Fig. 7.2.
R> multigraph(sgn, scope = scps)
```

```
# balance semiring with paths and semipaths of the triad as Signed class object
R> noquote(cbind(
+    semiring(signed(sgn), type = "balance", symclos = FALSE, k = 1)$Q
+    , "    ",
+    semiring(signed(sgn), type = "balance", symclos = FALSE, k = 2)$Q
+    , "    ",
+    semiring(signed(sgn), type = "balance", symclos = TRUE, k = 2)$Q
+    , "    ",
+    semiring(signed(sgn), type = "balance", symclos = TRUE, k = 3)$Q
+    ))

  1 2 3      1 2 3      1 2 3      1 2 3
1 o o p      p o o      p a a      a a a
2 n o a      a o n      a a n      a a a
3 p o o      o o p      a n a      a a a
```

7.8.3   Structural Balance in Incubator Network A, $\mathscr{X}_A$

```
# load data
R> require(multiplex)
R> data(incA)

# create signed network 'netsA' from first and third relation type in 'incA'
R> netsA <- incA$net[,,c(1,3)]

# record the first (that is the largest) component in the signed structure
R> com <- comps(netsA)$com[[1]]

# from 'netsA' select the relational system of the recorded component
R> nsA <- rel.sys(netsA, type = "toarray", sel = com)

# from 𝒳_A select the recorded component of the two types of attributes
R> nsAa <- rel.sys(incA$net[,,c(1,3,4:5)], type = "toarray", sel = com)
```

```
## Fig. 7.3.

# redefine node / edge / graph characteristics
R> scp <- list(directed = TRUE, signed = TRUE, pos = 0, cex = 3, lwd = 2, fsize = 9,
+     ecol = c("#00C000","#CD0000"), bwd = .5, vcol0 = "#808080", swp = TRUE)

# plot multigraph
R> multigraph(nsA, layout = "force", seed = 123, scope = scp, vcol = "#3399FF")
```

7.8.4   Balance Structures in Table 7.4

```
# balance structure with semicycles
R> semiring(signed(nsA), type = "balance", k = 4, symclos = TRUE)$Q

     5  9 14 30 46 48 50 59 60 63 64 71 72 75 76 84 90
5    a  a  a  a  a  a  a  a  a  a  a  a  a  a  a  a  a
9    a  a  a  a  a  a  a  a  a  a  a  a  a  a  a  a  a
14   a  a  a  a  a  a  a  a  a  a  a  a  a  a  a  a  a
30   a  a  a  a  a  a  a  a  a  a  a  a  a  a  a  a  a
46   a  a  a  a  a  a  a  a  a  a  a  a  a  a  a  a  a
48   a  a  a  a  a  a  a  a  a  a  a  a  a  a  a  a  a
50   a  a  a  a  a  a  a  a  a  a  a  a  a  a  a  a  a
59   a  a  a  a  a  a  a  a  a  a  a  a  a  a  a  a  a
60   a  a  a  a  a  a  a  a  a  a  a  a  a  a  a  a  a
```

```
63 a a a   a   a   a   a   a   a   a   a   a   a   a   a   a   a
64 a a a   a   a   a   a   a   a   a   a   a   a   a   a   a   a
71 a a a   a   a   a   a   a   a   a   a   a   a   a   a   a   a
72 a a a   a   a   a   a   a   a   a   a   a   a   a   a   a   a
75 a a a   a   a   a   a   a   a   a   a   a   a   a   a   a   a
76 a a a   a   a   a   a   a   a   a   a   a   a   a   a   a   a
84 a a a   a   a   a   a   a   a   a   a   a   a   a   a   a   a
90 a a a   a   a   a   a   a   a   a   a   a   a   a   a   a   a

attr(,"class")
[1] "Rel.Q"   "balance"   "semipaths"
```

```
# balance structure with cycles
R> semiring(signed(nsA), type = "balance", k = 4, symclos = FALSE)

...

$Q
     5  9 14 30 46 48 50 59 60 63 64 71 72 75 76 84 90
5    a  a  o  o  o  o  o  o  o  o  o  o  o  a  a  o  o
9    a  a  o  o  o  o  o  o  o  o  o  o  o  a  a  o  o
14   a  a  a  o  a  a  o  a  o  a  o  a  a  a  a  a  a
30   a  a  a  o  a  a  o  a  o  a  o  a  a  a  a  a  a
46   a  a  a  o  a  a  o  a  o  a  o  a  a  a  a  a  a
48   a  a  a  o  a  a  o  a  o  a  o  a  a  a  a  a  a
50   a  a  a  o  a  a  o  a  o  a  o  a  a  a  a  a  a
59   a  a  a  o  a  a  o  a  o  a  o  a  a  a  a  a  a
60   a  a  o  o  o  o  o  o  o  o  o  o  o  a  a  o  o
63   a  a  a  o  a  a  o  a  o  a  o  a  a  a  a  a  a
64   o  o  o  o  o  o  o  o  o  o  o  o  o  o  o  o  o
71   a  a  a  o  a  a  o  a  o  a  o  a  a  a  a  a  a
72   a  a  a  o  a  a  o  a  o  a  o  a  a  a  a  a  a
75   a  a  o  o  o  o  o  o  o  o  o  o  o  a  a  o  o
76   a  a  o  o  o  o  o  o  o  o  o  o  o  a  a  o  o
84   a  a  a  o  a  a  o  a  o  a  o  a  a  a  a  a  a
90   a  a  a  o  a  a  o  a  o  a  o  a  a  a  a  a  a

...

attr(,"class")
[1] "Rel.Q"   "balance"   "paths"
```

7.8.4.1   Signed Graph in Figure 7.3.

```
# define actors clustering
R> clus <- c(1, 1, 2, 3, 2, 2, 3, 2, 4, 2, 5, 2, 2, 1, 1, 2, 2)

# plot with clustering information and actor attributes
R> multigraph(nsA, layout = "force", seed = 123, scope = scp, vcol = 2:6,
+     clu = clus, att = nsAa)
```

8

Affiliation Networks

8.1 Structural Analysis of Affiliation Networks

Social systems in which there are two different sets of elements related with each other constitute *affiliation networks*, also known as *two-mode* and *bipartite* networks. Breiger (1974) points out that in affiliation networks there is a duality between actors and their groups, where the term *duality* refers to the "invariance in models of social structure and processes" (White, 2013).

Certainly, the notion of "groups" is extended to other types of entities related to the actors such as events, actor attributes, etc. These represent a separate domain from the actor set. As a result, affiliation networks represent two-mode data within a domain and a co-domain for the two types of sets, and this is in contrast with the usual one-mode social networks where there is just one set of relations on a single domain of social actors.

An example of an affiliation network is given in Table 8.1 where a cross table records the memberships of the G20 (Group of 20) countries or **Group of Twenty** according to Wikipedia (2019). Here the country names are specified in ISO 3166-1 alpha-3 codes in the row names (International Organization for Standardization, 2019), and the countries of the abbreviations are given in Appendix A. To the right of the table is specified the economic classification according to International Monetary Fund – IMF. Appendix A and Appendix C provide further details on this network with economic and socio-demographic indicators, and trade data of a pair of commodities among these nations.

To perform a network analysis of $\mathscr{X}_{G20}^{B}$, we treat Table 8.1 as a *rectangular* adjacency matrix with cross-domain relations; however, the last column with the economic classification of the countries is a vector of "factors" with two "levels", and it does not have a binary format. In order to be part of the adjacency matrix, the column corresponding to the economic classification needs a transformation first, which is possible by replacing the entries by two vectors with binary data where an actor belongs *either* to an "Advanced" or to an "Emerging" economy.

	ARG	AUS	BRA	CAN	CHN	DEU	FRA	GBR	IDN	IND	ITA	JPN	KOR	MEX	RUS	SAU	TUR	USA	ZAF
Advanced	0	1	0	1	0	1	1	1	0	0	1	1	1	0	0	0	0	1	0
Emerging	1	0	1	0	1	0	0	0	1	1	0	0	0	1	1	1	1	0	1

Algebraic Analysis of Social Networks: Models, Methods and Applications using R,
First Edition. J. A. R. Ostoic. Companion website: www.wiley.com/go/ostoic/algebraicanalysis.

Table 8.1 Affiliation network of G20 countries, $\mathscr{X}_{G20}^B$. *The last column after the affiliation matrix gives the International Monetary Fund economic classification of the countries.*

	P5	G4	G7	BRICS	MITKA	DAC	OECD	Cwth	N11	IMF-EC
ARG	0	0	0	0	0	0	0	0	0	Emerging
AUS	0	0	0	0	1	1	1	1	0	Advanced
BRA	0	1	0	1	0	0	0	0	0	Emerging
CAN	0	0	1	0	0	1	1	1	0	Advanced
CHN	1	0	0	1	0	0	0	0	0	Emerging
DEU	0	1	1	0	0	1	1	0	0	Advanced
FRA	1	0	1	0	0	1	1	0	0	Advanced
GBR	1	0	1	0	0	1	1	1	0	Advanced
IDN	0	0	0	0	1	0	0	0	1	Emerging
IND	0	1	0	1	0	0	0	1	0	Emerging
ITA	0	0	1	0	0	1	1	0	0	Advanced
JPN	0	1	1	0	0	1	1	0	0	Advanced
KOR	0	0	0	0	1	1	1	0	1	Advanced
MEX	0	0	0	0	1	0	1	0	1	Emerging
RUS	1	0	0	1	0	0	0	0	0	Emerging
SAU	0	0	0	0	0	0	0	0	0	Emerging
TUR	0	0	0	0	1	0	1	0	1	Emerging
USA	1	0	1	0	0	1	1	0	0	Advanced
ZAF	0	0	0	1	0	0	0	1	0	Emerging

Treating the economic classification of the countries—and other such kinds of "groups" in networks—as actor attributes, as with the Florentine Families network for example is just one possibility in the analysis (cf. Chapter 5). In this case, however, it is better to use the economic classification information to facilitate the visualization of the network, and this is since we seek to emphasize the distinction between the domain and the co-domain sets for the analysis of affiliation networks.

8.1.1 Visualization and Partition of Two-mode Data

Before focusing on algebraic methods to analyze affiliation network data, we start with the visualization of two-mode structures. We can depict affiliation networks in different ways, and the visualization techniques allow having flavors of the type of structure this data have. Visualizing two-mode networks gives us clues of their internal mechanisms.

For instance, the first graph in Figure 8.1 depicts the "classic" bipartite graph with two columns, one for the domain of actors and other for the co-domain of events. In this case, the graph is rotated 90 degrees clockwise, and undirected edges represent the affiliation relation. Bipartite graphs provide us with a rough sensation of the network structure; however, even for medium size two-mode networks it is difficult to get further insights about affiliation patterns. Certainly, it is possible to have more columns either for the actor set or for the set of events, and a second graph below the picture shows another type of bipartite graph with two columns for the actors and three columns for the events.

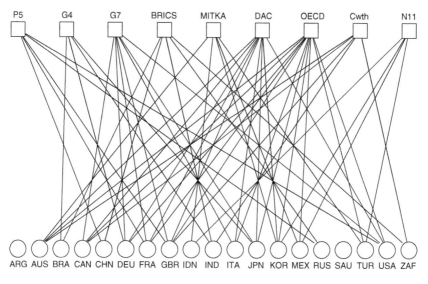

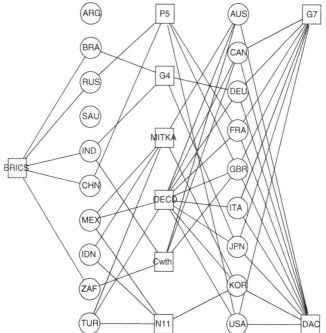

Figure 8.1 Bipartite graphs of the two-mode network $\mathscr{X}_{G20}^{B}$ with G20 countries and their affiliations. *Above: Bipartite graph with two rows. Below: Clustered bipartite graph.*

The second graph of Figure 8.1 is a *clustered bipartite graph* since it is based on the cluster information of the given economic classification of the countries to produce. Hence, each column represents countries either with Advanced economies or having Emerging countries. In the case of the "events", the organizations are split in three categories, one for each class of actors according to their economic classification, and the third category is for supranational organizations that are common for both classes of countries. That is, events where both classes of actors are affiliated with.

In affiliation networks, the co-domain relationships already provides some information for different forms for partitions of the system. This means that the partition is based on patterns of relations between domain and co-domain, and for the G20 countries, for example, we can use the clustering information of the bipartite graph with five columns given in Figure 8.1, rather than "merely" apply the country's economic classification in $\mathscr{X}_{G20}^{B}$. For instance, this approach makes explicit that BRICS organization has members only from emergent economies, while DAC and G7 group advanced economies where DAC groups all countries in this category.

The partition of two-mode networks—and one-mode configurations as well—is a form of distinction among the network members. In this sense, the five-column division made in the bipartite graph implies that countries of both emergent and advanced economies are split in those either belonging to the (politically rival) organizations G7 and BRICS or not belonging to them.

As a result, there are four classes of actors in network $\mathscr{X}_{G20}^{B}$ based on actor attributes and their patterns of relationships as well, and these are Advanced, Advanced-G7, Emerging, and Emerging-BRICS. Most of the graphs of this affiliation network are going to be based on this four-category model in the rest of the Chapter, and for convenience, the categories "Advanced-G7" and "Emerging-BRICS" will be sometimes shorten to "A-G7" and "E-BRICS".

8.1.2 Binomial Projection

Besides bipartite graphs, there are other ways to visualize affiliation networks, which allow making other kinds of partitions of these structures as well. For instance, for the visualization of affiliation networks it is also possible to consider both the actors and the events as different classes of the same "set of entities". This method is regarded as a *binomial projection* to two-mode data, as in the one algorithm described in Borgatti (2012, cf also Everett and Borgatti (2013)). This is a "direct" approach—rather than a "conversion" (Borgatti and Halgin, 2011)—and it makes viable to apply the layout algorithms used for the visualization of one-mode network structures such as a force-directed or stress majorization layouts.

Apart from the fact that the binomial projection allows plotting the of bipartite data as a one-mode network graph, another advantage of this approach is that it is a more convenient way to get insights into the network configuration than the bipartite graphs from before. In particular, one advantage of the binomial projection over bipartite graphs is that it makes easier to distinguish the different components and isolated actors especially in medium and large network configurations, which is an important part of the analysis of two-mode network data through visualization.

A "multilevel"-like multigraph with the binomial projection of the G20 countries affiliation network $\mathscr{X}_{G20}^{B}$ is given in Figure 8.2, and this configuration is made with the force-directed layout used for one-mode networks in previous chapters. Thanks to the repealing and attracting

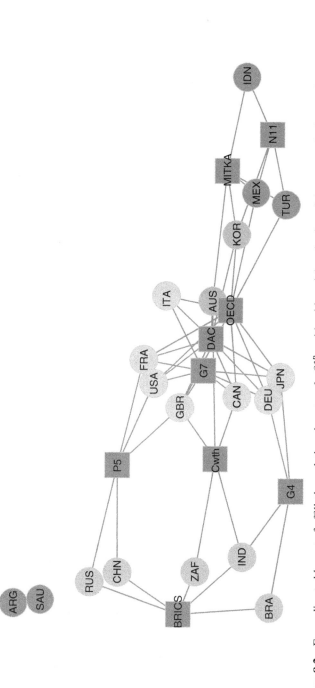

Figure 8.2 Force-directed layout of affiliation relations in network $\mathscr{P}_{G20}^{-B}$ with a binomial projection of the two mode data. *Color node orange is for organizations, plus olivegreen for Advanced-G7, khaki for Advanced, aquamarine for Emerging-BRICS, and azure for Emerging economies.*

forces among the actors and the events that are playing in this algorithm, actors tend to be "grouped" around the organizations they are affiliated with, and then we can take a look at some characteristics inherent among these network members. For instance, with the binomial projection of the data it is more evident from the graph that advanced economies are more integrated then emerging countries. Perhaps such aspect may be expected from theory in such institutionalized setting like the G20 countries network, but with other two-mode social systems the visualization of the binomial projection of two-mode data can result a very useful manner to perform the analysis.

The bipartite graph with clustered information constitutes a partition of the affiliation network $\mathscr{X}^B_{G20}$, and in the same manner we can get further insights into other actors and events from this network with the graphs of the binomial projection of the data. For instance, it is easier to notice in the multigraph with a force-directed layout in Figure 8.2 that RUS and CHN have identical affiliations in network $\mathscr{X}^B_{G20}$, which means that the two countries are regarded as structurally equivalent in the entire network $\mathscr{X}^B_{G20}$, and if we take the countries affiliated to MITKA, only TUR and MEX are structurally equivalent in network $\mathscr{X}^B_{G20}$. However, there are other forms to make a positional analysis of affiliation networks, and we look at some options in the next section.

8.2   Common Affiliations

A positional analysis of affiliation networks can be based on shared or common affiliations among the network members. Since two-mode network structures are made of two different types of domain, the term "common affiliations" in bipartite networks refers to either/or the number of:

- actors the events have in common
- events the actors attend in common

and these are two perspectives corresponding to the network actors and network events, respectively. We start the analysis of common affiliation in $\mathscr{X}^B_{G20}$ taking the actor perspectives first and then we pursue with the events perspective.

8.2.1   Actors Perspective

An important aspect in the analysis of two-mode networks is the value of the tie between actors in the domain and their common links in the co-domain. The resulting structure is known as a *co-membership* or *co-affiliation network*, which is represented by a one-mode and symmetric adjacency matrix where the entries are the number of common affiliations or common membership among the actors. Hence, in the graphical representation of co-affiliation networks the nodes represent the actors in the domain while the edges stand for the affiliations in the co-domain.

This co-affiliation network is in principle a valued network, although it is possible to have a dichotomous structure where a pair of actors either have or not have a common membership. Besides, edges are typically undirected, and this is because a "common affiliation" implies a mutual type relationship, which also means that the matrix representing the co-affiliation network results being symmetric.

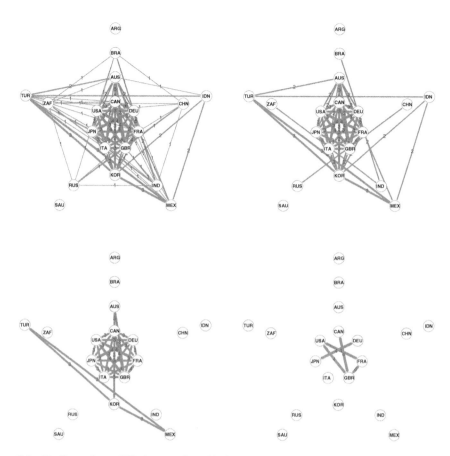

Figure 8.3 Undirected co-affiliation graphs with four salient structures of the G20 countries network $\mathscr{X}_{G20}^B$. *Edges are valued with number of common memberships between countries.*

For example, Figure 8.3 provides four different configurations with different strengths of the tie among G20 countries representing their common affiliations in the supranational organizations from the Wikipedia article. The plots are valued graphs with colored edges with a concentric layout having 4 radii where each radius represents a specific class of actors in $\mathscr{X}_{G20}^B$ as explained in the previous section. In these plots, the two radii at the center correspond to the advanced economies where the members of G7 make their own class, and—in the same manner—the emerging economies are located in the two peripheral radii where the members of BRICS make their own class as well.

We clearly notice by looking at the pictures that advanced economies are more heavily linked than emerging countries, and this is if we account for the supranational organizations included in $\mathscr{X}_{G20}^B$. By dropping the edges with a lower weight than an established cutoff value, it is possible to obtain a "skeleton" of the network structure that is based on similarities among actors. In fact, all four graph versions of the G20 co-affiliation network represent a particular salient structure of the configuration based on similarity; however, the salient structure of this kind for the G20 co-affiliation network differs from the Pathfinder semiring approach used in Chapter 9 where the criterion is rather based on dissimilarities among the network members.

8.2.2   Events Perspective

Common affiliation structures in bipartite networks are also from the *events perspective*, and this is because of the dual nature of this kind of network.

For instance, if we take a look at the bipartite graph with clustered information of $\mathscr{X}_{G20}^{B}$ in Figure 8.1, then we can see that there are some organizations acting as "political bridges" between countries with an emerging and an advanced economy. From an event perspective, the relations of the political *bridge organizations* that do not overlap with each other within the G20 countries network is represented by the extract matrix transpose.

	ARG	AUS	BRA	CAN	CHN	DEU	FRA	GBR	IDN	IND	ITA	JPN	KOR	MEX	RUS	SAU	TUR	USA	ZAF
P5	0	0	0	0	1	0	1	1	0	0	0	0	0	0	1	0	0	1	0
G4	0	0	1	0	0	1	0	0	0	1	0	1	0	0	0	0	0	0	0
MITKA	0	1	0	0	0	0	0	0	1	0	0	0	1	1	0	0	1	0	0

This is by keeping in mind, for example, that all countries related with N11 are related with MITKA as well in $\mathscr{X}_{G20}^{B}$. Because the above three supranational organizations P5, G4, and MITKA do not have common relations with the countries, it means that the co-affiliation structure among organizations in this case is empty. However, from the actors perspective such co-affiliation structure is represented as a binary and symmetric adjacency matrix where the 1 entries for each actor are with the actors having a common affiliation in the network.

Another other possibility of non-overlapping organizations from an event perspective comes from considering the organizations that draw together the advanced and emergent economies in $\mathscr{X}_{G20}^{B}$. These politically "rival" organizations are G7 and BRICS, plus MITKA, which has both advanced and emerging economies as members, and this is without overlapping the other two organizations. Certainly, this constitutes another criterion for the classification of the actors in the $\mathscr{X}_{G20}^{B}$ network with the following event perspective.

	ARG	AUS	BRA	CAN	CHN	DEU	FRA	GBR	IDN	IND	ITA	JPN	KOR	MEX	RUS	SAU	TUR	USA	ZAF
G7	0	0	0	1	0	1	1	1	0	0	1	1	0	0	0	0	0	1	0
BRICS	0	0	1	0	1	0	0	0	0	1	0	0	0	0	1	0	0	0	1
MITKA	0	1	0	0	0	0	0	0	1	0	0	0	1	1	0	0	1	0	0

The information provided by the event perspective allows performing a classification of the actors in a given affiliation network, and it contributes to a form of network reduction. In the case of $\mathscr{X}_{G20}^{B}$, however, it seems more informative to look at the structure made of the countries belonging to the political bridges rather than the system made of organization countries belonging to a separate economic classification.

Hence, if we take the network actors in $\mathscr{X}_{G20}^{B}$ that are either affiliated to the three political bridges, P5, G4, MITKA, and (in some cases or) to the politically rival organizations G7 and BRICS, then we obtain another G20 countries affiliation network with "bridge organizations". This latter system is denoted by $\mathscr{X}_{G20b}^{B}$, and has fewer of members than the full G20 countries affiliation network. While the affiliation matrix for $\mathscr{X}_{G20}^{B}$ has size 19×9, the affiliation matrix for the network with bridge organizations $\mathscr{X}_{G20b}^{B}$ has size 14×5.

Figure 8.4 presents the clustering version of the bipartite graph of the affiliation network with bridge organizations $\mathscr{X}_{G20b}^{B}$ with and without reduction of the structurally equivalent events.

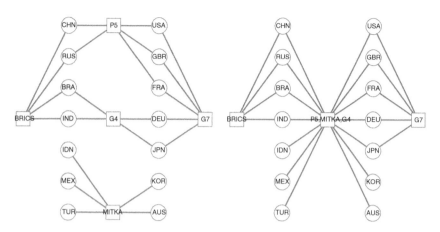

Figure 8.4 Clustered bipartite graph with and without reduction of structurally equivalent events in affiliation network $\mathscr{X}_{G20b}^{B}$. *The class of events corresponds to the bridge organizations in G20 affiliation countries network* $\mathscr{X}_{G20}^{B}$.

The placement of the countries is according to their economic classification, and the events are of three kinds: the two types of organizations for emerging and advanced economies, and the political bridges between these two classes of actors. First to the left of the picture are the political bridges given as separate entities, and to the right of the figure the bridge organization makes a single class.

The bipartite graph as a representation form evidences in a clear manner that the involved organizations in this network structure do not overlap with each other. Moreover, with this disposition of the actors and events it is easier to see in this network configuration that there are actors with identical affiliations and hence they can also make a class in terms of Structural equivalence.

It is important to mention that it is also possible to take both the actors and the events perspective at the same time in the modeling. In such case, the algebraic realm provides methods such as the Galois derivations within the Formal concept analysis framework to make the connection between actors and events in two-mode network structures.

8.2.3 Affiliation Network with Bridge Organizations $\mathscr{X}_{G20b}^{B}$

At this point, we performed analyses of the entire affiliation network $\mathscr{X}_{G20}^{B}$ with the G20 countries, and also with the reduced version with bridge organizations in $\mathscr{X}_{G20b}^{B}$. Recall that this network is a smaller part of this system made of the countries belonging to the three political bridges P5, G4, and MITKA, and/or to the politically rival organizations G7 and BRICS.

If we take a look at the network structure of $\mathscr{X}_{G20b}^{B}$, the connections between the members of this network produce a configuration that is given as a graph format in Figure 8.5. The plot of this affiliation network has a forced directed layout using the binomial projection approach to two-mode data, and this graph representation is isomorphic to the bipartite graph given to the left of Figure 8.4.

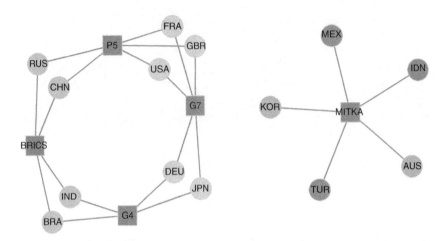

Figure 8.5 Co-affiliation network $\mathcal{X}_{G20b}^B$ with non-overlapping G20 bridge organizations. *Graph with binomial projection and force-directed layout with color nodes as in Figure* 8.2.

The two differences lie first in the layout applied, and second in the fact that the vertices in the graph of Figure 8.5 representing the co-affiliation structure of $\mathcal{X}_{G20b}^B$ with a binomial projection have distinguished colors to differentiate the four classes of actors, and which are based on their economic classification. That is, the color node «olivegreen» is for Advanced-G7, «khaki» for Advanced, «aquamarine» for Emerging-BRICS, and «azure» for Emerging economies, whereas for the network events the organizations in $\mathcal{X}_{G20b}^B$ are also represented by another shape and the color «orange» to distinguish these entities from the actors as we did before with the bipartite graphs.

For networks with a two-mode data, actors with identical affiliations in the system are structurally equivalent. Often it is enough to count with the binomial projection of the two-mode data to detect structurally equivalent actors and events, particularly when the network is not too large. As we can see in Figure 8.5, the depicted system $\mathcal{X}_{G20b}^B$ has two components, and in this case it is relatively easy to establish which countries are structurally equivalent in this political network.

In the same manner as with one-mode networks, structurally equivalent actors and events in affiliation networks are likely to behave in the same manner. In the case of the G20 countries, this constitutes a very important aspect when it comes to make a substantial analysis about how and why the nations acts as such in the international political arena.

8.3 Formal Concept Analysis

An algebraic approach for the analysis of affiliation networks is found in *formal concept analysis* (Ganter and Wille, 1996) where the cross table representation of data is mathematically defined. As the authors mention, the adjective "formal"—in concept analysis—simply means that we are dealing with mathematical terms, and it is usually omitted from the narrative.

In terms of this analytical framework, the domain and co-domain of an affiliation network, respectively, are characterized as a set of *objects G* and a set of *attributes M*. The mathematical definition of a *formal context* given by Ganter and Wille (1996) is

$$\mathbb{K} = (\, G, \, M, \, I \,).$$

This means that $\mathbb{K}$ is obtained with an incidence relation $I \subseteq G \times M$ between the sets G and M, which is the triple represented by the cross table representing the data. From such data of the formal context arises another fundamental notion in formal concept analysis: that of formal concepts.

A *formal concept* $\mathbb{C}$ of a formal context $\mathbb{K}$ is a pair of sets of objects A and attributes B that is maximally contained on each other; that is, columns in the cross table representing the attribute set that help to cover the most entries in I.

Formally, this is defined as

$$\mathbb{C} = (A, B)$$

where $A \subseteq G, B \subseteq M, A' = B$, and $B' = A$, where A and B are said to be the *extent* and *intent* of the formal concept, respectively. A' and B' implies that the concept is determined by both parts.

The set of all concepts of a given context is denoted by $\mathbf{B}(G, M, I)$, and the adjective "formal", which emphasizes the mathematical aspect of the "context" and "concept" notions, is sometimes omitted in the original literature and here as well.

Visually, a concept $\mathbb{C}$ of $\mathbb{K}$ with an extent an intent A and B is highlighted in the cross table below as

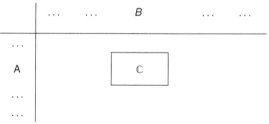

A formal concept (A, B) is then a "maximal rectangle" of the formal context where the incidence relation I ties extent A with intent B by the incidence relation I. In other words, formal concepts are a pair of sets of maximally contained objects and attributes, and the connection between these two parts of the formal concepts produce other structures. Additional definitions and the application of the Galois derivation notion to empirical two-mode networks are given in the next section.

8.4   Formal Concepts and Galois Derivations

Formal concept analysis makes it possible to make the connection between actors and events in affiliation networks and other kinds of two-mode data in the investigation. Such aspect is at the heart of this mathematical framework, and we introduce now a key concept that makes it possible the "connection" between the domain and the co-domain in a bipartite network.

A *Galois derivation* or Galois connection between the power sets of sets G and M is derived for any subsets $A \subseteq G$ and $B \subseteq M$ by the set of attributes common to all the objects in the intent, A', and the set of objects possessing the attributes in the extent, B':

$$A' = m \in M \mid (g, m) \in I \quad \text{(for all } g \in A)$$
$$B' = g \in G \mid (g, m) \in I \quad \text{(for all } m \in B)$$

where g and m are rows and columns in the derivation operation, and G and M result being two closed systems dually isomorphic to each other.

As a result, Galois derivations between the objects in the intent and the attributes in the extent produce the set of formal concepts in the formal context or the bipartite network in this case. The output of the Galois connections is not simply a set of elements from the formal context, but it is actually a *family* of concepts where the order of the elements is significant.

8.4.1 Concepts in the G20 Affiliation Network

After looking at the Galois derivation notion, we are able to count with a set of formal concepts from the context $\mathbb{K}$ with a full and a reduced format. This allows us to continue the algebraic analysis of the G20 affiliation network $\mathscr{X}^B_{G20}$ and its reduced version with bridge organizations $\mathscr{X}^B_{G20b}$ within the Formal Concept Analysis framework in the remaining part of this chapter.

8.4.1.1 Concepts in G20 Affiliation Network

The idea of applying Formal concept analysis to a co-affiliation network such as nation's membership to supranational is not new. Ganter and Wille (1996, pp. 28−30) constructed the Concept lattice of the context for developing nations in supranational groups. However, this time we are going to apply now the Galois derivations to the affiliation network $\mathscr{X}^B_{G20}$ with two different labeling format. For instance, a truncated version of the Galois connections between sets G and M of actors and events in network $\mathscr{X}^B_{G20}$ with a *full labeling* of the elements in this formal context is the following:

$$
\begin{aligned}
&\mathbb{C}_1 : \{P5\} && \{CHN, FRA, GBR, RUS, USA\} \\
&\mathbb{C}_2 : \{G4\} && \{BRA, DEU, IND, JPN\} \\
&\mathbb{C}_3 : \{DAC, G7, OECD\} && \{CAN, DEU, FRA, GBR, ITA, JPN, USA\} \\
&\mathbb{C}_4 : \{BRICS\} && \{BRA, CHN, IND, RUS, ZAF\} \\
&\quad \cdots && \quad \cdots \\
&\mathbb{C}_{25}
\end{aligned}
$$

In the above example, for instance, the G7 nations are also members of DAC and OECD, and some of these countries are connected to the organizations previously derived. We notice also that some countries appear more than once with the full labeling of the Galois derivations between actors and events in $\mathscr{X}^B_{G20}$, and this is because such countries have multiple affiliations. The organizations are also repeated in different concepts in this case since they have different countries as members. In fact with the full labeling of the Galois derivations there is one concept that groups all actors and another concept that groups all events at once, and these two particular concepts constitute the "ends" of the family of concepts in the context.

However, as said before, the output of a Galois derivation is not simply a set of elements, but actually a family of concepts where the order of the elements is significant. This is more evident with the *reduced labeling* of objects and attributes, which in most cases provides a more informative structure than the full labeling. With the reduced option the repeated objects discard the previous ones, whereas the recurrent attributes in the listing are also discarded but afterwards in the derivation.

Although the condensed labeling of the concepts is more appropriate for the analysis, the ordering of the concepts even with the reduced labeling is based, however, on the full derivation of the context. This is because with the reduced labeling there can be several concepts that are empty, and empty sets of extents and intents prevent producing an ordering structure that characterizes a family of elements, which in this case are formal concepts.

With the assurance that a full Galois derivation of the context precedes a reduced one, we have these connections in network $\mathscr{X}_{G20}^{B}$ with the reduced labeling below, and we bear in mind that the order of the concepts is the same as with the "full labeling" given above.

$$\mathbb{C}_1 : \{P5\} \qquad \{\varnothing\}$$
$$\mathbb{C}_2 : \{G4\} \qquad \{\varnothing\}$$
$$\mathbb{C}_3 : \{G7\} \qquad \{ITA\}$$
$$\mathbb{C}_4 : \{BRICS\} \qquad \{\varnothing\}$$
$$\ldots \qquad \ldots$$
$$\mathbb{C}_{25}$$

In this extract of the Galois derivations in context $\mathscr{X}_{G20}^{B}$, it is only the third concept that retains elements from both the extent and intent of the context, and hence members from both domains of the network. This means that this single country is related just to the three attributes of the full labeling without belonging to any other concept in the reduced labeling. The remaining countries and organizations belong to the other 21 concepts located afterwards.

8.4.1.2   Concepts in G20 Affiliation Network with Bridges

Another version of the affiliation network $\mathscr{X}_{G20}^{B}$ is made just with the countries associated with the bridge organizations in the system. This reduced formal context is denoted by $\mathscr{X}_{G20b}^{B}$, and below we look at the complete family of concepts with a reduced labeling that is product of a Galois derivation of intents and extents.

$$\mathbb{C}_1 : \{P5\} \qquad \{\varnothing\}$$
$$\mathbb{C}_2 : \{G4\} \qquad \{\varnothing\}$$
$$\mathbb{C}_3 : \{G7\} \qquad \{\varnothing\}$$
$$\mathbb{C}_4 : \{BRICS\} \qquad \{\varnothing\}$$
$$\mathbb{C}_5 : \{MITKA\} \qquad \{AUS, IDN, KOR, MEX, TUR\}$$
$$\mathbb{C}_6 : \{\varnothing\} \qquad \{\varnothing\}$$

$$\mathbb{C}_7 : \{\varnothing\} \qquad \{FRA, GBR, USA\}$$
$$\mathbb{C}_8 : \{\varnothing\} \qquad \{CHN, RUS\}$$
$$\mathbb{C}_9 : \{\varnothing\} \qquad \{DEU, JPN\}$$
$$\mathbb{C}_{10} : \{\varnothing\} \qquad \{BRA, IND\}$$
$$\mathbb{C}_{11} : \{\varnothing\} \qquad \{\varnothing\}$$

Hence, the set of formal concepts has a single portion where neither the intent nor the extent is empty when considering a reduced labeling of $\mathscr{X}_{G20b}^{B}$, which is concept 5. The rest of the concepts have at least one part empty with a reduced labeling, and there are a pair of concepts where both the intent and the extent are empty. However, rather than just enumerate the different formal concepts that are the product of a Galois derivation from a given context, it gives useful results to connect these concepts with an ordering relation, and in the next point we are going to look at the creation of a hierarchy structure of the different concepts.

8.5 Concept Lattice and Ordering of Concepts

8.5.1 Partial Ordering of the Concepts

The set of concepts product of the Galois derivations in the context serves to establish the partial ordering of the concepts. For the G20 formal context, such information is represented by object $\mathbb{C}$ that contains 25 formal concepts in both the full and reduced labeling of objects and attributes as in the previous example.

With a reduced labeling of objects and attributes in $\mathbb{C}$, recurrent objects in the family of concepts discard the previous ones produced in the derivation. This implies that these latter objects are covered for the one printed in the output, and such type of relation applies to all objects, and dually to the attributes as well. Hence, the set of inclusions of the concepts serves to clarify the disposition of the Galois derivations in the formal context.

The *partial ordering of the concepts* represents the hierarchy of the totality of all concepts of a context and is produced by the relation subconcept–superconcept, (A_1, B_1) and (A_2, B_2), of extents and intents. This is formally expressed as:

$$(A, B) \leq (A_2, B_2) \quad \Leftrightarrow \quad A_1 \subseteq A_2 \quad (\Leftrightarrow \quad B_1 \subseteq B_2).$$

That is, the partial ordering of the concepts corresponds to set of inclusions among the maximal rectangles in the formal context.

Table 8.2 presents the partial ordering of the 25 formal concepts in the context $\mathscr{X}_{G20}^B$ corresponding to G20 countries. For convenience, the row and column labels in the partial order table matches the number of concepts in the G20 affiliation network X_{G20}^B. However, the ordering relations are more precisely among the pairs of sets when the different concepts have the labels of both intents and extents.

By looking at the partial order table, we can see for instance that concept 10 is contained by all elements of the structure, whereas concept 25 includes the rest of the concepts resulting from the Galois derivations. Apart from these, other inclusion relations are difficult to observe with the matrix format and, as with the partially ordered semigroup from Chapter 5), the visualization of the partially ordered structure of inclusions among concepts results most of times being more informative.

8.5.2 Concept Lattice of the Context

The *Concept lattice* of the formal context, which is also known as *Galois lattice* in social network analysis (e.g. Freeman and White, 1993), is established based on the system of partially ordered elements corresponding to the set of all concepts. As seen before, the set of concepts in the formal context are derived through Galois connections among them, and hence the three concepts are intimately linked to each other.

Table 8.2 Partial ordering of the formal concepts in the G20 affiliation network $\mathscr{X}_{G20}^{B}$. *The same hierarchy applies to concepts having a full or a reduced format.*

≤	1	2	3	4	5	6	7	8	9	10	11	12	13	14	15	16	17	18	19	20	21	22	23	24	25
1	1	0	0	0	0	0	0	0	0	0	0	0	0	0	0	0	0	0	0	0	0	0	0	0	1
2	0	1	0	0	0	0	0	0	0	0	0	0	0	0	0	0	0	0	0	0	0	0	0	0	1
3	0	0	1	0	0	1	1	0	0	0	0	0	0	0	0	0	0	0	0	0	0	0	0	0	1
4	0	0	0	1	0	0	0	0	0	0	0	0	0	0	0	0	0	0	0	0	0	0	0	0	1
5	0	0	0	0	1	0	0	0	0	0	0	0	0	0	0	0	0	0	0	0	0	0	0	0	1
6	0	0	0	0	0	1	1	0	0	0	0	0	0	0	0	0	0	0	0	0	0	0	0	0	1
7	0	0	0	0	0	0	1	0	0	0	0	0	0	0	0	0	0	0	0	0	0	0	0	0	1
8	0	0	0	0	0	0	0	1	0	0	0	0	0	0	0	0	0	0	0	0	0	0	0	0	1
9	0	0	0	0	1	0	0	0	1	0	0	0	0	0	0	0	0	0	0	0	0	0	0	0	1
10	1	1	1	1	1	1	1	1	1	1	1	1	1	1	1	1	1	1	1	1	1	1	1	1	1
11	1	0	1	0	0	1	1	0	0	0	1	0	0	0	0	0	0	0	0	0	0	0	0	0	1
12	1	0	0	1	0	0	0	0	0	0	0	1	0	0	0	0	0	0	0	0	0	0	0	0	1
13	1	0	1	0	0	1	1	1	0	0	1	0	1	0	0	0	1	0	0	0	0	1	0	0	1
14	0	1	1	0	0	1	1	0	0	0	0	0	0	1	0	0	0	0	0	0	0	0	0	0	1
15	0	1	0	1	0	0	0	0	0	0	0	0	0	0	1	0	0	0	0	0	0	0	0	0	1
16	0	1	0	1	0	0	0	1	0	0	0	0	0	0	1	1	0	1	0	0	0	0	0	0	1
17	0	0	1	0	0	1	1	1	0	0	0	0	0	0	0	0	1	0	0	0	0	1	0	0	1
18	0	0	0	1	0	0	0	1	0	0	0	0	0	0	0	0	0	1	0	0	0	0	0	0	1
19	0	0	0	0	1	1	1	1	0	0	0	0	0	0	0	0	0	0	1	1	0	0	0	0	1
20	0	0	0	0	1	0	1	0	0	0	0	0	0	0	0	0	0	0	0	1	0	0	0	0	1
21	0	0	0	0	1	1	1	1	1	0	0	0	0	0	0	0	0	0	1	1	1	1	0	0	1
22	0	0	0	0	0	1	1	1	0	0	0	0	0	0	0	0	0	0	0	0	1	1	0	0	1
23	0	0	0	0	1	1	1	0	1	0	0	0	0	0	0	0	0	0	1	1	0	0	1	1	1
24	0	0	0	0	1	0	1	0	1	0	0	0	0	0	0	0	0	0	1	0	0	0	1	1	1
25	0	0	0	0	0	0	0	0	0	0	0	0	0	0	0	0	0	0	0	0	0	0	0	0	1

Formally, in the Concept lattice of the context, the greatest lower bound of the meet and the least upper bound of the join are defined in terms of objects and attributes and an index set T by the *Basic Theorem of Concepts Lattices* (Ganter and Wille, 1996).

$$\bigwedge_{t\in T} (A_t, B_t) = \left(\bigcap_{t\in T} A_t, \left(\bigcup_{t\in T} B_t \right)'' \right)$$

$$\bigvee_{t\in T} (A_t, B_t) = \left(\left(\bigcup_{t\in T} A_t \right)'', \bigcap_{t\in T} B_t \right).$$

As a result, the Galois connections among the concepts in a given formal context serve to create a poset structure with the inclusion relations that reflects the labeling of the derivations made on the context.

Depending on the labeling of the Galois derivations, the Concept lattice can have either a full or a reduced format. Recall that with a full format of the concepts both actors and events can have multiple entries in C, and in most cases it is more convenient to make explicit the sets of each concepts with a reduced format in Concept lattices. This is not only to avoid redundant information in the diagram, but more importantly to highlight the formal concepts in the formal context.

8.5.3   Concept Lattice of Network $\mathscr{X}_{G20}^B$

In the case of the G20 affiliation network $\mathscr{X}_{G20}^B$, the partial order structure, which was given in Table 8.2, constitutes the basis for the construction of the Concept lattice, which means that the Concept lattice of the context is the inclusion diagram of this poset. If this structure takes a reduced representation of the context, both objects and attributes are given just once in the concepts of the partial order structure. Otherwise attributes and objects are located in several places within the Concept lattice structure, as with the output we have seen before with the Galois derivations in $\mathscr{X}_{G20}^B$.

Figure 8.6 depicts the set of inclusions among the concepts in the formal context $\mathscr{X}_{G20}^B$ and also in the smaller version with the "cartels" and bridge organizations $\mathscr{X}_{G20b}^B$. These two representations have a reduced labeling of the concepts, and if a concept does not have a label—which often happens with reduced contexts—then the respective number of the concept is placed instead of the node rather than leaving the node unlabeled. The numbering of the concepts is arbitrary, and it does not indicate any placement in the ordering structure.

In the Concept lattice of the context, objects having more attributes are located downwards and hence covered by the objects with less attributes. Conversely, the reverse is true for the other part of the duality, which means that the most popular attributes are located more upwards than the less popular ones. This implies that the "levels" of the lattice somewhat reflect the covering in the two instances, but this is not necessarily true in the depiction.

In fact the placement of the elements in the inclusion diagram corresponds in this case to the layout algorithm applied by the *Rgraphviz* package (Hansen et al., 2018) that is an R (R Core Team, 2015) implementation of the *graphviz* program (AT&T Labs Research, 2019), and which has been incorporated into the *multiplex* package (Ostoic, 2019*b*). The placement of the elements in the layout can be rather arbitrary depending on particular cases of the data.

For instance, although in the network depicted in Figure 8.6 ZAF has fewer affiliations than ITA, the former country is in a lower level than the latter nation. Hence, to have precise information about the ordering of the countries, we need to count with the set of inclusion relations in the Concept lattice, particularly in context structures having a large number of objects and attributes.

Another significant aspect has to do with the labeling of the formal concepts. For example, in the Concept lattice for $\mathscr{X}_{G20}^B$, concepts 10, 19, 20, and 22 have both sets of actors and events empty with a reduced labeling, but this is not true with the full labeling. In fact, the hierarchy given in the partially ordered structure is based on the full labeling of the concepts, even if the presentation in the poset table and in the Concept lattice is with a reduced labeling.

It is worth saying that the smaller version of the Concept lattice corresponding to $\mathscr{X}_{G20b}^B$ is a *complete sublattice* of the Concept lattice for $\mathscr{X}_{G20}^B$. This is because the partial order structure is closed both under a suprema and under an infima concept element, which happen to be empty in this particular case. If there are any inclusions among either objects or attributes in this later

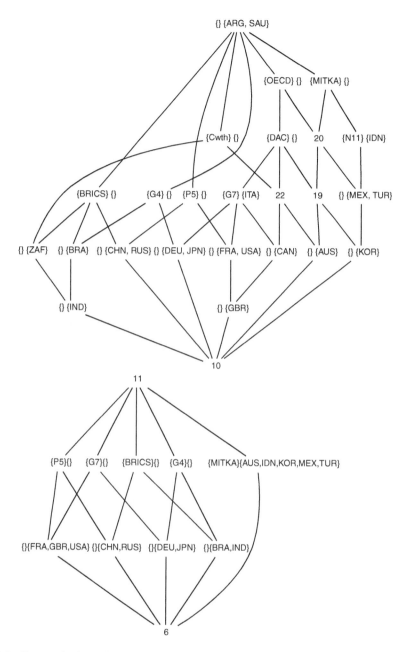

Figure 8.6 Concept lattices of the G20 affiliation network with reduced labeling of intents and extents. *Above: Entire affiliation network $\mathscr{X}_{G20}^B$. Below: Affiliation network with bridges $\mathscr{X}_{G20b}^B$.*

partial order, it is easy to see in the lattice diagram corresponding to $\mathscr{X}^B_{G20b}$ the structurally equivalent actors and events by their identical affiliations in the network. In fact, the structure of $\mathscr{X}^B_{G20b}$ provides a more effective categorization of these G20 countries than of the network $\mathscr{X}^B_{G20}$ above.

Finally, an example of a Concept lattice with a full format derived from the G20 countries affiliation network is given to the right of Figure 10.6 in Chapter 10 Multilevel Networks. In such case, the Concept lattice takes the positional system of classes of countries and organizations from $\mathscr{X}^B_{G20}$, and hence it constitutes a reduced affiliation network structure (details of the modeling are given in the corresponding chapter). With this representation form, we evidence that the set of all actors and the set of all event in the two-mode network correspond to the maximum and the minimum elements in the complete Concept lattice structure.

8.6   Order Filters and Order Ideals

For relatively large formal contexts, the corresponding poset structure of the concepts typically results is too complex, and the construction of the Concept lattice of the context allows us to gain significant insights—or at least information—about the affiliation network structure. However, often it would be useful to count with the set of implications among objects and attributes in the Concept lattice structure. In this sense, algebraic notions based on *upsets* and *downsets* are of fundamental importance in uncovering implication subsets in an ordered structure. These implications are ordered sets that are made, for any formal concept or set of formal concepts in the formal context, of all elements either in the lower or in the greater bound of the Concept lattice structure.

Formally, an *ordered set* is represented by the pair $(P, \leq)$ where a and b are some elements in P. A non-empty subset U [resp. D] of P is an upset [resp. downset] called a *order filter* [resp. *order ideal*] iff, for all $a \in P$ and $b \in U$ [resp. D]:

$$b \leq a \text{ implies } a \in U \qquad [\text{ resp. } a \leq b \text{ implies } a \in D]$$

For a specific element $x \in P$, the upset $\uparrow x$ formed for all the upper bounds of x is called a *principal order filter* generated by x. Dually, $\downarrow x$ is a *principal order ideal* with all the lower bounds of x. Order filters and order ideals not coinciding with P are called *proper*. In this sense, both order filters and order ideals of particular elements of the context are based on the partial ordering of the concepts, which is graphically expressed by the Concept lattice.

8.6.1   Principal Order Filters

With the notions of order filters and order ideals in place we try to find subsets of concepts in the partial ordered structure of a given formal context. In order to better understand these concepts, we are going to take the Concept lattice of the affiliation networks $\mathscr{X}^B_{G20}$ and $\mathscr{X}^B_{G20b}$, which are depicted in Figure 8.6, and where the latter structure is a subset of the G20 countries affiliation network.

If, for example, we take the contexts $\mathscr{X}^B_{G20}$ corresponding to the G20 affiliation network and the smaller structure $\mathscr{X}^B_{G20b}$, the principal order filter set of the third concept of intents and extents with reduced format, which is {G7} {ITA} in $\mathscr{X}^B_{G20}$, then the following subset of concepts arise:

$$\mathscr{X}_{G20}\ \mathbb{C}_3 \qquad\qquad\qquad \mathscr{X}_{G20b}\ \mathbb{C}_3$$

$\mathbb{C}_3$:	{G7}	{ITA}
$\mathbb{C}_6$:	{DAC}	{Ø}
$\mathbb{C}_7$:	{OECD}	{Ø}
$\mathbb{C}_{25}$:	{Ø}	{ARG, SAU}

| $\mathbb{C}_3$: | {G7} | {Ø} |
| $\mathbb{C}_{11}$: | {Ø} | {Ø} |

The output shows us that the principal order filter is a proper one, and starts with the concept itself. In both cases, it then takes the different inclusions to this concept until the maxima element of the partial order structure, which are $\mathbb{C}_{25}$ for $\mathscr{X}^B_{G20}$ and $\mathbb{C}_{11}$ for $\mathscr{X}^B_{G20b}$. The numbering of the concepts is somewhat arbitrary as well, and it does not represent either a linear order or a partial ordering of the elements.

8.6.2 Order Ideals and Principal Order Ideals

Likewise, order ideals [and proper order ideals] of one or more concepts are obtained as the dual part of the order filter [and proper order filter] notion. For example, the principal order ideals of the first concept in $\mathscr{X}^B_{G20}$ and $\mathscr{X}^B_{G20b}$ are given below. In this case, the concept that corresponds to the minimal element in the Concept lattice of $\mathscr{X}^B_{G20}$ is $\mathbb{C}_{10}$, and for $\mathscr{X}^B_{G20b}$ the minimal element in is concept $\mathbb{C}_6$.

$$\mathscr{X}_{G20}\ \mathbb{C}_1 \qquad\qquad\qquad \mathscr{X}_{G20b}\ \mathbb{C}_1$$

$\mathbb{C}_1$:	{P5}	{Ø}
$\mathbb{C}_{10}$:	{Ø}	{Ø}
$\mathbb{C}_{11}$:	{Ø}	{FRA, USA}
$\mathbb{C}_{12}$:	{Ø}	{CHN, RUS}
$\mathbb{C}_{13}$:	{Ø}	{GBR}

$\mathbb{C}_1$:	{P5}	{Ø}
$\mathbb{C}_6$:	{Ø}	{Ø}
$\mathbb{C}_7$:	{Ø}	{FRA, GBR, USA}
$\mathbb{C}_8$:	{Ø}	{CHN, RUS}

It is also possible to derive the order ideal of the intents within more than one concept in the partial order structure. For instance, the order ideal of the intents within two concepts in these partial order structures which G7 and BRICS are part of produce their affiliates in the G20 network.

$$\mathscr{X}_{G20}\ \{\mathbb{C}_3, \mathbb{C}_4\} \qquad\qquad \mathscr{X}_{G20b}\ \{\mathbb{C}_3, \mathbb{C}_4\}$$

$\mathbb{C}_3$:	{G7}	{ITA}	$\mathbb{C}_3$:	{G7}	{Ø}
$\mathbb{C}_4$:	{BRICS}	{Ø}	$\mathbb{C}_4$:	{BRICS}	{Ø}
$\mathbb{C}_{10}$:	{Ø}	{Ø}	$\mathbb{C}_6$:	{Ø}	{Ø}
$\mathbb{C}_{11}$:	{Ø}	{FRA, USA}	$\mathbb{C}_7$:	{Ø}	{FRA, GBR, USA}
$\mathbb{C}_{12}$:	{Ø}	{CHN, RUS}	$\mathbb{C}_8$:	{Ø}	{CHN, RUS}
$\mathbb{C}_{13}$:	{Ø}	{GBR}	$\mathbb{C}_9$:	{Ø}	{DEU, JPN}
$\mathbb{C}_{14}$:	{Ø}	{DEU, JPN}	$\mathbb{C}_{10}$:	{Ø}	{BRA, IND}

where $\mathbb{C}_{15}$–$\mathbb{C}_{18}$ in $\mathscr{X}^B_{G20}$ correspond to {Ø} with {BRA}, {IND}, {CAN}, and {ZAF}.

The minimal elements in the Concept lattices here are the same as with the previous example, and—again—the placement of the different concepts in the output does not correspond automatically to the ordering of concepts in the partial order structure neither in this case.

Before concluding the order filters and order ideals section in the partial ordering of the concepts and Concept lattice, there are a couple of things worth mentioning:

1. The partial order structure produced by the Galois connections of concepts in a given context produce always a complete lattice, which means that both the maximal and minimal element in the inclusion diagram are part of any set of order filters, order ideals, or principal order filters or else principal order ideals.
2. It is easy to see that the longest principal order filter or the longest principal order ideal departs either from the minimal or from the maximal element in the Concept lattice of the partial structure of the concepts in the formal context. Atoms and co-atoms in the Concept lattice, on the other hand, will generate the shortest order filters and order ideals, respectively.

8.7   Affiliation Networks: Summary

In this Chapter, we have looked at different representation forms for affiliation or two-mode networks, which are types of configuration that combine two different domains. Affiliation networks are themselves special types of multiplex networks and this is because they carry an intrinsic difference in their structures. Affiliation networks differ from one-mode networks in the presence of two domains, while there is a single domain in the one-mode structure.

Bipartite graphs having two or more columns represent affiliation networks in a graphical format where domain and co-domain share affiliation ties, and is possible to cluster actors and events in these types of graphs. Another approach is the binomial projection of two-mode data that allows the depiction of two-mode structures with the two visualization layouts employed for one-mode networks, which are force-directed and stress majorization algorithms.

Formal concept analysis provides an algebraic framework for the study of affiliation networks as formal contexts of formal concepts with intents and extents representing the two domains. We have also looked at Galois derivations that connect formal concepts with the two network domains, and the Concept lattice of the context together with order filters and order ideals serve to make inferences about features of the affiliation network structure.

8.8   Learning Affiliation Networks by Doing

8.8.1   G20 Affiliation Network

```
# construct 'G20' data set with events and actors
R> G20 <- data.frame(
+    P5     = c(0,0,0,0,1,0,1,1,0,0,0,0,0,0,1,0,0,1,0),
+    G4     = c(0,0,1,0,0,1,0,0,0,1,0,1,0,0,0,0,0,0,0),
+    G7     = c(0,0,0,1,0,1,1,1,0,0,1,1,0,0,0,0,0,1,0),
+    BRICS  = c(0,0,1,0,1,0,0,0,0,1,0,0,0,0,1,0,0,0,1),
+    MITKA  = c(0,1,0,0,0,0,0,0,1,0,0,0,1,1,0,0,1,0,0),
+    DAC    = c(0,1,0,1,0,1,1,1,0,0,1,1,1,0,0,0,0,1,0),
+    OECD   = c(0,1,0,1,0,1,1,1,0,0,1,1,1,1,0,0,1,1,0),
+    Cwth   = c(0,1,0,1,0,0,0,1,0,1,0,0,0,0,0,0,0,0,1),
+    N11    = c(0,0,0,0,0,0,0,0,1,0,0,0,1,1,0,0,1,0,0) )

R> rownames(G20) <- c("ARG","AUS","BRA","CAN","CHN","DEU","FRA","GBR","IDN",
+     "IND","ITA","JPN","KOR","MEX","RUS","SAU","TUR","USA","ZAF")
```

```
# event clustering information
R> ec <- c(1, 1, 2, 0, 1, 2, 1, 1, 1)

# actor clustering (IMF economic classification of countries)
R> ac <- c(0, 1, 0, 1, 0, 1, 1, 1, 0, 0, 1, 1, 1, 0, 0, 0, 0, 1, 0)
R> ac <- replace(ac, ac == 0, "Emerging")
R> ac <- replace(ac, ac == 1, "Advanced")

 [1] "Emerging" "Advanced" "Emerging" "Advanced" "Emerging" "Advanced" "Advanced"
 [8] "Advanced" "Emerging" "Emerging" "Advanced" "Advanced" "Advanced" "Emerging"
[15] "Emerging" "Emerging" "Emerging" "Advanced" "Emerging"
```

8.8.2   Bipartite Graphs in $\mathscr{X}^B_{G20}$

```
# bipartite graph with a horizontal layout
R> bmgraph(G20, rot = 90, mirrorX = TRUE, mirrorY = TRUE, cex = 3)

# bipartite graph with a vertical layout with clustering information
R> bmgraph(G20, layout = "bipc", clu = list(ac, ec), cex = 4)
```

```
# four classes of actors as factor with explicit levels
R> acc <- factor(ac, levels = c("A-G7", "Advanced", "E-BRICS", "Emerging"))
R> acc[which(G20[,3] == 1)] <- "A-G7"
R> acc[which(G20[,4] == 1)] <- "E-BRICS"


 [1] Emerging Advanced E-BRICS  A-G7      E-BRICS A-G7     A-G7     A-G7     Emerging
[10] E-BRICS  A-G7     A-G7      Advanced Emerging E-BRICS Emerging Emerging A-G7
[19] E-BRICS
Levels: A-G7 Advanced E-BRICS Emerging
```

```
# scope binomial projection for edges and vertices
R> sce <- list(ecol = "#FF7F24", lwd = 2)
R> scv <- list(vcol = c("#BCEE68","#BDB76B","#66CDAA","#838B8B","#FF7F00"),
+      vcol0 = 8, cex = 4, pos = 0, fsize = 8, ffamily = "mono")
```

```
# plot 'G20' with a force-directed layout and vertex clustering information
R> bmgraph(G20, layout = "force", seed = 11, rot = 45, scope = c(scv,sce),
+      vclu = list(acc, rep(1,ncol(G20))))
```

8.8.3   Co-affiliation Network of G20 Network

```
# "multilevel" structure from actors common membership
R> G20cn <- mlvl(y = G20, type = "cn")
```

```
# redefine scopes for graph, edges, and vertices
R> scg <- list(directed = FALSE, valued = TRUE)
R> sce <- list(ecol = "orange", values = TRUE )
R> scv <- list(cex = 5, vcol = "white", vcol0 = "gray51", pos = 0, fsize = 10,
+      fsize2 = 11, fcol2 = "blue", ffamily = "mono", fstyle = "bold")
```

```
# plot co-affiliation graphs with concentric layout and dropping edge values
R> multigraph(G20cn, layout = "conc", nr = acc, scope = c(scg,scv,sce), drp = 0)
R> multigraph(G20cn, layout = "conc", nr = acc, scope = c(scg,scv,sce), drp = 1)
R> multigraph(G20cn, layout = "conc", nr = acc, scope = c(scg,scv,sce), drp = 2)
R> multigraph(G20cn, layout = "conc", nr = acc, scope = c(scg,scv,sce), drp = 3)
```

8.8.4   Positional System of $\mathscr{X}_{G20b}^{B}$ with Events Classes

```
# event "bridges" in 'G20'
R> acb <- factor(ac, levels = c("P5", "G4", "MITKA", "none"))
```

```
R> acb[which(G20[,1] == 1)] <- "P5"
R> acb[which(G20[,2] == 1)] <- "G4"
R> acb[which(G20[,5] == 1)] <- "MITKA"
R> acb[which(is.na(acb))] <- "none"
```

```
 [1] none  MITKA G4    none  P5    G4    P5    P5    MITKA G4    none  G4    MITKA
[14] MITKA P5    none  MITKA P5    none
Levels: P5 G4 MITKA none
```

```
# extract of the G20 bipartite network
R> G20b <- G20[which(acb!="none"), c(1:5)]
```

```
     P5 G4 G7 BRICS MITKA
AUS  0  0  0     0     1
BRA  0  1  0     1     0
CHN  1  0  0     1     0
DEU  0  1  1     0     0
FRA  1  0  1     0     0
GBR  1  0  1     0     0
IDN  0  0  0     0     1
IND  0  1  0     1     0
JPN  0  1  1     0     0
KOR  0  0  0     0     1
MEX  0  0  0     0     1
RUS  1  0  0     1     0
TUR  0  0  0     0     1
USA  1  0  1     0     0
```

```
# events clustering
R> clue <- c(3, 3, 1, 2, 3)
```

```
# reduction of 'G20b' with events clustering information (valued)
R> G20e <- reduc(G20b, clu = clue, col = TRUE, valued = TRUE,
+    lbs = c("G7", "BRICS", "P5,MITKA,G4"))
```

```
     G7 BRICS P5,MITKA,G4
AUS  0     0           1
BRA  0     1           1
CHN  0     1           1
DEU  1     0           1
FRA  1     0           1
GBR  1     0           1
IDN  0     0           1
IND  0     1           1
JPN  1     0           1
KOR  0     0           1
MEX  0     0           1
RUS  0     1           1
TUR  0     0           1
USA  1     0           1
```

8.8.5 Clustered Bipartite Graph and Binomial Projection of $\mathcal{X}_{G20b}^{B}$

```
# Figure 8.4: clustered bipartite graph

# actor clustering used for permutation
R> clua <- c(14, 3, 12, 6, 5, 4, 2, 8, 9, 10, 7, 11, 13, 1)

# plots
R> bmgraph(G20b, layout = "bipc", scope = c(scv, ecol = "gray50", mirrorX = TRUE,
+    cex = 6, fsize = 12, lwd = 3), clu = list(ac[which(acb!="none")],c(2,2,1,3,2)),
+    perm = list(clua,1:5) )

R> bmgraph(G20e, layout = "bipc", scope = c(scv, cex = 6, fsize = 12, mirrorX = TRUE,
+    lwd = 3, ecol = "gray50"), clu = list(ac[which(acb!="none")],c(1,3,2)),
+    perm = list(clua,1:3) )
```

```
# Figure 8.5: binomial projection of 'G20b'

# scope graph, edges, and vertices
R> scp <- list(rot = 45, cex = 3, fsize = 9, ecol = "#FF7F24",
+    vcol = c("#BCEE68","#BDB76B","#66CDAA","#838B8B","#FF7F00"))

# plot bipartite graph with force-directed layout
R> bmgraph(G20b, layout = "force", seed = 132, scope = scp,
+    vclu = list(acc[which(acb!="none")], rep(1,ncol(G20b))))
```

8.8.6 Formal Concept Analysis

```
# Galois derivations of 'G20' with full format
R> galois(G20)

$P5
[1] "CHN, FRA, GBR, RUS, USA"

$G4
[1] "BRA, DEU, IND, JPN"

$`DAC, G7, OECD`
[1] "CAN, DEU, FRA, GBR, ITA, JPN, USA"

$BRICS
[1] "BRA, CHN, IND, RUS, ZAF"
...
```

```
# Galois derivations with reduced format of 'G20' (left) and 'G20b' (right)
R> G20_gc <- galois(G20, labeling = "reduced")
R> G20b_gc <- galois(G20b, labeling = "reduced")


...                                          ...
$gc                                          $gc
$`reduc`                                     $`reduc`
$`reduc`$`P5`                                $`reduc`$`P5`
character(0)                                 character(0)

$`reduc`$G4                                  $`reduc`$G4
character(0)                                 character(0)

$`reduc`$G7                                  $`reduc`$G7
[1] "ITA"                                    character(0)

$`reduc`$BRICS                               $`reduc`$BRICS
character(0)                                 character(0)

$`reduc`$MITKA                               $`reduc`$MITKA
character(0)                                 [1] "AUS, IDN, KOR, MEX, TUR"
...                                          ...
```

```
# 'G20' partial order table in 8.2
R> partial.order(G20_gc, type = "galois", lbs = paste0("c", seq(1, 25)))

# record the partial order tables of Galois derivations in 'G20' and 'G20b'
R> G20_gc_P <- partial.order(G20_gc, type = "galois")
R> G20b_gc_P <- partial.order(G20b_gc, type = "galois")
```

8.8.6.1   Concept Lattices in Figure 8.6

```
# plot Concept lattice of 'G20_gc_P'
R> diagram(G20_gc_P)

# Concept lattice of Galois connections in 'G20b'
R> diagram(G20b_gc_P, lwd = 2)
```

8.8.7   Order Filters and Order Ideals

```
# principal order filters third concept PO GC 'G20' (left) and 'G20b' (right)

R> fltr(3, PO = G20_gc_P)                    R> fltr(3, PO = G20b_gc_P)


$`3`                                         $`3`
[1] "{G7} {ITA}"                             [1] "{G7} {}"

$`6`                                         $`11`
[1] "{DAC} {}"                               [1] "11"

$`7`
[1] "{OECD} {}"

$`25`
[1] "{} {ARG, SAU}"
```

```
# principal order ideals Concept lattice of 'G20' (left) and 'G20b' (right)
R> fltr(c("G7", "BRICS"), PO = G20_gc_P, ideal = TRUE)
R> fltr(c("G7", "BRICS"), PO = G20b_gc_P, ideal = TRUE)


$`3`                                         $`3`
[1] "{G7} {ITA}"                             [1] "{G7} {}"

$`4`                                         $`4`
[1] "{BRICS} {}"                             [1] "{BRICS} {}"

$`10`                                        $`6`
[1] "10"                                     [1] "6"

$`11`                                        $`7`
[1] "{} {FRA, USA}"                          [1] "{} {FRA, GBR, USA}"

$`12`                                        $`8`
[1] "{} {CHN, RUS}"                          [1] "{} {CHN, RUS}"

$`13`                                        $`9`
[1] "{} {GBR}"                               [1] "{} {DEU, JPN}"

$`14`                                        $`10`
[1] "{} {DEU, JPN}"                          [1] "{} {BRA, IND}"

$`15`
[1] "{} {BRA}"

$`16`
[1] "{} {IND}"

$`17`
[1] "{} {CAN}"

$`18`
[1] "{} {ZAF}"
```

9

Valued Networks

A significant type of social configuration is that of a **valued network**, which is a system of relations among entities where ties have different values or "weights" for the intensity of the relationship. Valued networks are particularly useful for analyzing systems of relations representing a directed *flow* of resources or information among a set of entities. A flow of resource are things like the amount of traffic in urban spaces or the volume of trade between nations, etc. and where traffic and trade correspond to a transportation and a financial network, respectively.

In this sense, there are different kinds of problems that can be addressed with valued networks, like the "maximum flow" and the "minimum cost flow" problems, which consider the edges capacities and sending costs respectively (cf. Ahuja et al., 1993, for a comprehensive analysis and algorithms to solve these problems). Although network flows are out of the scope of this book, it is worth mentioning that what is important to take into account in the analysis of any kind of valued network is the *quantity* in the flow relation. This is because if quantity in the flow relation is not retained for the study of valued networks, we are at risk of ending analyzing (only) dichotomous data rather than a weighted structure.

In this Chapter, we are going to concentrate here on the algebraic treatment of valued systems, and we will perform structural analyses of real-life valued networks that are also multiplex. First, we are going to look at valued paths in semigroup structures where the attached operation considers the weight of the ties. Then we consider two-mode networks that are valued. The analysis of valued two-mode networks are within the Formal Concept Analysis framework that was introduced in Chapter 8 Affiliation Networks, where we focused on the notion of conceptual scaling. At this end, we will see two reduction strategies for valued structures; first by applying an algebraic approach, called "Pathfinder network analysis" by their authors, for the analysis of weighted ties that are undirected. Then we apply the triangle inequality principle for the reduction of valued networks that are directed, which is especially useful for most valued networks.

Pathfinder and the triangle inequality are two approaches for the reduction of valued networks, which correspond to the combination of two operations among the valued ties. As a result, the algebraic structure of semiring once more serves as the setting for performing analysis of valued network data.

Algebraic Analysis of Social Networks: Models, Methods and Applications using R,
First Edition. J. A. R. Ostoic. Companion website: www.wiley.com/go/ostoic/algebraicanalysis.

9.1   Relational Structure of Valued Networks

As with simple networks, valued networks can have a multiplicity character as well. This means that it is possible to combine the different types of relations occurring in the valued system, almost in the same way as we did in Chapter 5 with the positional systems of multiplex networks $\mathscr{X}_A$ and $\mathscr{X}_F$. However, the composition of valued relations takes another operation than the relational composition of dichotomous data we have seen before with the construction of semigroup of relations.

As Pattison (1993) notices, value assignment to labeled valued paths in multiplex networks can be performed with at least two rules: the "ordinary" product of the links, and the max − min composition of relational ties. Hence, for multiplex networks that are binary, the *product* of two matrix relations is based on Boolean multiplication, whereas for multiplex networks that are valued another method to produce the composition of a pair of matrix relations is the max − min product.

With the ordinary composition the value on the path between i and j is based on the dot product

$$w_{i,j} = \sum_k w_x(i,k) \cdot w_y(k,j)$$

where w represents the path value made of relations x and y for nodes i, j, and $k \in \mathscr{G}^V$.

And with the max − min product the valued path is defined as

$$w_{i,j} = \max_k \{\min(w_x(i,k), w_y(k,j))\}$$

That is, it takes the minimum value of the sending and receiving scores with intermediary nodes, and then the maximum value of these outcomes.

Note that the order of semigroup structures from valued networks with the max − min product approach is very depending to the number of actors in the network and the "amount of tie values". Typically, valued networks with same size and order than dichotomous structures yield to much larger semigroups than with dichotomous networks, and this is because the max − min product is often more sensitive to the range of the weight scores. Therefore one must try to find an "appropriate" range and scale to produce a manageable relational structure, and this is by limiting the different score values of the network especially when such range is large. Making categories of the scores is clearly one strategy to follow.

9.1.1   Valued Paths in the G20 Trade Network

Now is time to produce the relational structure of a valued network that corresponds to the G20 Trade network. The data is found in Appendix C where valued matrices represent the amount of trade of two kinds of commodities—Fresh milk and honey—among the *Group of 20* network, which is the system of countries affiliated to the G20 supranational organization.

The affiliation part corresponds to the network $\mathscr{X}_{G20}^B$, and this structure has been the subject of an extensive analysis in Chapter 8 Affiliation Networks. The G20 Trade network, however, is a one-mode system made of these two commodities is which a valued configuration.

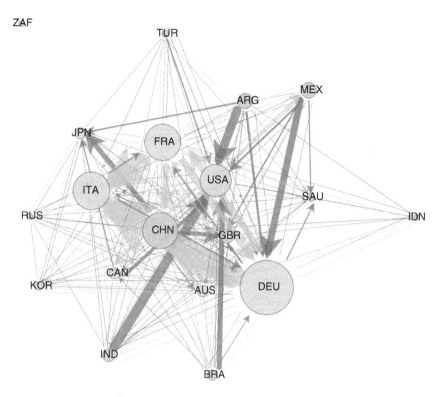

Figure 9.1 Trade network $\mathcal{X}_{G20}^V$ of milk & honey among the G20 countries. *Arc weights represent amount of trade between nations where edge colors deepskyblue and coral stand, respectively, for milk and honey. Vertex size reflects countries overall international trade (all in year 2017).*

This particular trade system can denoted as $\mathcal{X}_{G20mh}^V$ with subscripts '*m*' and '*h*' that stand for milk and honey, but for convenience we abbreviate such notation in the rest of the chapters just as $\mathcal{X}_{G20}^V$.

Figure 9.1 presents the valued multigraph of the Trade network of fresh milk & honey among the G20 countries in year 2017 or $\mathcal{X}_{G20}^V$. The plot is made with a force-directed layout, and again we notice that the network has 19 actor members, and this is because the European Union and other affiliated countries to the G20 group are disregarded in the analysis for practical reasons. There is just one component in this network and a single isolated actor, which is ZAF.

The nodes in the multigraph of Figure 9.1 are colored, and the advanced economies within the G20 organization are depicted with a «darkolivegreen2» color, whereas the emergent economies are depicted with «aquamarine3». Besides, the weights in the arcs reflect the amount of trade between two countries with a commodity where a «coral» color correspond to honey and a «deepskyblue» is to fresh milk.

At this point, it is important to mention that there can also be a multilevel perspective in the analysis with a combination of one- and two-mode systems into a single structure as we will see in Chapter 10 Multilevel Networks. For now, however, we are going to focus just on the multiplicity and in the value of the different ties in $\mathcal{X}^V_{G20}$, and this in order to illustrate the construction of a relational structure of this valued network.

For example, the "smallest" multilevel structure in the Trade network $\mathcal{X}^V_{G20}$ is a positional system with two kinds of relations among three classes of actors plus their affiliation ties. Table 10.1 in Chapter 10 provides the tie values in the multilevel structure of this Trade network that is valued system, and for the construction of its relational structure we concentrate just on the two types of commodities, which are fresh milk M and honey H. This means that the network configuration is not only valued but it is also a multiplex structure. Since this configuration represents a positional system, we denote this valued network as $\mathcal{S}^V_{G20}$.

Figure 9.2 provides different aspects of the structure of this network positional system, which is denoted by $\mathcal{S}^V_{G20}$ (rather than $\mathcal{S}^V_{G20mh}$), and we can see, for example, that the network configuration is a triad made of two bundle classes and self-relations that commonly appear in reduced structures such as this one. The *bundle census* that enumerates the different bundle classes occurring in $\mathcal{S}^V_{G20}$ is given at the top left of the picture, and here the heads of the arcs

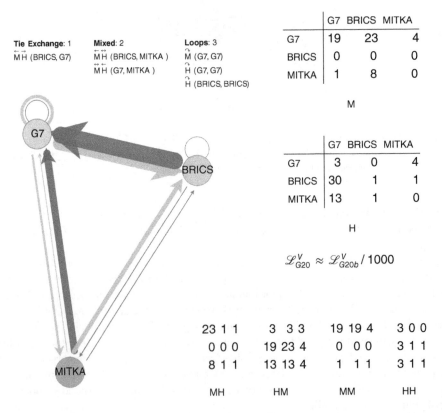

Figure 9.2   Positional system of the valued Trade network $\mathcal{S}^V_{G20}$ of fresh milk M and honey H commodities between classes of G20 countries. *Left top and bottom: Bundle patterns and valued multigraph where colors are as in Figure 9.1. Right top and bottom: Generator relations and compounds of generators with max-min composition.*

on the top of each type of relation point the direction of the tie where mixed bundles have a two-headed arc.

Below these bundle classes is depicted the valued multigraph of this particular network, and the layout in this case corresponds to a stress majorization scheme (Gansner et al., 2005). In this graph, the thickness of the loop edges has a different scale than the thickness of the directed edges, and this is just for practical reasons.

The adjacency matrices at the top right of the picture record the values of the two kinds of role relations, and it is a rounded version of the thousandth values of the role relations in the positional system given in Chapter 10. This transformation allows having a smaller number of values than the original structure, and for example, relation M counts with 5 different values where 1 and 4 occur in H as well.

In Table 10.1, M has more values and none of them coincides with the values in H. In other words, the two generators in $\mathscr{S}^V_{G20}$ are the flow values of primitives in $\mathscr{S}^V_{G20bmh}$ or rather $\mathscr{S}^V_{G20b}$, which is a smaller version of $\mathscr{S}^V_{G20}$ with only bridge organizations, divided by 1000 and rounded as well. One consequence of this transformation is that network structure becomes simpler, and we can expect a simpler role structure as well. This is since there are "less different values" to take into account in the computation of the compounds. As with $\mathscr{X}^V_{G20}$, we simplify the notation of the positional system of bridge organizations and related structures as $\mathscr{S}^V_{G20b}$.

Figure 9.2 at the bottom right provides also the valued adjacency matrices with all valued paths of length two that are product of the primitives with a max − min procedure. Next with the semigroup construction of valued path relations are the details of this procedure in the computing of the compounds including where these value scores came from. For this particular network, there are 4 compounds with length 2 in the relational structure, and 13 compounds with larger width where the longest string has five elements.

9.1.2 Constructing Valued Paths

As with multiplex networks made of dichotomous data, both the relational and role structures of a given valued network can be made in terms of a semigroup object. As said before, for valued systems the endowed operation to the semigroup structure is based on the max − min product composition of the values representing the tie strengths in valued networks rather than the Boolean multiplication of dichotomous data. For instance, the composite role relation MH among the G7 countries product of the max − min procedure is computed from the values of M this position "sends" to the rest of the actors in the system, and the values of H this actor "receives" from the rest of the positions in the network.

The composition of string relations through the max − min procedure for the valued Trade network $\mathscr{S}^V_{G20}$ in Figure 9.2 involves the two generators M and H. In a generic way, the max − min product for these two relations for actors i, j and $k \in \mathscr{X}^V_{G20}$ are

$$\mathsf{M} \circ \mathsf{H} = \max_k \{\min(w_\mathsf{M}(i, k),\ w_\mathsf{H}(k, j))\} = \mathsf{MH}$$

where w denotes the "weight" entry in the valued matrix of the generator in $\mathscr{S}^V_{G20}$. The same rules apply for other compound valued relations in this positional system or in larger configurations such as $\mathscr{X}^V_{G20}$.

For instance, the max − min product for the first listed position in $\mathscr{S}_{G20}^{V}$ is:

$$
\begin{matrix}
19 & 23 & 4 \\
3 & 30 & 13
\end{matrix}
\qquad
\begin{matrix}
3 & 23 & 4
\end{matrix}
\qquad\qquad
23
$$

And for compound HM, the role relations among the G7 countries is

$$
\begin{matrix}
3 & 0 & 4 \\
19 & 0 & 1
\end{matrix}
\qquad
\begin{matrix}
3 & 0 & 1
\end{matrix}
\qquad\qquad
3
$$

Next, we apply the same rules to the rest of the cells in the adjacency matrices for these compounds; that is, find the minimum values in row and column vectors of the corresponding actors in the adjacency matrices for the involved relations first and then pick up the maximum value of such outcomes. Certainly, in a case where the outgoing tie of a given actor is zero, such as the case of BRICS for MH in $\mathscr{S}_{G20}^{V}$, then the min vector belonging to this actor is empty, as is the max product.

Longer compounds are created by the right multiplication of composite relations and generators with the max − min product composition. As a result, for the valued path MHM between MITKA and BRICS, for instance, the following values are part of the computation (cf. Figure 9.2):

$$
\begin{matrix}
8 & 1 & 1 \\
23 & 0 & 8
\end{matrix}
\qquad
\begin{matrix}
8 & 0 & 1
\end{matrix}
\qquad\qquad
8
$$

And the rest of the weights in the valued matrix for this compound with $k = 3$ are

$$
\begin{matrix}
19 & 23 & 4 \\
0 & 0 & 0 \\
8 & 8 & 4
\end{matrix}
$$
$$
\text{MHM}
$$

In a similar way, the compounds combined with the primitive ties generate larger compound relations, which in turn are operands for further right multiplication with max − min composition until the algebraic structure is closed. Such algebraic system is a complex structure represented by the semigroup of weighted relations, which serves to represent the relational structure of valued networks.

9.1.3 Semigroup and Equations of Valued Relations

As with multiplex networks having dichotomous relations, the set of unique generators and compound valued relations make the algebraic structure of semigroup. However, in the case of multiplex networks that are valued, the associated operator to the semigroup is the max-min product of valued string relations rather than Boolean multiplication. The Axiom of Quality allows having a closed algebraic system as well, where some of the representative valued strings (if not all) are part of the set of equations.

Figure 9.3 serves to illustrate with different Cayley graphs the semigroup construction process of the valued positional system $\mathscr{S}^V_{G20}$ for this particular network. As we can see in the picture, in the case of the positional system of the valued Trade network $\mathscr{S}^V_{G20}$ the longest representative compound is at $k = 5$ when the semigroup structure gets closed. This means that longer compounds product of max-min composition are equated to one of these representative strings, and the order of the semigroup remains at 19 unique representative strings.

There is a set of equations in Figure 9.3 as well, and we can see that the two primitive relations generate a pair compounds each first. This means that there is a total of four compound relations with length 2, and the matrices of these compound relations were already given in Figure 9.2.

The multiplication table below represents all compounds of length 2 in $\mathscr{S}^V_{G20}$ where the composition of two relations in terms of the max − min product is:

○	M	H
M	MM	MH
H	HM	HH

9.1.4   First Letter Law in Semigroup Structure

Recall from Chapter 5 Role Structure in Multiplex Networks that if the first constituent letter of the compounds in the algebraic structure coincide, then the associated pattern corresponds to the First Letter law. In the case of the semigroup structure for $\mathscr{S}^V_{G20}$, we can evidence just by looking at the Cayley graphs that the First Letter law applies, and this is case. Besides, as longer compounds are created afterwards from these string relations through right multiplication of the max − min operation, we notice in Figure 9.3 that each of the two primitive relations in this valued system generates compounds belonging to its own component.

The fact that each primitive relation has its own component suggests that these structures constitute at least two "sub-semigroups" of the algebraic system; one for M and one for H. Moreover, each component of generators for the semigroup structure has their components as well. For instance, commodity honey H differentiates the idempotent H = HH from the compounds where M is present as well. In the case of commodity milk, there are four sub-components in the semigroup systems where both H and M are interweaved.

The following multiplication tables are extracts of the semigroup structure for the two commodities with one- and two-path compounds, and in both cases are the longest strings involved in the max-min composition:

○	M	MM	MH	...
M	MM	MM	MMH	...
MM	MM	MM	MMH	...
MH	MHM	MHMM	MH	...
...	...	...	...	...

○	H	HM	HH	...
H	HH	HHM	HH	...
HM	HMH	HM	HMHH	...
HH	HH	HHM	HH	...
...	...	...	...	...

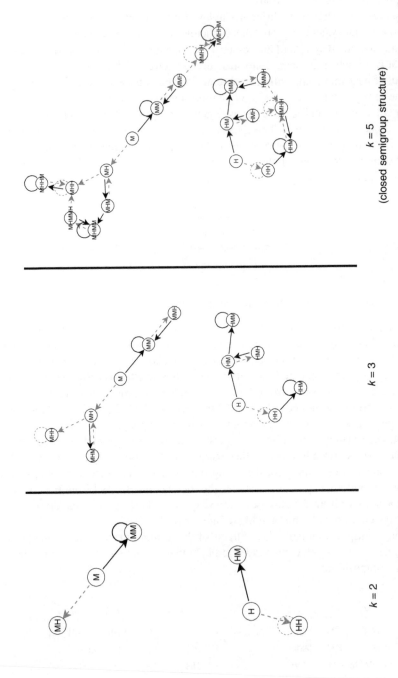

Figure 9.3 Cayley graphs of the valued positional system $\mathscr{S}_{G20}^{V}$ with increasing path length. *Solid and dashed arcs are M and H, respectively. Equations of compounds with length 2 are:* MM = MMM = MMHM = MMHMM, MH = MHMH, HM = HMHM, HH = HHH.

Once more, we evidence the First Letter law pattern in these configurations that is going to last in the rest of the semigroup structure. This First Letter table is therefore associated with interlock between the two commodities or flow relations of this valued network.

As a result, the compound relations having milk or honey as the starting relation operate like the respective commodity in the valued path. This is because with the First Letter law the composition of any two elements always results in an element that is in the same class as the first element of the composition (Wasserman and Faust, 1994, p. 447).

9.2   Many-valued Contexts

As with one-mode structures, affiliation networks can be valued as well, and the analysis of this kind of data requires its own techniques. Although we are able to plot valued two-mode network data with the binomial approach as weighted multigraphs, it would be nice to take a further step from the visualization and perform an algebraic analysis of affiliation networks that are also valued.

The Formal Concept Analysis framework from Chapter 8 Affiliation Networks constitutes an algebraic approach to two-mode data, and in this case, the attributes of the different contexts represent either the presence or the absence of a relationship of the attribute with the object where values are either 1 or 0. Attributes, however, can have different values other than one and zero in order to reflect the intensity in the relationship or the weights of the different attributes. If there are different weights in the attribute information, the context is considered as "valued" or "many-valued" rather than a "one-valued" context as with dichotomous data.

A *many-valued context* $\mathbb{K}^V$ (Wille, 1982; Ganter and Wille, 1996) is defined as

$$\mathbb{K}^V = (G, \ M, \ W, \ I).$$

where M stands for the many-valued attributes in $\mathbb{K}^V$, W represents the "weights" or the attribute values, and I the incidence relation. If W has k elements, then the context is said to be an k-valued context.

In many-valued contexts, the incidence relation is between the three sets in $\mathbb{K}^V$; i.e. $I \subseteq G \times M \times W$, and it can represented as $m(g) = w$, which reads as "the attribute m has a value w for the object g", for $m \in M$, $g \in G$, and $w \in W$. Typically the possible values of w include 0 as well, and this can be expressed as $m(g) = \varnothing$, meaning that with this notation m acts as a partial map from g in w.

Many-valued contexts are useful settings to represent contexts where the objects have different properties that are important to take into account in the analysis. Such properties usually correspond to categorical data with values that are ordinal like "grades", or nominal values such as "color", or among others, the dichotomic scale of "yes"/"no". In the rest of this section, we are going to look at scaling techniques to analyze a real-world many-valued context in algebraic terms with the different steps in the analytical process.

9.2.1   Conceptual Scaling

The concept system of a many-valued context depends on different *scales* depending on the nature of the data in the formal context. For instance, the different forms for partition of networks refers to a nominal scale, whereas a particular ordinal scale serves to represent structures of hierarchy and rank orders. We have seen hierarchy and rank orders before both with one-valued contexts and with partially ordered semigroups, while the partition of objects into extents relates to a nominal scale within the formal context analysis approach.

In this sense, the notion of *conceptual scaling* (Ganter and Wille, 1996) refers to a context where each attribute of a many-valued context is interpreted in the process of scaling, and this methodological procedure yields to a *derived context with respect to* the scaling chosen for the establishment of dichotomous data. Hence, the conceptual scaling notion constitutes a method of interpretation where both the scale and the transformation of the values of a many-valued context are part of the analytical process that is a matter of interpretation and not mathematically compelling.

For instance, the values *high, low, very-high,* and *very-low* can be represented as two well-known scales represented by the following tables:

	very-high	high	low	very-low
very-high	1	0	0	0
high	0	1	0	0
low	0	0	1	0
very-low	0	0	0	1

	$\leq$very-high	$\leq$high	$\geq$low	$\geq$very-low
very-high	1	1	0	0
high	0	1	0	0
low	0	0	1	0
very-low	0	0	1	1

Nominal $\mathbb{N}_n = \langle n, n, = \rangle$ Inter-ordinal $\mathbb{I}_n = \langle n,n, \leq | n,n, \geq \rangle$

Hence, in the *nominal* scale the identity matrix implies that each attribute value correspond only to the value itself, and its basic meaning is the partition of the set. On the other hand, with the one-dimensional *inter-ordinal* scaling the superlative "very" to the value implies in this case the value as well, and it represents a linear "betweenness" relation among the elements in the set. In fact the inter-ordinal scale $\mathbb{I}_n$ is the union of two *ordinal* scales where $\mathbb{O}_n = \langle n, n, \leq \rangle$ represents two rank orders dual to each other of a chain of extents. Ganter and Wille (1996, p. 57) designate a formal context where the set of values is ordered in a natural way as *ordinal context* for each attribute. Besides, they define other elementary scales of ordinal type that serve to represent different types of attribute set within the Formal concept framework.

9.2.2   Conceptual Scaling of $\mathscr{X}_{G20}^{B}$

An example of a conceptual scaling of a many valued two mode structure is found in Table 9.1, which provides a four-valued context of the affiliation network $\mathscr{X}_{G20}^{B}$ with economic and socio-demographic indicators. The ordinal values of these indicators correspond to the raw data given in Appendix C, where TRADE, nominal gross domestic product (NOM._GDP), and gross domestic product per capita (GDP_PC) are the three economic indicators. On the other hand, human development index (HDI), POPULATION, and AREA correspond to the three socio-demographic indicators of the countries in the G20 network.

In order to study a given many-valued context in terms of the formal concept analysis framework, it is required to perform a *transformation* of the many-values of the context into dichotomous data made by 1's and 0's. This yields to a one-valued context where each attribute value has its own column. As a result, derived contexts increase in size comparing with their respective many-valued contexts, and this constitutes a backward for the analysis. However, a derived context allows performing Galois derivations between extents and intents, which in turn allow constructing the partially ordered structure of these connections where the interpretation of the data typically takes place in terms of inclusion.

In the case of the many-valued context of Table 9.1, for example, the transformation of the "highs" and the "lows" correspond respectively to scores above and below the arithmetic mean of the economic and socio-demographic indicator for the G20 countries in this network. Certainly, it is possible to use other measures of central tendency such as the median rather than the mean for establishing the ordinal values. In such case, the low, middle, and upper quartiles or another measure of relative standing can serve to define the attribute ordinal values.

Hence, the scale chosen in Table 9.1 for the many-valued context $\mathscr{X}_{G20}^V$ is inter-ordinal, and the ordinal values are based on the arithmetic mean of the two kinds of indicators for the G20 countries. The derived context becomes a one-valued context with a transformation process where the category *high* represent the scores higher or equal to the mean in the respective indicator, and *low* are the lower scores to the indicator mean. On the other hand, category values *very-high* and *very-low* correspond to the scores higher and lower to the arithmetic means of the

Table 9.1 A four-valued context of $\mathscr{X}_{G20}^B$ for economic and socio-demographic indicators and its derived context with respect to an inter-ordinal scaling. *The establishment of the ordinal scale values is based on the arithmetic mean of the scores given in Appendix C.*

	Economic			Socio-demographic		
	TRADE	NOM._GDP	GDP_PC	HDI	POPULATION	AREA
ARG	very-low	very-low	low	low	very-low	low
AUS	very-low	very-low	very-high	very-high	very-low	high
BRA	very-low	low	low	low	low	high
CAN	low	low	very-high	very-high	very-low	high
CHN	very-high	very-high	very-low	very-low	very-high	high
DEU	high	high	very-high	very-high	low	very-low
FRA	low	low	high	high	very-low	very-low
GBR	low	low	very-high	very-high	very-low	very-low
IDN	very-low	very-low	very-low	very-low	high	low
IND	low	low	very-low	very-low	very-high	low
ITA	low	low	high	high	very-low	very-low
JPN	high	high	high	high	low	very-low
KOR	low	very-low	high	high	very-low	very-low
MEX	low	very-low	low	low	low	low
RUS	low	very-low	low	low	low	very-high
SAU	very-low	very-low	low	high	very-low	low
TUR	very-low	very-low	very-low	low	very-low	very-low
USA	very-high	very-high	very-high	very-high	high	high
ZAF	very-low	very-low	very-low	very-low	very-low	very-low

Table 9.1 (contd.).

	T.vh	T.h	T.l	T.vl	N.vh	N.h	N.l	N.vl	G.vh	G.h	G.l	G.vl	H.vh	H.h	H.l	H.vl	P.vh	P.h	P.l	P.vl	A.vh	A.h	A.l	A.vl
ARG	0	0	1	1	0	0	1	1	0	0	1	0	0	0	1	0	0	0	1	1	0	0	1	0
AUS	0	0	1	1	0	0	1	1	0	1	0	0	1	1	0	0	0	0	1	1	0	1	0	0
BRA	0	0	1	1	0	0	1	0	0	0	1	0	0	0	1	0	0	0	1	0	1	0	0	0
CAN	0	0	1	0	0	0	1	0	0	1	0	0	1	1	0	0	0	0	1	1	1	0	0	0
CHN	1	1	0	0	1	0	0	0	0	1	0	1	0	0	1	0	1	1	0	0	1	0	0	0
DEU	0	1	0	0	0	1	0	0	0	1	0	0	0	1	0	0	0	0	1	1	0	0	1	1
FRA	0	0	1	0	0	0	1	0	0	1	0	0	0	1	0	0	0	0	1	1	0	0	1	1
GBR	0	0	1	0	0	0	1	0	0	1	0	0	0	1	0	0	0	0	1	1	0	0	1	1
IDN	0	0	1	0	0	0	1	1	0	0	1	1	0	0	1	0	0	1	0	0	0	1	0	0
IND	0	0	1	0	0	0	1	0	0	0	1	1	0	0	1	1	1	1	0	0	0	1	0	1
ITA	0	0	1	0	0	0	1	0	0	1	0	0	0	1	0	0	0	0	1	1	0	0	1	1
JPN	0	1	0	0	0	1	0	0	0	1	0	0	0	1	0	0	0	0	1	0	0	0	1	1
KOR	0	0	1	0	0	0	1	0	0	1	0	0	0	1	0	0	0	0	1	1	0	0	1	0
MEX	0	0	1	0	0	0	1	1	0	1	1	0	0	0	1	0	0	0	1	0	0	1	0	0
RUS	0	0	1	0	0	0	1	1	0	1	0	0	0	0	1	0	0	0	1	0	1	0	0	0
SAU	0	0	1	1	0	0	1	0	0	1	0	0	0	1	0	0	0	0	1	1	0	0	1	0
TUR	0	0	1	1	0	0	1	1	0	1	0	0	0	1	0	0	0	0	1	0	0	1	0	1
USA	1	1	0	0	1	1	0	0	0	1	0	0	0	1	0	0	0	1	0	0	1	0	0	0
ZAF	0	0	1	1	0	0	0	1	0	0	1	1	0	0	0	1	0	0	1	0	0	1	0	1

Attribute ordinal values in the derived context very-high=vh, high=h, low=l, very-low=vl are preceded by their respective indicators. Economic: T=overall Trade, N=Nom. GDP (nominal gross domestic product), G=GDP_PC (gross domestic product per capita). Socio-demographic: H=HDI (human development index), P=Population, A=Area.

scores in the two categories defined previously. In this way, each scale value for every attribute type has now its own column in the formal context with the record of the ties with the respective objects, and where *very-high* and *very-low* imply *high* and *low*, respectively.

9.2.3 Concept Lattices of Many-valued Contexts

The conceptual scaling constitutes a methodological procedure to reduce the many-valued concepts into the basic type through a scale transformation. In the case of the valued context $\mathscr{X}_{G20}^{V}$, each attribute yields four categories with an inter-ordinal scale. This means that after the transformation within the conceptual scaling procedure, the derived context with respect to the scale ends up having 24 attributes types where half of them correspond to economic indicators and the other half are of a socio-demographic character. Because of the high amount of attributes in the derived context, the distinction between the two categories of indicators allows a better interpretation of the many-valued formal context corresponding to this network.

In this sense, Figure 9.4 provides two Concept lattices of the many-valued context $\mathscr{X}_{G20}^{V}$ with an inter-ordinal scale: one for economic indicators and the other for socio-demographic indicators of the G20 countries. The lattice with economic indicators to the left of the picture has 26 concepts, and the lattice of the G20 countries with socio-demographic indicators to the right has 44 concepts. Since there are fewer elements in the Concept lattice with economic indicators than the diagram with socio-demographic ones, it appears that there is a larger correlation among countries and the economic indicators than the socio-demographic ones.

One way to analyze the partial order structure of the concepts having the two kinds of indicators is by considering order filters and order ideals as with the partial order structures of dichotomous formal contexts in Chapter 8. However, for many-valued contexts with an inter-ordinal scale like the four-valued context $\mathscr{X}_{G20}^{V}$, we already know that the *highs* and *lows* of the different indicators are included in their respective "*verys*".

Recall that the inclusion among the attributes in formal contexts goes in opposite direction to the inclusion among the objects, as well as the interpretation of these. Hence, the most common ordinal scale values are close to the maximal element of the Concept lattice while countries having fewer values are closer to the minimal than those countries with more ordinal scale values. In the case of economic indicators in $\mathscr{X}_{G20}^{V}$ the Concept lattice to the left of Figure 9.4 already shows that the greatest lower bounds of the maximal element correspond to highs and lows of the three indicators. For the three socio-economic indicators, the highs and lows are also the greatest lower bounds of the maximal element of the Concept lattice, and in all cases, these cover the very-highs and the very-lows value attributes.

9.2.3.1 Order Filters, Order Ideals

Because the scale is inter-ordinal, by looking at the order filters or principal order filters of the extents do not bring further insights to the analysis in this case. On the other hand, order ideals and principal order ideals will produce a set of concepts where the "membership" of the object to those categories of the attributes values applies. Hence, countries with a very high or very low value of economic indicators such as trade, nominal gross domestic product, and gross domestic product per capita are also in the highs and lows of these indicators.

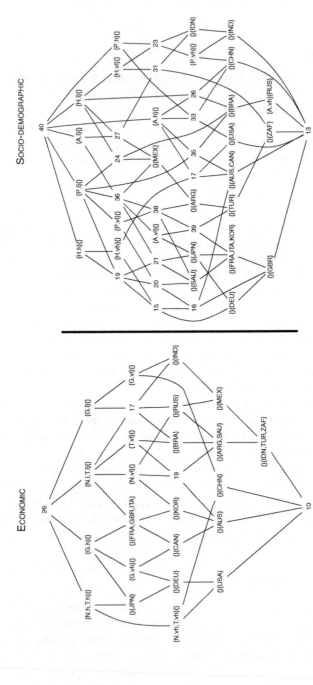

Figure 9.4 Concept lattices representing a four-valued context of the Group of Twenty network $\mathscr{R}_{G20}^V$ with an inter-ordinal scale for economic and socio-demographic indicators. *Attribute ordinal values of indicators are given in Table 9.1.*

Just for illustration purposes, below are the principal order ideals of the two extreme values in the inter-ordinal scale corresponding to the overall TRADE in $\mathscr{X}_{G20}^V$. These formal concepts comprise the *very-low* and the *very-high* level of commerce among the G20 countries.

$$\mathscr{X}_{G20}^V \ T.vl \qquad\qquad\qquad \mathscr{X}_{G20}^V \ T.vh$$

$$
\begin{array}{ll}
\mathbb{C}_4 : \{T.vl\} & \{\varnothing\} \\
\mathbb{C}_{10} : \{\varnothing\} & \{\varnothing\} \\
\mathbb{C}_{19} : \{\varnothing\} & \{\varnothing\} \\
\mathbb{C}_{20} : \{\varnothing\} & \{AUS\} \\
\mathbb{C}_{21} : \{\varnothing\} & \{BRA\} \\
\mathbb{C}_{22} : \{\varnothing\} & \{IDN,TUR,ZAF\} \\
\mathbb{C}_{27} : \{\varnothing\} & \{ARG, SAU\}
\end{array}
\qquad
\begin{array}{ll}
\mathbb{C}_1 : \{N.vh, T.vh\} & \{\varnothing\} \\
\mathbb{C}_{10} : \{\varnothing\} & \{\varnothing\} \\
\mathbb{C}_{11} : \{\varnothing\} & \{USA\} \\
\mathbb{C}_{12} : \{\varnothing\} & \{CHN\}
\end{array}
$$

Certainly, since the type of data is ordinal, such information could have been deduced from Table 9.1 without the need of the Concept lattice or the partial order structure. However, order filters and order ideals in the Concept lattice allows having combinations of scale values in the many-valued formal context, and this is very difficult to deduce from tables that represent a given formal context.

9.3   Pathfinder Network Analysis

A special type of semiring structure that is useful for valued network reduction is found in "Pathfinder network analysis" (Schvaneveldt et al., 1989). Pathfinder analysis is based on symmetric adjacency matrices representing valued networks where the values reflect the "proximity" between pairs of network members. Then there is a series of transformations on the network data that produces a distance structure of the system that is obtained by removing unimportant ties.

The computation of a Pathfinder network analysis is based on the Pathfinder algorithm that takes the network data where the adjacency matrix of this data is regarded as a *proximity matrix*. The criteria for the computation depend on two parameters, q and r, which take different values according to the desired level of reduction of the Pathfinder network.

The r parameter in the Pathfinder algorithm is a distance function influencing the weights to give to the different links until just the largest connection determines weight of a path. The links are called "components" in this context, and in psychological terms, this distance function means that perceived difference between entities is determined by the dissimilarity of the most dissimilar relations connecting the entities.

On the other hand, the q parameter places an upper limit on the number of ties in paths used to establish the minimum distance between actors in the proximity matrix. (Schvaneveldt et al., 1989, pp. 257–8). In this sense, the r and q parameters are generalizations of the definition of network distance, and the Pathfinder algorithm serves to obtain the *salient structure*, which is the "skeleton" or "backbone" of the valued network by producing a dissimilarity measure from the proximity data. This means that the resultant structure, which is called a *Pathfinder network*, becomes "r-metric" and "q-triangular" or more precisely "q-angular".

The Pathfinder network equals the proximity matrix when q equals one, and the minimum number of ties in the valued network is obtained with Pathfinder analysis when both parameters have their maximum values. That is, when $q = n - 1$, with n being the number of elements in the proximity matrix, and $r = \infty$.

In some cases, the salient structure with Pathfinder Network Analysis results in the *minimal value spanning tree* where the sum of tie weights is minimal over the set of all possible trees in the system.

9.3.1   Pathfinder Semiring

Batagelj et al., (2014, p. 97) assert that the two parameters in the Pathfinder algorithm, r and q, correspond to a semiring structure where q represents the length of all walks computed over the semiring in a dissimilarity matrix $\mathbf{W}^{(q)}$, and r is a distance measure, the Minkowski distance.

The *Pathfinder semiring* is formally defined for $r \geqslant 1$ as

$$\langle\ \mathbb{R}_0^+,\ \bigoplus,\ \boxed{r},\ \infty,\ 0\ \rangle$$

where the first operator is the min product

$$a \bigoplus b = \min(a, b)$$

and the second one corresponds to the *Minkowski operation*

$$a\ \boxed{r}\ b = \sqrt[r]{a^r + b^r}.$$

The two special elements in the semiring structure, ∞ and 0, are particular cases where the *r-metric* only applies in the transformation of the proximity data when $r < \infty$.

Increasing the value of r implies that the larger components in the network receive greater weight in establishing the paths of the salient structure after applying the Pathfinder algorithm.

For instance, if $r = 1$, then $a\ \boxed{r}\ b = a + b$, which means that all ties have equal weight in establishing the value of a path. As r increases then the ties with greater value receive greater weight until r reaches infinity, which means that $a\ \boxed{r}\ b = \max(a, b)$; that is, only the largest component matters.

The familiar Euclidean distance, which is the straight-line distance between two points in Euclidean space, occurs when $r = 2$, which means that

$$a\ \boxed{r}\ b = \sqrt{a^2 + b^2}.$$

Thanks to the absorption property with $a^\star = 0$, the Pathfinder semiring becomes a closed structure, which is expressed as

$$a \bigoplus (a\ \boxed{r}\ b) = a \bigoplus (b\ \boxed{r}\ a) = a.$$

9.3.2 Pathfinder Algorithm

To understand how the Pathfinder semiring algorithm works, let us assume a symmetric valued network $\mathscr{X}^V$ with the proximity matrix $\mathbf{W}$ between n actors in the network. In order to transform the proximity matrix by removing unimportant ties, the q parameter establishes the "number of iterations" minus 1 applied to $\mathbf{W}$, and the r parameter defines the "distance" between the network members. This means that we need to clone the proximity matrix of the valued network $\mathbf{W}$ into two matrices $\mathbf{Q}$ and $\mathbf{D}$ for the two parameters.

Initially, both $\mathbf{D}$ and $\mathbf{Q}$ equal the original proximity matrix, and the first iteration round corresponds to 2-paths. Then both $\mathbf{Q}$ and $\mathbf{D}$ are transformed through a number of iterations from 2 to the value of parameter q according to the entry values of $\mathbf{W}$ and r.

As a result, first the values in $\mathbf{Q}$ are updated either by max product or by the Minkowski distance $\boxed{r}$ depending on whether $r = \infty$ or $r \neq \infty$. The max product applies to each entry of the sending row a and the receiving column b in $\mathbf{W}$, and the r-metric is computed on these vectors as well. Then we apply the $\oplus$ operation to the outcome of the previous transformation and choose the minimum value.

After such computations, the entries in $\mathbf{D}$ are updated by values in $\mathbf{Q}$ when corresponding weights in $\mathbf{Q}$ are lower than $\mathbf{D}$, and this is made through $q - 1$ iterations. Afterwards, there is another round $1, \ldots, n$ where the entries in $\mathbf{Q}$ are updated by values from the proximity matrix $\mathbf{W}$ when corresponding weights in $\mathbf{D}$ are lower than the original proximity matrix. Otherwise the corresponding entry becomes ∞, and the link is removed from the matrix $\mathbf{Q}$, which represents the salient structure of the valued network as a proximity matrix.

The pseudo-code given below illustrates the computation process of the Pathfinder algorithm. The procedure is based on a proximity matrix and the two parameters q and r, which produces the dissimilarity matrix $\mathbf{W}^{(q)}$ as $\mathbf{Q}$ that results being an updated version of $\mathbf{W}$.

```
 1: procedure PFNET(W, q, r)                    ▷ proximity matrix and two parameters
 2:     Q ← W
 3:     D ← W
 4:     for k ← 2, q do
 5:         for i ← 1, n do
 6:             for j ← 1, n do
 7:                 if r = ∞ then
 8:                     Q[i,j] ← min(max(W[i, ], W[,j]))
 9:                 else
10:                     Q[i,j] ← min(W[i, ]ʳ + W[,j]ʳ)^(1/r)
11:                 if D[i,j] > Q[i,j] then
12:                     D[i,j] ← Q[i,j]
13:     for i ← 1, n do
14:         for j ← 1, n do
15:             if D[i,j] < W[i,j] then
16:                 Q[i,j] ← ∞                                     ▷ entry is removed
17:     return Q                                               ▷ salient structure of W
```

9.4   Pathfinder Semiring to Co-affiliation Network in $\mathscr{X}^B_{G20}$

This section is to illustrate the salient structure of a valued network through the rules of the pathfinder semiring. In this case, we are going to take the affiliation network of the group of 20 nations from Chapter 8 Affiliation Networks and that is represented by $\mathscr{X}^B_{G20}$. More concretely, the two networks correspond to the actors and then the events co-affiliations. Hence, co-occurrences are regarded as proximity data, and therefore they represent undirected relations among the entities in both configurations.

For instance, Figure 9.5 above provides valued graphs representing the co-affiliation configuration for the advanced economies in $\mathscr{X}^B_{G20}$, and below are the co-occurrences for the events as well. We can see in the first configuration for example that every actor is connected to every other actor, which means that such type of structure corresponds to a *clique* made of

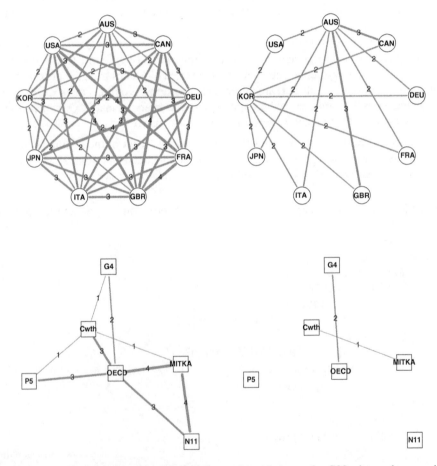

Figure 9.5   Co-affiliation networks and Pathfinder semiring skeletons for G20 advanced economies and bridge organizations. *Values represent the number of common membership and affiliates among actors and events, respectively. Top-left and right: Undirected network with actor co-affiliation as proximity data, and Pathfinder semiring with $q = 2$ and $r = \infty$. Bottom-left and right: Same as above for events.*

valued ties. On the other hand, the edge density for the network of common affiliate members is lower.

Since ties are proximity information, the graphs are undirected and with valued edges where the thickness reflects either the number of common membership among the actors in the first case or the amount of "common affiliate members" of the bridge organizations in $\mathscr{X}_{G20}^B$ for the second configurations. Besides, there are different values in the edges of Figure 9.5 with the exact number of co-occurrences between connected network members in both cases. This proximity data is given as the usual adjacency matrix format in Appendix C, and these quantities constitute the basis to perform the analysis with the Pathfinder semiring arrangement.

Each of the two networks in Figure 9.5 has its corresponding Pathfinder semiring skeleton to the right of the picture. In both cases, the parameters are $q = 2$ and $r = \infty$, which means that there is still room for reduction in the edge density by increasing the value of the q parameter in the Pathfinder semiring. For example, if the value of q equals to 3 or more up to $(n - 1)$, which is the maximum possible value, then the two edges with the largest value in the latter graph will be removed. However, the reduction occurs as long the value of r is larger than one; otherwise the original structure remains unchanged in this particular network.

Nevertheless, we evidence that the edge density in both co-affiliation networks are reduced dramatically where almost all ties equal or higher than 3 are removed from the graph. In the case of the co-affiliation network for events, the Pathfinder semiring implies that for (closed) 2-chains only the relations between $\{OECD, G4\}$ and $\{Cwth, MITKA\}$ prevail in the salient structure, and this information is valuable for a substantial interpretation of the system.

It is important to mention that the Pathfinder semiring approach does not necessarily reduce the edge density of all valued networks. If, for example, we consider in the Pathfinder analysis also the G20 countries with emerging economies (cf. Table 8.1 in Chapter 8 Affiliation Networks), then all edges will be removed no matter what are the parameter values of q and r.

Pathfinder network analysis was originally designed for symmetric proximity matrices corresponding to undirected networks. In the case of directed valued network structures, then the principle of triangle basic inequality applies as a counterpart to the Pathfinder network analysis.

9.5   Triangle Inequality

Pathfinder semirings are useful in producing the salient structure of valued networks made of undirected relations. However, when the valued network is directed as typically occurs with systems of flow relations, the salient structure is based on the notion of "triangle basic inequality of side lengths" or just triangle inequality. This form for reduction for directed valued structures is rather based on the strict Pathfinder algorithm as with undirected networks.

Triangle inequality is one of the three axioms of a metric space in a Euclidean plane, say $\mathbb{R}^2$, which is a function d defined on pairs of points (x, y) from a set X; i.e.,

$$d(x, y) = \sqrt{(y_1 - x_1) + (y_2 - x_2)}$$

that denotes the distance between x and y.

The *triangle inequality* axiom says that "the sum of the lengths of any two sides must be greater than or equal to the length of the third side." This statement is expressed in formal terms as

$$d(x, y) + d(y, z) \geqslant d(x, z) \quad \text{for } x, y, \text{ and } z \in X \qquad \text{where } d \in \mathbb{R}^2$$

For the sake of completeness, the other two axioms of the metric space are given:

1. "the distance between two points is always positive except when the two points are the same" and
2. "the distance from point x to point y is the same as the distance from y to x"

Formally, these two statements are expressed as

$$d(x, y) = 0 \quad \text{iff } x = y$$

and

$$d(x, y) = d(y, x).$$

The principle of triangle inequality for valued networks means that given a valued tie there is a path with two ties connecting its endpoints with lower values sum, then the tie is removed (Batagelj et al., 2014). In this way, the concept of triangle inequality resembles the Pathfinder analysis of the co-membership network from the previous section, and this is since it produces the salient structure of the valued network as well. In this case, however, the triangle inequality notion is based on the "dissimilarity" among the actors in the network rather than proximity data, as with the Pathfinder analysis of undirected systems.

9.5.1   Application of Triangle Inequality to a Valued Configuration

Now it is time to apply the principles of triangle inequality to a valued structure, and we take as example the triadic configuration in Figure 9.2, which corresponds to the positional system of the valued Trade network of milk and honey commodities between classes of the G20 countries network $\mathscr{X}_{G20}^V$. Since the triangle inequality axiom do not take into account loops, we ignore self-relationships when computing triangle inequality to valued networks. This is even though the semirings objects with paths and semipaths for signed networks from Chapter 7 consider self-relations in the computations of structural balance. Certainly, the application of triangle inequality by considering loops would be the next research task.

Nonetheless, we start the analysis with the amount of milk trade between the three classes of countries in the mentioned year (the "blue" arcs in the graph of Figure 9.2). From the information given in the adjacency matrix, we can represent the valued ties with weights above the arrow-ties as follows:

$$G7 \xrightarrow{4} MITKA, \quad MITKA \xrightarrow{8} BRICS, \quad G7 \xrightarrow{23} BRICS$$

In this case, the application of triangle inequality means that given the sum of the weights in ties from G7 to MITKA and from MITKA to BRICS is lower than the weight of the tie from G7 to BRICS, then this latter tie is removed from the salient structure of the network. Since this is true for this triadic configuration, the tie between G7 and BRICS does not form part of the network backbone since the "triangle inequality principle" accounts for the dissimilarity or distance among the actors reflected in tie weights.

For the trade of honey in this valued Trade network there are two weighted triads where we also apply the triangle inequality principle:

$$\text{BRICS} \xrightarrow{1} \text{MITKA}, \; \text{MITKA} \xrightarrow{13} \text{G7}, \; \text{BRICS} \xrightarrow{30} \text{G7}$$
$$\text{MITKA} \xrightarrow{1} \text{BRICS}, \; \text{BRICS} \xrightarrow{30} \text{G7}, \; \text{MITKA} \xrightarrow{13} \text{G7}$$

As we can see that as with the previous case with, the trade of milk, the tie with more weight, which is the trade of honey from BRICS to G7, is removed from the salient structure.

In a similar way, the same principle applies to other triads belonging to larger configurations to obtain a reduced structure of the entire network. It is important to mention, however, that the computation is performed on all triads simultaneously, which means that often the 2-paths can belong to different triads where the tie may remain for one triad and removed for the other configuration

9.5.2   Triangle Inequality in Multiplex Networks

For valued networks that are also multiplex, the principle of triangle inequality is applied for each type of relation separately. This is even though it is possible to collapse the different levels in the relationship to produce the skeleton of the valued network with triangle inequality as a monoplex structure. However, in the case of the valued configuration of $\mathscr{X}_{G20}^{V}$, it is clear from the bundle census and the multigraph in Figure 9.2 that by collapsing the tie levels we end up with a triadic configuration having three reciprocated relations among its members. These reciprocated bonds then have a combination of the two kinds of ties, and the valued ties in this case are these six relations:

$$\text{G7} \xrightarrow{23} \text{BRICS}, \quad \text{BRICS} \xrightarrow{1} \text{MITKA}, \; \text{MITKA} \xrightarrow{14} \text{G7}, \; \text{BRICS} \xrightarrow{30} \text{G7}, \; \text{G7} \xrightarrow{4} \text{MITKA}$$
$$\text{MITKA} \xrightarrow{9} \text{BRICS}$$

Hence, the tie value from MITKA to G7 is the sum of the tie values in the two commodities $1 + 13$, and from MITKA to BRICS is $8 + 1$, etc.

The salient structure of this monoplex configuration then removes the reciprocal bond between BRICS and G7 as with the previous case, but now in both directions. In this case, however, the trade flow between these two political blocks is much higher than with the MITKA countries and this is by counting either multiplex or monoplex structures.

For multiplex networks, the application of the triangle inequality principle without losing the multiplicity in the relationships constitutes a challenge, and this is not even considering valued loops that characterize positional systems, as mentioned before. This is because by collapsing the different types of tie in one there will be an ambiguity in the direction of the ties whenever the multiplex bundle is not tie entrainment. Certainly, since the multiplex network in this case is valued as well, the weight of the particular ties can define the direction of the bond or bundle relation.

For instance, in the case of the positional system $\mathscr{S}_{G20}^{V}$ (cf. Figure 9.2), the triangle with collapsed ties becomes

$$\text{BRICS} \xrightarrow{30+23} \text{G7}, \; \text{MITKA} \xrightarrow{13+1+4} \text{G7}, \; \text{MITKA} \xrightarrow{1+1+8} \text{BRICS}$$

However, since the two bundle classes of tie exchange and mixed that are occurring in this positional system have a mutual character, the directionality of the ties are lost. This means that the principle of triangle inequality does not strictly apply here, but instead the Pathfinder algorithm. In such case, the bond to be removed will be from BRICS to G7 with the default parameters $q = 2$ and $r = \infty$.

9.6 Trade Network $\mathscr{X}_{G20}^{V}$ with Triangle Inequality

Now we are going to apply a triangle inequality principle to the valued directed structure of $\mathscr{X}_{G20}^{V}$, which represents the entire Trade network of milk and honey among the G20 countries in year 2017. The triadic configuration represented as a valued multigraph in Figure 9.1 is the positional system of this network, which means that we scale up the application of the triangle inequality principle.

Tables C.1 and C.1 in Appendix C provide the valued matrices representing the amount of trade of these two commodities. The matrices in each commodity as given in Appendix C provide score values fro the whole trade flow in USD $\times 10000$. On the other hand, the arrays in Table C.2 drop small values from the matrices, those that are less than 250 in the trade commerce for the respective commodity.

The triangle inequality norm is going to be applied on these two second matrices, which are specified at the end of Appendix C for both types of valued relation. The reason for such transformation of the data is that triangle inequality is very sensitive to small weights, which will prevail over the rest of the ties even if these are just above their values. This means that the outcome for this particular data resembles a configuration by dropping values greater than a relative small cutoff value, and this is something we want to avoid in the illustration.

Our purpose is thus to make a distinction in the salient structure of a valued network between merely dropping values and the reduction through the triangle inequality approach, which has a relational character. However, when producing the salient structure of the valued network by applying the triangle inequality principle, we bear in mind that an alternative method to obtain the skeleton is to consider the multiplex valued network as a monoplex configuration.

Figure 9.6 provides two salient structures of the Trade network of milk and honey among the G20 countries in year 2017 or $\mathscr{X}_{G20}^{V}$. First, there is the skeleton of the network by dropping values less or equal to 25000 USD to the left of the picture. Then to the right is the salient structure of this configuration in terms of the triangle inequality principle, and this is since this trade system is made with directed and valued relations.

Because $\mathscr{X}_{G20}^{V}$ has two types of relations and hence it is a multiplex network, there is actually a pair of skeletons in the valued multigraph representing the salient structure of the network, one for *each* type of relation or commodity of the G20 Trade network ξ_{G20}^{V}.

There are a number of relations in $\mathscr{X}_{G20}^{V}$ that have been removed from the valued network by means of triangle inequality. Appendix C provides the trade flow data of the two commodities as matrices where the removed ties are in cells with a colored background. Such ties are removed from the Trade network because the sum of the trade values of any two relations in a triad of $\mathscr{X}_{G20}^{V}$ is greater than or equal to the trade value of the remaining relation of the triangle as the triangle inequality principle states.

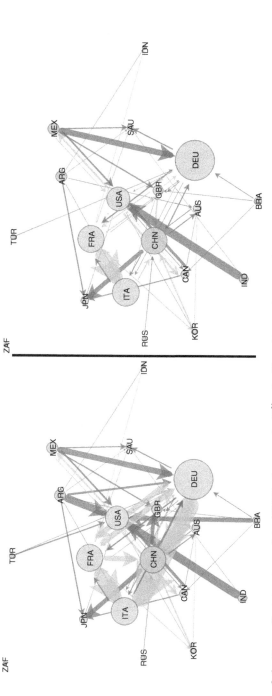

Figure 9.6 Two salient structures of the G20 Trade network $\mathscr{R}_{G20}^{-V}$ of milk and honey in year 2017. *Left: Skeleton by dropping values ≈ 25000 USD or less. Right: Salient structure of the skeleton with triangle inequality applied. Arcs and vertices as in Figure 9.1.*

We use the notion of triangle inequality as a method for producing the salient structure of the multiplex configuration, and this is despite we can reduce dramatically the network density by dropping large score values. We also recall that for networks the triangle inequality takes each type of relation separately, avoiding the ambiguity of monoplex structures, but at the price of losing the multiplicity in the ties in the process.

9.7   Valued Networks: Summary

Valued networks have ties that carry a value or a weight to denote the intensity in the relationship between the network actors, and the quantity of the flow relation is an important piece of information for the study of these type of networks. It is important to retain the tie values; otherwise, we are at risk of ending analyzing dichotomous data rather than a weighted structure.

The concatenation of relations in valued networks produced valued paths, which are part of the semigroup structure of multiplex valued network. The attached operation in this case was the max − min operation rather than the Boolean as with multiplex dichotomous networks. We constructed the relational structure of the Trade network of the G20 countries, which is valued, and where the semigroup structure followed the First Letter law principle.

It was also possible to study two-mode structures that are valued with the Formal concept analysis framework where many-valued contexts and conceptual scaling allow the integration of weights in the analysis of cross-domain relations. The affiliation network comprised the same actors as the previous example and we applied an inter-ordinal scale of economic and socio-demographic indicators of the G20 countries.

Finally, we have seen different forms for reduction of valued structures through Pathfinder network analysis in the case of undirected valued networks, and the application of the triangle inequality principle for directed valued networks where we were able to produce the salient structure of the Trade network among the G20 countries. In the case of Pathfinder semiring, we produced skeletons of co-affiliation networks both from the perspective of the actors and from the perspective of the events.

9.8   Learning Valued Networks by Doing

9.8.1   *Valued Network*

```
# load Trade network positional system data from public repository

R> load(url("www.wiley.com/go/ostoic/algebraicanalysis"))

# 'vnet' is a valued network from positional system 𝒮ᵛ_{G20b}
# (drop thousand rounded and actor labels renamed)
R> vnet <- round(SG20bmh/1000)
R> dimnames(vnet)[[1]] <- dimnames(vnet)[[2]] <- c("G7","BRICS","MITKA")


, , M

      G7 BRICS MITKA
G7    19    23     4
BRICS  0     0     0
MITKA  1     8     0


, , H

      G7 BRICS MITKA
G7     3     0     0
BRICS 30     1     1
MITKA 13     1     0
```

```
# bundle classes with loops
R> summaryBundles(bundles(vnet, loops = TRUE))

                              Bundles
Txch        <-{M} ->{H}  (BRICS, G7)
Mixd1 <-{M} <->{H}  (BRICS, MITKA)
Mixd2       <->{M} <-{H}  (G7, MITKA)
Loop1             o{M}  (G7, G7)
Loop2             o{H}  (G7, G7)
Loop3       o{H}  (BRICS, BRICS)
```

```
# define different types of scopes
# graph
R> scg <- list(valued = TRUE, loops = TRUE, rot = 90, mirrorX = TRUE)
# vertices and text
R> scv <- list(cex = 9, vcol = c("#BCEE68","#66CDAA","#A0A17B"), vcol0 = "#C0C0C0",
+    pos = 0, fsize = 14)
# edges
R> sce <- list(ecol = c("#00BFFF","#CD5B45"), bwd = .5)
```

```
# plot valued network with stress majorization layout
R> multigraph(vnet, layout = "stress", seed = 4, scope = c(scg, scv, sce))
```

9.8.2   Semigroup of Valued Network with max-min Product

We continue the analysis with the valued structure in vnet.

```
# obtain the semigroup of 'vnet'
R> Svnet <- semigroup(vnet, valued = TRUE)

...
$st
 [1]  "M"    "H"    "MM"   "MH"   "HM"   "HH"   "MMH"  "MHM"  "MHH"  "HMM"
[11]  "HMH"  "HHM"  "MMHH" "MHMM" "MHHM" "HMMH" "HMHH" "MMHHM" "MHMMH"

$S
      1  2  3  4  5  6  7  8  9 10 11 12 13 14 15 16 17 18 19
 1    3  4  3  7  8  9  7  3 13 14  4 15 13  3 18 19  9 18  7
 2    5  6 10 11 12  6 16  5 17 12 17 12 17 10 12 17 17 12 16
 3    3  7  3  7  3 13  7  3 13  3  7 18 13  3 18  7 13 18  7
 4    8  9 14  4 15  9 19  8  9 15  9 15  9 14 15  9  9 15 19
 5   10 11 10 16  5 17 16 10 17 10 11 12 17 10 12 16 17 12 16
 6   12  6 12 17 12  6 17 12 17 12 17 12 17 12 12 17 17 12 17
 7    3 13  3  7 18 13  7  3 13 18 13 18 13  3 18 13 13 18  7
 8   14  4 14 19  8  9 19 14  9 14  4 15  9 14 15 19  9 15 19
 9   15  9 15  9 15  9  9 15  9 15  9 15  9 15 15  9  9 15  9
10   10 16 10 16 10 17 16 10 17 10 16 12 17 10 12 16 17 12 16
11    5 17 10 11 12 17 16  5 17 12 17 12 17 10 12 17 17 12 16
12   12 17 12 17 12 17 17 12 17 12 17 12 17 12 12 17 17 12 17
13   18 13 18 13 18 13 13 18 13 18 13 18 13 18 18 13 13 18 13
14   14 19 14 19 14  9 19 14  9 14 19 15  9 14 15 19  9 15 19
15   15  9 15  9 15  9  9 15  9 15  9 15  9 15 15  9  9 15  9
16   10 17 10 16 12 17 16 10 17 12 17 12 17 10 12 17 17 12 16
17   12 17 12 17 12 17 17 12 17 12 17 12 17 12 12 17 17 12 17
18   18 13 18 13 18 13 13 18 13 18 13 18 13 18 18 13 13 18 13
19   14  9 14 19 15  9 19 14  9 15  9 15  9 14 15  9  9 15 19

attr(,"class")
[1] "Semigroup" "numerical" "valued"
```

```
# Cayley graph of 'Svnet' with customized labels and layout
R> ccgraph(Svnet, cex = 4, seed = 123, bwd = .5, rot = 45, pos = 0, col = 8, lwd = 2,
+    lbs = Svnet$st)
```

9.8.2.1   Equations with max-min Product

```
# strings equations with valued option until longest compound in semigroup
R> strings(vnet, valued = TRUE, equat = TRUE)$equat


$equat
$equat$`MM`
[1] "MMM"    "MMHM"   "MMHMM"

$equat$MH
[1] "MHMH"

$equat$HM
[1] "HMHM"

$equat$HH
[1] "HHH"

$equat$MMH
[1] "MMMH"

$equat$MHH
[1] "MHMHH"

$equat$HMM
[1] "HMHMM"

$equat$HHM
[1] "HHHM"   "HHHM"   "HMHHM"

$equat$MMHH
[1] "MMMHH"

$equat$HMHH
[1] "HHMH"   "HMMHH" "HHMMH" "HHMHH"


attr(,"class")
[1] "Strings" "valued"
```

9.8.3   Many-valued Contexts

9.8.3.1   Conceptual Scaling of G20 Affiliation Network

```
# load object 'G20mv' from a public repository

R> load(url("www.wiley.com/go/ostoic/algebraicanalysis"))

# Table 9.1 with economic and socio-demographic indicators
R> G20mv
```

```
      T   N   G   H   P   A
ARG  vl  vl  l   l   vl  l
AUS  vl  vl  vh  vh  vl  h
BRA  vl  l   l   l   l   h
CAN  l   l   vh  vh  vl  h
CHN  vh  vh  vl  vl  vh  h
DEU  h   h   vh  vh  l   vl
FRA  l   l   h   h   vl  vl
GBR  l   l   h   vh  vl  vl
IDN  vl  vl  vl  vl  h   l
IND  l   l   vl  vl  vh  l
ITA  l   l   h   h   vl  vl
JPN  h   h   h   h   l   vl
KOR  l   vl  h   h   vl  vl
MEX  l   vl  vl  l   l   l
RUS  l   vl  l   l   l   vh
SAU  vl  vl  l   h   vl  l
TUR  vl  vl  vl  l   vl  vl
USA  vh  vh  vh  vh  h   h
ZAF  vl  vl  vl  vl  vl  vl
```

```
# define inter-ordinal scale with four "levels"
R> scl <- data.frame(c(1, 0,0,0), c(1,1,0,0), c(0,0,1,1), c(0,0,0,1))
R> colnames(scl) <- rownames(scl) <- c("vh", "h", "l", "vl")
```

```
     vh  h  l  vl
vh   1   1  0  0
h    0   1  0  0
l    0   0  1  0
vl   0   0  1  1
```

```
# apply conceptual scaling to 'G20mv' with 'scl' and customized separator
R> G20mvsc <- cscl(G20mv, scl, sep = ".")
```

	T.vh	T.h	T.l	T.vl	N.vh	N.h	N.l	N.vl	G.vh	G.h	G.l	G.vl	H.vh	H.h	H.l	H.vl	P.vh	P.h	P.l	P.vl	A.vh	A.h	A.l	A.vl
ARG	0	0	1	1	0	0	1	1	0	0	1	0	0	0	1	0	0	0	1	1	0	0	1	0
AUS	0	0	1	1	0	0	1	1	1	1	0	0	1	1	0	0	0	0	1	1	0	1	0	0
BRA	0	0	1	1	0	0	1	0	0	1	0	0	0	1	0	0	0	0	1	0	0	1	0	0
CAN	0	0	1	0	0	0	1	0	1	1	0	0	1	1	0	0	0	0	1	1	0	1	0	0
CHN	1	1	0	0	1	1	0	0	0	0	1	1	0	0	1	1	1	1	0	0	0	1	0	0
DEU	0	1	0	0	0	1	0	0	1	1	0	0	1	1	0	0	0	0	1	0	0	0	1	1
FRA	0	0	1	0	0	0	1	0	0	1	0	0	0	1	0	0	0	0	1	1	0	0	1	1
GBR	0	0	1	0	0	0	1	0	0	1	0	0	1	1	0	0	0	0	1	1	0	0	1	1
IDN	0	0	1	1	0	0	1	1	0	0	1	1	0	0	1	1	1	0	0	0	0	0	1	0
IND	0	0	1	0	0	0	1	0	0	0	1	1	0	0	1	1	1	1	0	0	0	0	1	0
ITA	0	0	1	0	0	0	1	0	0	1	0	0	0	1	0	0	0	0	1	1	0	0	1	1
JPN	0	1	0	0	0	1	0	0	0	1	0	0	0	1	0	0	0	0	1	0	0	0	1	1
KOR	0	0	1	0	0	0	1	1	0	1	0	0	0	1	0	0	0	0	1	1	0	0	1	1
MEX	0	0	1	0	0	0	1	1	0	0	1	1	0	0	1	0	0	0	1	0	1	1	0	0
RUS	0	0	1	0	0	0	1	1	0	0	1	0	0	0	1	0	0	0	1	1	0	0	1	0
SAU	0	0	1	1	0	0	1	1	0	0	1	0	0	1	0	0	0	0	1	1	0	0	1	0
TUR	0	0	1	1	0	0	1	1	0	0	1	1	0	0	1	0	0	0	1	1	0	0	1	1
USA	1	1	0	0	1	1	0	0	1	1	0	0	1	1	0	0	0	1	0	0	0	1	0	0
ZAF	0	0	1	1	0	0	1	1	0	0	1	1	0	0	1	1	0	0	1	1	0	0	1	1

Each indicator has four columns now, one for each scale level.

9.8.3.2   Galois Connections with Conceptual Scaling

```
# conceptual scaling of economic indicators and customized separator
R> G20mv.ec <- galois(G20mvsc[, 1:12], labeling = "reduced", sep = ",")

# conceptual scaling for socio-demographic indicators
R> G20mv.sd <- galois(G20mvsc[, 13:24], labeling = "reduced", sep = ",")
```

9.8.3.3   Concept Lattices in Figure 9.4

```
# economic indicators
R> diagram(partial.order(G20mv.ec, type = "galois"), lwd = 2, ecol = "#333333")

# socio-demographic indicators
R> diagram(partial.order(G20mv.sd, type = "galois"), lwd = 2, ecol = "#333333")
```

9.8.4   Pathfinder Semiring

G20 countries affiliation network (extract, data from Chapter 8)

```
R> G20

    P5 G4 G7 BRICS MITKA DAC OECD Cwth N11
ARG  0  0  0     0     0   0    0    0   0
AUS  0  0  0     0     1   1    1    1   0
BRA  0  1  0     1     0   0    0    0   0
CAN  0  0  1     0     0   1    1    1   0
CHN  1  0  0     1     0   0    0    0   0
DEU  0  1  1     0     0   1    1    0   0
...
```

```
# co-membership relations in 'G20' as a "multilevel structure" with mlvl()
R> G20c <- mlvl(y = G20, type = "cn")

# look at the "structure" in object 'G20c'
R> str(G20c)

List of 3
 $ mlnet: num [1:19, 1:19] 0 0 0 0 0 0 0 0 0 0 ...
 ..- attr(*, "dimnames")=List of 2
 .. ..$ : chr [1:19] "ARG" "AUS" "BRA" "CAN" ...
 .. ..$ : chr [1:19] "ARG" "AUS" "BRA" "CAN" ...
 $ lbs  :List of 2
 ..$ dm : chr [1:19] "ARG" "AUS" "BRA" "CAN" ...
 ..$ cdm: chr [1:9] "BRICS" "Cwth" "DAC" "G4" ...
 $ modes: chr "1M"
 - attr(*, "class")= chr [1:2] "Multilevel" "cn"
```

```
# from Chapter8 we get actors clustering with explicit levels in factors
R> acc

 [1] Emerging Advanced E-BRICS  A-G7     E-BRICS  A-G7      A-G7      A-G7
 [9] Emerging E-BRICS  A-G7      A-G7     Advanced Emerging E-BRICS Emerging
[17] Emerging A-G7      E-BRICS
Levels: A-G7 Advanced E-BRICS Emerging
```

```
# select "Advanced" and "A-G7" countries from multilevel structure in 'G20c'
R> G20c2 <- rel.sys(G20c$mlnet, sel = which(acc %in% c("Advanced", "A-G7")),
+     type = "toarray")

    AUS CAN DEU FRA GBR ITA JPN KOR USA
AUS   0   3   2   2   3   2   2   3   2
CAN   3   0   3   3   4   3   3   2   3
DEU   2   3   0   3   3   3   4   2   3
FRA   2   3   3   0   4   3   3   2   4
GBR   3   4   3   4   0   3   3   2   4
ITA   2   3   3   3   3   0   3   2   3
JPN   2   3   4   3   3   3   0   2   3
KOR   3   2   2   2   2   2   2   0   2
USA   2   3   3   4   4   3   3   2   0
```

```
## redefine scopes
# graph
R> scg <- list(directed = FALSE, valued = TRUE)
# vertices and text
R> scv <- list(cex = 4, pos = 0, vcol0 = 1, vcol = 0, fsize2 = 10, fcol2 = 1)
# edges
R> sce <- list(ecol = "orange", values = TRUE)
```

```
# Fig. 9.5 co-affiliation network for G20 advanced economies.
R> multigraph(G20c2, scope = c(scg, scv, sce))
```

```
# check if 'G20c2' is symmetric data
R> isSymmetric(G20c2)

[1] TRUE
```

```
# plot pathfinder skeleton for co-affiliation network with q = 2 and r = ∞
R> multigraph(pfvn(G20c2, q = 2), scope = c(scg, scv, sce))
```

9.8.5   Triangle Inequality

```
# load Trade network data from a public repository

R> load(url("www.wiley.com/go/ostoic/algebraicanalysis"))

# the "structure" in 'G20mh'
R> str(G20mh)

 num [1:19, 1:19, 1:2] 0 0 0 0 0 246 801 0 0 0 ...
 - attr(*, "dimnames")=List of 3
 ..$ : chr [1:19] "ARG" "AUS" "BRA" "CAN" ...
 ..$ : chr [1:19] "ARG" "AUS" "BRA" "CAN" ...
 ..$ : chr [1:2] "M" "H"
```

```
# drop 10 thousand from Trade network 'G20mh' and then round it
R> g20mh <- round(G20mh/10000)

# drop values less or equal to 250
R> g20mh250 <- replace(g20mh, g20mh<= 250, 0)
```

```
# make the G20 Trade network of milk & honey as a monoplex valued structure
R> g20mhmono <- mnplx(g20mh250, dichot = FALSE)

# this is the G20 overall trade
R> G20ot <- rowSums(g20mhmono) + colSums(g20mhmono)
```

```
## redefine scopes
# graph
R> scg <- list(valued = TRUE)
# vertices
R> scv <- list(cex = G20ot, vcol = c("#BCEE68","#BCEE68","#66CDAA","#66CDAA"))
# edges
R> sce <- list(ecol = c("#00BFFF","#CD5B45"), bwd = .5, alpha = c(.9,.7))
# text
R> sct <- list(ffamily = "mono", fsize = 9, fstyle = "bold", fcol = "black", pos = 0)
```

```
# plot Fig.9.1: 'G20mh' with clustering information
R> multigraph(G20mh, layout = "force", seed = 0, scope = c(scg,scv,sce,sct),
+     clu = acc)
```

```
# plotting 'g20mh250' with clustering information and coordinates from 'G20mh'

# get a set of coordinates for 'G20mh' with a force-directed layout
R> cG20 <- frcd(G20mh, seed = 0, maxiter = 100)

# plot Fig.9.6a: skeleton by dropping values
R> multigraph(g20mh250, coord = cG20, scope = c(scg, scv, sce, sct), clu = acc)
```

```
# check if 'g20mh250' is symmetric data
R> isSymmetric(g20mh250)

[1] FALSE
```

```
# plot Fig.9.6b: skeleton with triangle inequality
R> multigraph(pfvn(g20mh250), coord = cG20, scope = c(scg,scv,sct,sce), clu = acc)
```

10

Multilevel Networks

10.1   Structural Analysis of Multilevel Systems

Multilevel networks are still today in the cutting edge of research in social network analysis, and the level of complexity these structures carry on exceeds the other kinds of networks we have seen so far. These—besides simple networks—are multiplex, signed, two-mode, and valued structures. For the G20 countries network (cf. Chapter 8 Affiliation Networks), nations are embedded within supranational organizations, and such organizations can even be nested within larger organizations such as the United Nations in the case of a system like this made of countries.

According to Brass and Borgatti (2019, pp. 187–188), a *cross-level research* will seek for appropriate ways to measure constructs at different "levels" of analysis established in the end by the social entities defined in the research design. Hence, a social entity can be an individual node, a dyad, a triad, a group of four or more nodes, and even networks, which means that there is an overlapping of entities.

In the typology of ties of Brass and Borgatti (2019, cf. Fig. 8.2), the Trade network $\mathscr{X}_{G20}^V$ is made of flows of merchandises, which in this case is an "event" type of tie. On the other hand, co-membership among nations in $\mathscr{X}_{G20}^B$ corresponds to a "state" type of tie, which is based on similarities among the actors. For our purpose, however, the individual nations and the organizations that for the G20 countries network are defined as actors and events in the two-mode network representation, make the multilevel structure for the analysis.

Since the configuration of the G20 countries' network is relatively large and complex for a multilevel representation, we illustrate the components in a multilevel system with a smaller configuration. For the introduction of multilevel structures, we start with the visual representation of a multilevel network data made of two kinds of social actors that are four clients and three attorneys with different types of relations between and within them.

The network made of *Clients and attorneys* illustrated as "simultaneous analysis of graphs and bipartite graphs" by Wasserman and Iacobucci (1991) is an early example of a multilevel structure, and Figure 10.1 depicts this network with different graph format options–described afterward–and the adjacency matrices that include cross-domain relations.

Algebraic Analysis of Social Networks: Models, Methods and Applications using R,
First Edition. J. A. R. Ostoic. Companion website: www.wiley.com/go/ostoic/algebraicanalysis.
© 2021 John Wiley & Sons Ltd. Published 2021 by John Wiley & Sons Ltd.

What makes the Clients and attorneys network a *multilevel* structure—and not "merely" a multiplex network—is first that there is more than one domain in the network structure, and second that the relational content across and within domains differs. Hence, the presence of a co-domain in the system is primarily what characterizes a multilevel network structure, but a second condition, however, is that one or both domains have within relations among its members. This means that a multilevel network "combines" the one- and the two-mode network structures into a single configuration.

In this particular network, the clients and the attorneys are two classes of individual human actors that connect by different types of relations within and across them. The two classes of actors then constitute different domain structures or *levels* in the network configuration even if none of them is an "event".

We could say that in the case of the G20 countries affiliation network $\mathscr{X}_{G20}^{B}$, for example, there are two classes of actors as well. These are the advanced and emerging economies according to the IMF economic classification, and one could specify from this feature the trade network $\mathscr{X}_{G20}^{V}$ of milk and honey among the G20 countries as a multilevel network structure with two classes of network members as well. It is because the ties occurring both between and within classes of actors in $\mathscr{X}_{G20}^{V}$ have the same relational content that this configuration has a single domain only, which means that the trade network is just a multiplex structure with attribute-based categories of actors.

When there is in the system a bundle class pattern with a multiplexity character such as tie entrainment, tie exchange, or one of the varieties of mixed bundles (cf. Chapter 3 Multiplex Network Configurations), then the multilevel structure constitutes a multiplex network as well. Both the Clients and attorneys network and the multilevel structure of the Group of 20 are multiplex systems, and this is because there are different types of ties in the cross-level relationship in the multilevel network and one of the levels or domains in the other system has a multiplicity of relations. Moreover, one can even imagine social systems having more than two domains, and where there are ties across and within some or all domains, but the study of multilevel networks with more than two domains is out of the scope here.

For a recent collection of articles about multilevel structures within social network analysis, refer to Lazega and Snijders (2016), although the works there are almost entirely from statistical points of view. We try in this chapter to perform algebraic analyses of multilevel network structures, and like for many types of analyses of complex configurations, the visualization plays a key role. Therefore we continue with the visual representation of multilevel structures.

10.2 Visual Representation of Clients and Attorneys Multilevel Network

Multilevel structures are complex in nature, and the ability to visualize them as graphs constitutes an important step in the analysis. This is because visualization not only gives us a flavor of the data, but also because it allows us to represent the information from the different domains in the multilevel network "at once" and hence we can infer structural phenomena in the social system.

For instance, if we take a look at the Clients and attorneys network from Figure 10.1, we can see that there is a domain with elements depicted as circles, and a co-domain with elements given

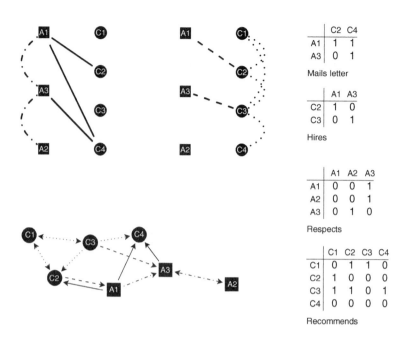

Figure 10.1 Three multilevel structures of the Clients and attorneys network data. *Left and middle above: Bipartite graphs with relations across and within domains. Left below: Directed multigraph of all relations with a force-directed layout of the binomial projection where "solid", "dashed", "dotted", and "dashed-dotted" arcs in the graphs correspond to Mails letter, Hires, Recommends, and Respects relations. Right: Adjacency matrices for relations where* A1-3 *and* C1-4 *stand for attorneys and clients, respectively.*

as squares. There are three different visual representations of the Clients and attorneys network multilevel structure in this case where the two cross-domain relations are "Mails a letter" and "Hires" from attorneys and clients. There is also in Figure 10.1 a pair of relation types within domains in the picture that are "Respects" among attorneys and "Makes a recommendation" among clients.

All these graphs have relations across domains, and the two graphs at the top of the picture correspond to a "classical bipartite graph" with two columns, which we used to represent affiliation networks in Chapter 8. The two bipartite graphs are undirected and they represent the distinct types of relations in this particular network. Obviously, each domain does not need to be depicted as a single column or a single row for multilevel network structures. Besides, it is also possible to have other bipartite graph options with three or more columns as before with affiliation networks.

The ties between a domain and a co-domain that characterizes affiliation networks are typically depicted as undirected, but in multilevel network structures cross-domain relations can be directed as well. The multilevel multigraph at the bottom of Figure 10.1, which includes the four kinds of relations, has directed edges to point the direction of the ties both within and across domains. The shape of the arcs correspond to the shapes used in the bipartite graphs above, and the graph depiction in this case is an application of a force-directed layout algorithm to the binomial projection of the two-mode data.

We can see in the multigraph of Figure 10.1 with the binomial projection the two types of relations across domains. Since there is a pair of network members, namely {A1, C2}, which have an exchange ties of different type, it means that the multilevel structure of the Clients and attorneys network is also multiplex. This is even if the only bundle with a multiple character occurs between domains, and the dyadic bundles within the two domains are just of a simple character. The other attorney that is hired is A3, who is the most respected in the network, by Client C1. However, neither of these clients receives a mail letter from the attorneys.

10.2.1   Additional Features

The Clients and attorneys network is a small network, which served to illustrate the principles of a multilevel structure. In this case, both domains have within and cross relations of a single type and a multiplex type, respectively. Hence, this system is a multilevel network that is multiplex as well, and this is because two network members exchange ties of different type.

A less complex multilevel network than the Clients and attorneys network will have within relations in just one domain while the cross-domain relations are still of a single type. In another extreme, a more complex multilevel structure will have multiplex ties across and within domains.

Whether or not a multilevel structure must have a relationship between the domain and the co-domain is a matter of definition, and we only assume that a cross-domain relation is a pre-condition for a multilevel arrangement. Otherwise, the different domains will constitute just a cascade of separate systems as with the case of multilayer networks.

Finally, the lack of ties within the domains is what characterizes bipartite networks, but one could consider an affiliation network as a special case of a multilevel structure. This is as long any multimode network structure has the possibility of having ties among the actors and/or the among events as a precondition.

10.3   Multilevel Structure of the G20 Network

In this section, we are going to look at the multilevel structure of the G20 countries network, taking both the trade network $\mathscr{X}_{G20}^V$ from Chapter 9 with valued relations, and the affiliation network $\mathscr{X}_{G20}^B$ where countries are related to supranational organizations. In this case, one of the domains is made of the one-mode structure found in $\mathscr{X}_{G20}^V$, whereas the two-mode structure provides the relations across domains.

The multilevel analysis starts with the structure with all nations in the G20 network, and later on we will reduce the number of actors and concentrate in the analysis only the multilevel structure of the G20 network with bridge organizations, $\mathscr{X}_{G20b}^M$. It is with this latter system that we construct, at the end of the Chapter, a relational algebra for a multilevel reduced structure, and this algebra considers both within and across domain relations in the network.

10.3.1   Multilevel structure of all G20 countries $\mathscr{X}_{G20}^M$

Indeed, it is also possible to construct a multilevel structure from the two systems of the Group of 20 countries from Chapters 8 and 9. In the case of the trade network of milk and honey of the G20 network $\mathscr{X}_{G20}^V$, for instance, this corresponds to a multiplex structure where the G20 countries have different political affiliations in network $\mathscr{X}^B_{G20}$. Hence, the system of relations in $\mathscr{X}^V_{G20}$ corresponds to the domain of the multilevel network structure, whereas the set of affiliations in $\mathscr{X}^B_{G20}$ constitutes the co-domain of the network.

However, since $\mathscr{X}^V_{G20}$ is a valued network with a high density in the flow of relations, it is more convenient to work in the analysis of the multilevel system with the salient structure of this network. In Chapter 9 Valued Networks, we produced the salient structure of $\mathscr{X}^V_{G20}$ with the triangle inequality of the Pathfinder semiring approach, and this skeleton of this valued trade network will serve as the one-mode part of the multilevel structure for the G20 network.

As a result, in this case the multilevel structure is made of the combination of the G20 affiliation network and the salient structure of the trade network of milk and honey, which represent, respectively, two-mode structure and a one-mode valued multiplex system. We denote the multilevel structure when considering all members in the G20 network as $\mathscr{X}^M_{G20}$, which is a combination of the salient configuration of $\mathscr{X}^V_{G20}$ and $\mathscr{X}^B_{G20}$.

In this particular multilevel configuration, there are within relations in the system domain, but there are any within relations in the co-domain, and this is because the supranational organizations are not directly linked. Certainly, it is also possible to have a system in the G20 countries network where there are within ties in both domains of the systems, such as the Clients and attorneys network mentioned at the beginning of this chapter.

Figure 10.2 presents two valued multigraphs with a force-directed layout that corresponds to the trade network in year 2015, however. The valued multigraphs encode information both from $\mathscr{X}^B_{G20}$ and the salient structure of $\mathscr{X}^V_{G20}$, which means that the two graphs are representation forms for a multilevel network structure even if the co-domain members are not represented by vertices in the first graph. Since there are not within relations in one domain of the network, then undirected edges with the co-membership relations stand for the information attached to the co-domain of the system. In other words, the information of the co-domain is incorporated as a co-affiliation tie in the multilevel structure.

The nodes in both multilevel multigraphs have distinguished colors to differentiate the four classes of actors as explained in the presentation of the two-mode data in Chapter 8. In the case of events, these are represented as well in both multigraphs, but the events are represented as edges in the first graph, and as vertices in the second multigraph. Besides, there is another color to distinguish the organizations either as a node or as an edge type in these multigraphs of the multilevel structures.

If we take a look at the multigraph in Figure 10.2, there is the salient structure of $\mathscr{X}^V_{G20}$ after applying the triangle inequality version of the Pathfinder semiring (cf. Chapter 9). This means that this configuration represents the backbone of this valued structure that is based on the dissimilarities among the network actors. The arcs in this case represent the presence of a trade flow of the two commodities between pair of nations, and the undirected edges link the countries when they have common affiliations.

On the other hand, the multigraph in Figure 10.3 provides a "multilevel binomial projection" where all members of $\mathscr{X}^B_{G20}$, in the domain and co-domain, are depicted in visualization device that can be called a valued *multilevel graph*. This representation form has different node shapes for the actors and the events in the network, and the presence of directed and undirected edges as well. As with the previous case, the directed edges correspond to the two types of trade flow in $\mathscr{X}^V_{G20}$, and the undirected ones to the co-membership relations in $\mathscr{X}^B_{G20}$.

Certainly, the graph of the salient structure of trade network $\mathscr{X}^V_{G20}$ with co-membership ties has fewer vertices than the one with both $\mathscr{X}^B_{G20}$ and the salient structure of $\mathscr{X}^V_{G20}$. However,

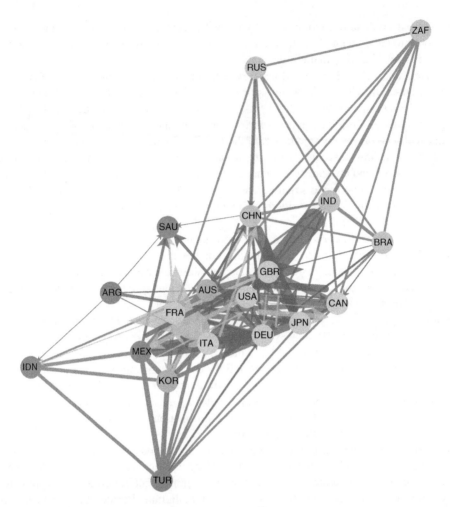

Figure 10.2   Multilevel valued multigraph of network $\mathscr{X}_{G20}^{M}$ with co-membership ties among countries in $\mathscr{X}_{G20}^{B}$. *This arrangement is the salient structure of trade network.*

having more vertices, and even more edges as well, can sometimes provide a better picture of the multilevel structure in the network than multilevel graphs with fewer vertices.

In the case of the multilevel multigraphs in Figure 10.2, it is possible to "see", for instance, that IDN has common membership and hence is tied with MEX thanks to MITKA (and N11 as well in $\mathscr{X}_{G20}^{M}$). These two countries are also linked to KOR, TUR, and AUS because of their common affiliations in the organization MITKA as well. Moreover, we can verify that some nations have important trade ties with these commodities without too much political power, and conversely some other nations are politically linked in the affiliation part of the multilevel system, but they are isolated in the particular trade network such as ZAF. Through the visualization of multigraphs then is possible to make similar analyses for other actors and relations that are occurring in the multilevel network.

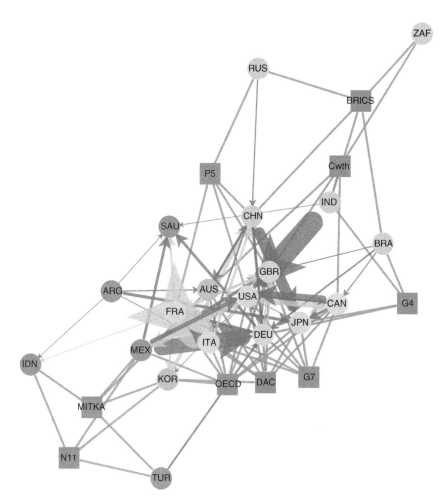

Figure 10.3 Binomial projection of the multilevel valued multigraph $\mathscr{X}_{G20}^{M}$. *As with Figure 10.2, this arrangement is the salient structure of trade network $\mathscr{X}_{G20}^{V}$ in terms of triangle inequality.*

10.4   Multilevel Positional System of G20 Network with Bridges

We continue the analysis of the multilevel structure by constructing the positional system of $\mathscr{X}_{G20}^{M}$ with bridge affiliations or $\mathscr{S}_{G20b}^{M}$. In this case, we take the multilevel network $\mathscr{X}_{G20b}^{M}$ that corresponds to the G20 countries that are affiliated to bridge affiliations, but also to the rival organizations G7 and BRICS.

We start with a visual interpretation of the multilevel network $\mathscr{X}_{G20b}^{M}$ and then we perform a positional analysis of the G20 countries with bridge affiliations that will produce a positional system made of categories of actors and events. As with multiplex networks, the goal of the positional analysis is to reduce the complexity in the system in order to facilitate a substantial interpretation of it.

10.4.1   Visual Interpretation of the Multilevel Structure in $\mathscr{X}^B_{G20b}$

The multilevel structure of the G20 countries network with bridges, represented as $\mathscr{X}^M_{G20b}$, corresponds to the G20 countries affiliated to the bridge organizations in $\mathscr{X}^B_{G20}$. That is, this structure disregards the organizations DAC, OECD, Cwth, and N11, and hence we take in this case just the actors in the trade network $\mathscr{X}^V_{G20}$ that are affiliated to the bridge organizations considered in Fig 8.5 in Chapter 8.

Figure 10.4 depicts two multilevel network structures of network $\mathscr{X}^B_{G20b}$ with a single force-directed layout. One graph corresponds to the trade network $\mathscr{X}^V_{G20b}$ with co-membership ties among actors in $\mathscr{X}^B_{G20b}$, and the other graph below encodes the co-affiliation relations through vertices with a square shape representing the elements of the network co-domain. The "star" graph to the right of the picture represents a multilevel structure where both edges for cross-domain relations and arcs for within-domain ties are of the same width. The "circle" graph to the left is represented as a valued multigraph where the thickness of the edges-directed or not-reflects the amount of trade in the within relations in the one-dimension part, and the number of common affiliations in the cross-domain part.

Both representation forms—where co-membership ties are represented by edges and by vertices—have their advantages and disadvantages. The salient structure of the trade network $\mathscr{X}^V_{G20b}$ with co-membership ties among actors has fewer nodes, and the multilevel structure of co-membership as vertices has not, however, fewer edges. On the contrary, the multilevel structure of co-membership vertices has the double of co-affiliation ties than the graph with co-membership ties among actors, and this is since they are "split" by a vertex, even though co-affiliation edges tend to be shorter in length.

Typically, the presence of the network events as nodes provides a more complete picture of the multiplex structure, which is an important aspect for the substantial interpretation of the system. Visually, the valued multigraph such as the graph with co-membership ties among nations to the left of Figure 10.4 can result, being a better representation of multilevel structures than a multigraph with arcs and edges with the same thickness, as with the graph to the right.

In the case of $\mathscr{X}^M_{G20b}$, for instance, the added information to the graph allows differentiate the arcs from the trade network $\mathscr{X}^V_{G20b}$ from the edges corresponding to the affiliation ties in $\mathscr{X}^B_{G20b}$. The fact that the edges have the same width in the multilevel graph with organizations as vertices makes sense because in this case a country can either be affiliated or not affiliated to a certain organization, but for depicting the co-membership relations among nations valued multigraphs are typically desirable.

It is easier to distinguish the trade flows having the co-affiliation relations as vertices in Figure 10.4, which in this case happens more among the G7 countries and CHN. Since this kind of layout ends up by "grouping" nations by their relations with organizations and, vice versa, organizations by their relations with countries, it is also easier to see the trade flow between nations with a different economic classification. This is even if organizations they are affiliated with are plotted in the two graphs.

The multilevel structures of $\mathscr{X}^B_{G20b}$ are smaller than the systems corresponding to $\mathscr{X}^B_{G20}$, and in all cases we used valued graphs. This is in order to reflect the intensity of the relationships and distinguish the amount of trade flows among the countries with the two kinds of commodities. However, it is certainly possible to count with a multilevel graph where the weights in the edges are equal, and this may constitute an advantage since many times larger thickness of the edges can make difficult to observe important ties such as affiliation and co-membership.

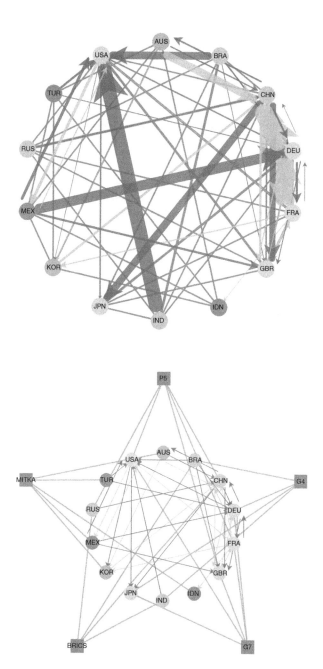

Figure 10.4 Two multilevel graph representations of G20 Trade network with bridges, $\mathscr{X}^M_{G20b}$. *Top: Valued multigraph with co-membership as edges. Bottom: Multilevel graph with organizations as vertices.*

10.4.2 Positional Analysis of $\mathcal{X}_{G20b}^{M}$

For the G20 countries network with bridges $\mathcal{X}_{G20b}^{M}$, a positional analysis takes the "non-overlapping affiliations" of nations to the organizations G7, BRICS, and MITKA. Hence, there are three *classes* of actors that correspond to their links to one of these three organizations, and we can represent the categories as 1 to 3.

AUS	BRA	CHN	DEU	FRA	GBR	IDN	IND	JPN	KOR	MEX	RUS	TUR	USA
3	2	2	1	1	1	3	2	1	3	3	2	3	1

The three different classes of actors in $\mathcal{S}_{G20b}^{M}$ that correspond to the non-overlapping organizations are going to be represented by G7.C, BRICS.C, and MITKA.C, and they are composed of: {DEU, FRA, GBR, JPN, USA}, {BRA, CHN, IND, RUS}, and {AUS, IDN, KOR, MEX, TUR}, respectively. As a result, in this positional system the classes of actors stand for the network members in the domain of the bipartite network with bridges $\mathcal{X}_{G20b}^{B}$, and these are affiliated with the respective organizations in the co-domain of the affiliation network $\mathcal{X}_{G20}^{B}$.

Figure 10.5 provides a multilevel valued structure of the positional system $\mathcal{S}_{G20b}^{M}$ as image matrices. In this case, the positional system comprises trade flow relations of the two commodities among G20 countries, and the affiliation to bridge and rival organizations. As Figure 10.5 shows it, all the image matrices are valued including the affiliation table, and the numbers in the two image matrices among actors are accumulated trade values of the valued data given in Appendix C.

	G7.C	BRICS.C	MITKA.C
G7.C	19249	22865	3538
BRICS.C	0	0	0
MITKA.C	752	7572	412

Fresh Milk

	G7.C	BRICS.C	MITKA.C
G7.C	3050	296	293
BRICS.C	29577	584	1187
MITKA.C	13477	860	0

Honey

	P5	G4	G7	BRICS	MITKA
G7.C	3	2	5	0	0
BRICS.C	2	2	0	4	0
MITKA.C	0	0	0	0	5

Affiliation of classes to bridge organizations

Figure 10.5   Multilevel valued image matrices of $\mathcal{S}_{G20b}^{M}$ made of $\mathcal{X}_{G20b}^{V}$ with class affiliations in $\mathcal{X}_{G20b}^{B}$.

The third table below constitutes a "valued bipartite matrix" where two bridge organizations are given along with the organizations of the three classes of actors. The values in this rectangular image matrix correspond to the number of nations affiliated to such organizations, and we can see, for instance, that the bridge organizations P5 and G4 have members both from classes G7.C and BRICS.C, which makes these countries a separate class from MITKA.C.

As a result, the positional system $\mathscr{S}^M_{G20b}$ made from network $\mathscr{X}^V_{G20b}$ with class affiliations in $\mathscr{X}^B_{G20b}$ is another multilevel valued structure of the G20 network. The three classes of actors belong to a system with two types of trade relations (milk & honey) that they have in common. Besides, the multilevel structure includes also the affiliation relations with the bridge organizations P5, G4, and MITKA, in addition to their membership with their respective "headquarters". The multilevel analysis in algebraic terms is then based on the positional system made by these three arrangements.

10.4.3   Depiction of Multilevel Positional System $\mathscr{S}^M_{G20b}$

Naturally, a multilevel valued structure such as the positional system of the G20 network can be represented graphically as well. The difference is that the set of vertices in the graph—for the domain in this case—is made of classes of nations rather than individual countries. There is also the possibility that categories of events correspond to the co-domain structure and this is when events are classified.

Figure 10.6 depicts the multilevel graph with a binomial projection of a positional system network from $\mathscr{X}^V_{G20}$ with class affiliations in $\mathscr{X}^B_{G20b}$. Here the configuration is made of three different classes of countries, G7.C, BRICS.C, and MITKA.C, and the network includes their affiliations with the respective organizations as well.

As with the valued networks in Chapter 9, the thickness of the edges among actors reflects the amount of trade flow of the respective commodity, which in this case is between categories of countries in $\mathscr{S}^M_{G20b}$. Valued loops in the multigraph represent the amount of trade among countries within the same category, and the affiliation ties that are weighted, as well as reflecting the number of nations in the category that are members of the organization the class is tied with. Since there are only three classes of actors, it is more likely that the valued graph provides more information that is not at the expense of other relations, and this is since cross-domain relations do not "cross" with the ties within domains.

If we look at the events in the multilevel network positional system $\mathscr{S}^M_{G20b}$, the "size" of the organization in terms of number of associates corresponds to the width of the edge connecting the event to the class of actors in Figure 10.6. In this sense, the multilevel multigraph is valued not only for the relations within the domain, but also with the ties across the two domains.

*
* *

Another reduction of the configuration in Figure 10.6 (not shown here) is the multilevel structure with co-affiliation of classes of actors depicted as edges. The graph of this reduced structure is then a triad made of categories of actors only, and this is since the event affiliations are represented by edges. In this case, classes G7.C and BRICS.C have a co-membership edge thanks to their common affiliation with G4 and P5. On the other hand, MITKA.C is just related by trade relations with the other two classes of actors, without affiliation edges.

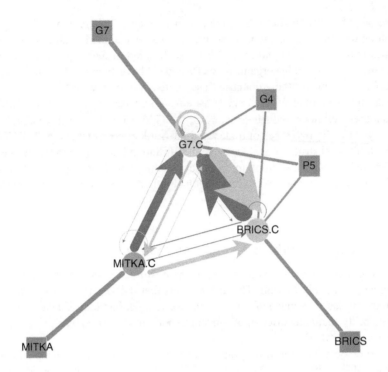

Figure 10.6   Valued multigraph of the multilevel positional system $\mathscr{S}^M_{G20b}$ derived from $\mathscr{X}^V_{G20b}$ and with class membership in $\mathscr{X}^B_{G20b}$. *Classes of nations with common affiliations are* G7.C, BRICS.C, *and* MITKA.C, *where the thickness of the edges reflect the strength of the tie. Valued loops and valued directed edges have different scales.*

In such a case, however, the form for graph reduction is as a result less informative than a multi-level graph where organizations are represented as nodes. This is even if this reduced graph has a smaller number of vertices than the multilevel valued graph of the positional system network with affiliations of $\mathscr{X}^V_{G20}$ with class affiliations from $\mathscr{X}^B_{G20b}$ as in $\mathscr{S}^M_{G20b}$.

10.5   Algebraic Approaches to Multilevel Networks

As stated earlier, in the introductory part of this chapter, multilevel networks are made of at least two "levels" or domain structures, and different kinds of algebraic objects can help us for the analysis. However, the visualization of the multilevel network is usually the step before the algebraic analysis itself. This is because the visualization of multilevel networks as multigraphs has extraordinary advantages; this is not only because graphs allow us to have a flavor of such network structure, but also because in multilevel graphs it is possible to "see" the different associations between the bundle patterns that are occurring at the various domain structures in the system.

Performing an algebraic analysis of the multilevel network after its visualization constitutes a next phase in the study of such complex structure, and we are going to perform algebraic analyses of the different types of relations in one (or both) of the domain structures. For an algebraic analysis, we first take a look at the multilevel structure of the Group of 20 network from Chapter 8 and from Chapter 9, where the multilevel system combines one- and two-mode network data. In this case, the pair of domain structures is in the form of a flow trade of two kinds of commodities among some G20 countries in the trade network $\mathscr{X}^V_{G20}$, and their affiliations to some other supranational organizations in the reduced system $\mathscr{X}^B_{G20}$, which includes bridges.

If we consider in the multilevel structure the relationship across different domains first, then we can apply Galois connections between the two sets of entities as we did previously in Chapter 8 with two-mode networks. In this way, we could obtain a partially ordered structure of both the domain and the co-domain within the Formal concept analysis framework. Since both affiliation and multilevel networks have a domain and a co-domain in their respective structures, the formal concept approach can certainly be useful for the study of multilevel structures.

Besides, the relational algebra expressed in the partially ordered semigroup can serve to model multilevel systems as well, and it constitutes another application of this algebraic object that has been used for the modeling of multiplex and valued multiplex networks. Recall from Chapter 5 Role Structure in Multiplex Networks that it was possible to model one-mode multiplex network structures such as $\mathscr{X}_A$ and $\mathscr{X}_F$ with attribute-based information of the actors, which were treated in a relational manner as self-relationships with the use of diagonal matrices. In the case of multilevel networks, we can also benefit from the use of diagonal matrices to model the cross-domain relations. Hence, rather than incorporating actor attributes in the modeling of multiplex networks, we can incorporate "actor affiliation" by using a separate diagonal matrix for each element or attribute in the co-domain of the multilevel network.

10.5.1 G20 Multilevel Network

Constructing the algebra of the G20 multilevel network make it possible to gain more insights about the configuration of $\mathscr{X}^M_{G20}$, as with both multiplex and affiliation networks. Networks $\mathscr{X}^V_{G20}$ and $\mathscr{X}^B_{G20}$ are, for instance, themselves multiplex systems, and the multilevel structure of these as domain structures adds another complexity "layer" that is product of the combination of the two levels.

In the case of the visualization of the multilevel structure $\mathscr{X}^M_{G20}$, which is made of $\mathscr{X}^V_{G20}$ and $\mathscr{X}^B_{G20}$ (and the smaller version of $\mathscr{X}^M_{G20b}$ or $\mathscr{X}^V_{G20b}$ plus $\mathscr{X}^B_{G20b}$), the analysis begins with the positional system of the affiliation network $\mathscr{X}^B_{G20}$. This is because the positional system is a condensed structure where we are able to see in a clearer manner the relationship between the two domains of the network.

Figure 10.7 provides a visual representation of a multilevel structure of the reduced positional system that will help us to cover the analytical process of representing the relational structure of multilevel networks. Now all edges and loops have the same scale, and there has been a surjection product of structural equivalence.

Hence, the starting point in the analysis of the multilevel network structure is in this case the positional system $\mathscr{S}^M_{G20b}$. This system is made of classes of actors and classes of events in $\mathscr{X}^V_{G20b}$ and $\mathscr{X}^B_{G20b}$, which means that the network members are related between and within the domains of the system. That is what characterizes a multilevel structure.

In fact, the multilevel structure is made of the ("skeleton" of) positional system $\mathscr{S}^M_{G20b}$ with collapsed bundles. This means that the two types of relations in $\mathscr{X}^V_{G20}$ are represented by a single

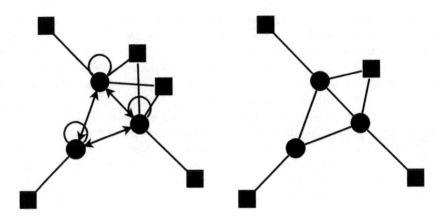

Figure 10.7 Multilevel graphs of positional system $\mathscr{S}^M_{G20b}$ with collapsed bundles. *A reduced positional system is obtained by removing redundant information in edges, and by applying* SE *to positions.*

edge that is directed for trade flows between classes of actors and undirected loops for relations within classes of actors.

On the other hand, the affiliation ties between classes of actors and categories of events are represented as custom by undirected edges. Both the direction of the ties and the loops in the graph to the right of Figure 10.7 are removed, and this simplify dramatically the multilevel structure. Since relational structures are typically large and complex, we are going to represent and analyze next this reduced system made of 3 connected entities in one domain and 5 cross-domain relations in algebraic terms.

One strategy for performing an algebraic analysis of multilevel systems is to build their structure of relations or relational structure or even better their structure of role relations or role structure for systems made of classes of actors. The process of the construction of the relational or role algebra for a multilevel network configuration is similar to the process of constructing the role structure of multiplex networks $\mathscr{X}_A$ and $\mathscr{X}_F$ as we did with the in Chapter 5 Role Structure in Multiplex Networks. That is, the first step is to construct the network positional system with classes of actors who are related through different types of collective role relations.

From a graphical perspective, the difference between the multiplex and multilevel networks we have seen so far, is that a "multilevel structure" graph is made of a combination of directed and undirected edges where directed relations are within the domain and the undirected edges stand for the cross-domain ties. This also means that there are two different domains to take into account for constructing the multilevel network relational structure while for multiplex networks the relational structure is just on a single domain.

In the case of the affiliation and trade networks $\mathscr{X}^B_{G20b}$ and $\mathscr{X}^V_{G20b}$, for instance, the flow relations of the two commodities in the trade system constitute two generator relations for the semigroup configuration. On the other hand, the countries' membership in organizations with the affiliation network provides "structural information" for additional generator relations in the algebraic structure of the multilevel network. Having the cross-domain relations as diagonal matrices makes it possible to pursue the construction of the partially ordered semigroup of the multilevel structure. Yet the Concept lattice that is built within the Formal concept analysis approach is still applicable to the cross-domain relationships in a multilevel network, and it also constitutes part of the *algebra* for multilevel network structures.

The construction of the algebra for a given multilevel structure requires first performing a homomorphic reduction of the network constituents. For the multilevel network $\mathscr{X}^M_{G20b}$, which is made of $\mathscr{X}^V_{G20b}$ and $\mathscr{X}^B_{G20b}$, the generating set of the semigroup structure corresponds to the two types of relations in the trade network $\mathscr{X}^V_{G20b}$ that are in the domain. On the other hand, the events in the co-domain are represented in a relational manner with the aid of diagonal matrices, as with the positional analysis with actor attributes made in Chapter 5, and then the different affiliation ties constitute additional generators of the semigroup structure. The semigroup structure is based on the positional system denoted as $\mathscr{S}^M_{G20b}$ that have three positions G7.C, BRICS.C, and MITKA.C, and which is partially depicted in Figures 10.6 and 10.4.

The generating set for the semigroup of this multilevel structure is

$$\Sigma_{\mathscr{G}_{G20b}} = \{m,\ h,\ G,\ B,\ A,\ P\}$$

where m and h stand for the two trade commodities milk and honey. With respect to the organizations, G, B, A stand for G7, BRICS, and MITKA, respectively. Since both P5 and G4 are structurally equivalent network members, these relations are equated and are represented by letter P.

10.5.2   Visualization of Multilevel Network Algebra

A visual representation of the algebra for the multilevel structure $\mathscr{S}^M_{G20b}$, which is the positional system for $\mathscr{X}^M_{G20b}$, is given in Figure 10.8. Here the multilevel positional system "role algebra" has two separate algebraic objects that stand for the relationships in the domain structures. To the left of the picture is the partially ordered semigroup that represents the role algebra of the multilevel network, and to the right of the picture there is a Concept lattice with the distinct associations between the elements in the domain and the co-domain of the multilevel structure. The string generators of the semigroup structure m, h, P, G, B, and A are depicted in the Cayley graph with "solid", "dashed", "dotted", "dotdash", "longdash", and "twodash" line types, respectively.

On the other hand, the Concept lattice is made of the network actors, and in this case for a multilevel network this set is made of the actors and the events. Typically the reduced labeling of the Concept lattice provides more useful information than the full labeling, but for illustration purposes this partial order structure is given with a full labeling of multilevel network. Hence, the Concept lattice represents the partial order structure of the Galois connections between intents and extents, which in this case is between the classes of G20 nations and organizations associated with these positions of categories of countries.

As a result, two partially ordered structures serve to represent the relational system for multilevel networks in Figure 10.8. One is the inclusion lattice of the string relations in the partially ordered semigroup, and the other partial order structure corresponds to the Concept lattice with the actor and the events in the multilevel system. The partially ordered semigroup representing the relational structure of multilevel network has a multiplication table as well, which in this case is represented graphically by a Cayley color graph. Following the convention established here, string relations that are equated are represented by a comma in lattice structures and by the equal sign in the vertices of the Cayley graph.

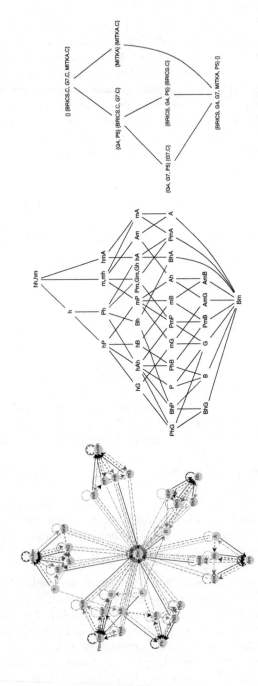

Figure 10.8 Visual representation of the multilevel role algebra for the "levels" in positional system $\mathscr{P}^M_{G20b}$. *Left and middle: Cayley graph and inclusion lattice of the semigroup. Right: Concept lattice where the reduced labeling of $\mathbb{C}$ is:* $\varnothing > (\{G4, P5\} \{\varnothing\} > (\{G7\} \{G7.C\}, \{BRICS\} \{BRICS.C\}\}),$ $\{MITKA\} \{MITKA.C\}) > \varnothing.$

10.5.3   *Substantial Interpretation*

If we start the substantial interpretation of the multilevel structure with the partially ordered semigroup, each equation given in Figure 10.8 for the role structure has its own story For instance, m is an idempotent element that is also equated with mh, and this means that for $\mathscr{X}^M_{G20b}$ all countries that trade fresh milk also trade honey first and then milk through a third partner. Besides, both G, B, A, and the bridge organizations P5 and G4 that are represented by P are idempotents as well. Substantially, such equations do not reveal something new, since they are just telling us that these organizations relate to themselves.

On the other hand, the equations G=PG=GP and B=PB=BP are telling us that all actors affiliated with either G or B are affiliated with P as well. P, which is representing the bridge organization, does not have a common actor with A or MITKA. This implies that the two strings PA and AP are empty and equated with Bm, which stands for the zero element in the representation of the semigroup structure.

The partially ordered semigroup has a universal relation as well, and this element has hh=hm (or hh,hm in the inclusion lattice) as representative strings. Hence, the inclusion lattice is complete with the universal and the empty relation in both extremes, and having the rest of the representative strings in the hierarchy of relations. Here, it is worth mentioning that not all classes of countries that trade fresh milk also trade honey and vice versa.

With respect to the Concept lattice, it is clear the interlock among the nations in G, B and P, whereas the MITKA countries lie in a separate set of inclusions in the partial order structure. The fact that the class of actors affiliated to this latter event follow a different path allows us to perform another analysis of a simpler multilevel structure where only G7 and BRICS countries are involved. Substantially, this makes a lot of sense since this network member does not have any common affiliation either with G7.C or with BRICS.C.

10.6   Reducing Complexity in $\mathscr{X}^M_{G20b}$

It was clear already from Figures 8.4 and 8.5 in Chapter 8 that the binomial projection of the two-mode data brought two components, one for the MITKA countries and the other component for those associated with G7 and BRICS. Hence, the reduced multilevel structure with only G7 and BRICS countries is the multilevel positional system $\mathscr{S}^M_{G20b}$ after dropping class MITKA.C. Apart from their respective organizations, these two classes of nations, G7 and BRICS, are associated with P5 and G4 as well, and this is not the case with the set of countries associated with MITKA.

There are several reasons for the reduction of complexity in the relational structure. If we take into the modeling just the G7 and the BRICS countries affiliated to P5 and G4, additional equations occur, which reduces the network structure. Certainly, there are few actors in the network now, and this implies, for instance, that generator P, which stands for organizations P5 and G4, becomes an identity relation without a structuring effect in the role structure. Besides, the two primitive relations representing the commodities have additional equations and hence the set of representative strings has fewer compounds. By disregarding position MITKA.C, the order of the partially ordered semigroup structure dramatically reduces from 33 elements in the Role structure (cf. Figure 10.8) to just 11 representative strings in the semigroup of role relations of $\mathscr{S}^M_{G20b}$. The same amount of reduction applies to the Concept lattice with cross-domain ties of the multilevel structure.

Figure 10.9 provides the role algebra of the reduced multilevel structure with G7 and BRICS countries only, and we clearly evidence the decrease in complexity of the role structure after dropping MITKA.C. First, in the upper part of the picture is given the partially ordered semigroup of this multilevel system in a graphical format where we evidence that the representative equation for the maxima element in the Concept lattice has changed slightly.

The line types in the Cayley graph to the left that stand for the generator relations of the semigroup follow the shapes used for the primitives in the Cayley graph of Figure 10.8. The only exception is that there is no need for generator A for the MITKA countries, which means that the "twodash" arcs representing this primitive relation are removed from the Cayley graph. To the right of the Cayley graph of semigroup structure in Figure 10.9 there are the string role relations that form a partially ordered structure by inclusion, which represented as a lattice diagram. These two structures stand for the partially ordered semigroup of the role relations.

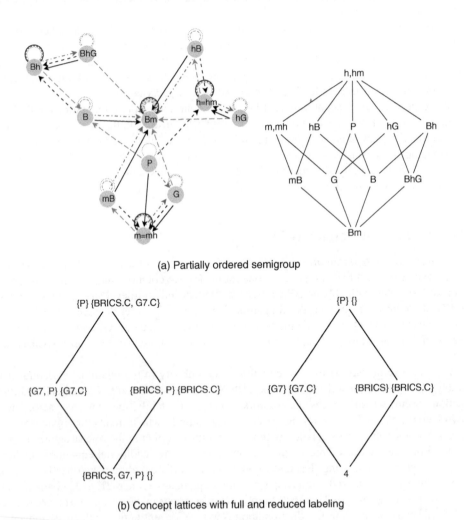

(a) Partially ordered semigroup

(b) Concept lattices with full and reduced labeling

Figure 10.9   Role algebra of the multilevel network $\mathscr{S}^{M}_{G20b}$ after dropping MITKA.C.

The lattice diagram is isomorphic to the inclusion lattice of network $\mathcal{Q}_B$ depicted in Figure 6.8 in Chapter 6, and which corresponds to the role structure of Incubator network B. Such isomorphism suggests that the structure in both positional systems have similar algebra, and this is not only because the order of the two semigroups are equal but also the set of inclusions of their strings. Naturally, the substantial interpretation of the role structures differ in this case, and this is since the two networks have ties with a different relational content.

Besides the partially ordered semigroup with the relationships among strings in a single domain, the cross-domain relations of the multilevel structure of G7 and BRICS countries is given in the lower panel of Figure 10.9. In this case, there is one representation of the Concept lattice with full labeling, and another representation with reduced labeling of the concepts. For instance, now it is easy to see in the Concept lattice with the full format that BRICS.C and G7.C are affiliated with P, which means that BRICS.C is affiliated with P and BRICS, whereas G7.C is affiliated with G7 and BRICS. Besides, there is any country in the system that is affiliated to the three organizations BRICS, G7, and P.

The set of inclusions in the Concept diagram given in Figure 10.8 for 3 classes of actors and 4 organizations prevails over the set of inclusions in the smaller Concept lattice, except that MITKA countries and organization are not part of the diagram of partial order structure, and organizations P5 and G4 are represented by P in the lattice. With respect to the reduced labeling of the Concept lattice, this partial order structure tells us nothing more than that the P organization involves BRICS and G7 nations, BRICS countries are associated with BRICS, and G7 countries are associated with G7.

On the other hand, the reading of the link structure in the semigroup of relations in this latter system is less difficult than the three-class positional system, and this is because of its reduced size thanks to the additional equations produced with the two-class positional system. Naturally, in this new system Bm is still an empty string, but relation h becomes an idempotent element in the semigroup of role relations as the same as m and as before the strings representing the organizations. That is, $m = mm = mmm = \dots$, $h = hh = hhh = \dots$, $P = PP = PPP = \dots$, $G = GG = GGG = \dots$, and $B = BB = BB = \dots$

10.7   Further Algebraic Representations of Multilevel Structures

It is desirable that the algebra to represent multilevel network structures is a more integrated ensemble; that is, not "merely" a set made of a partially ordered semigroup and the Concept lattice. This is a current challenge and some clues lie in the following directions:

- One possibility is to construct the relational structure of the one and two-mode structures with a binomial projection, which means that the "dimension" of the matrix generators equals the number of actors and events in the network. With this other option, however, there is a higher risk of producing a larger semigroup object than before, and we need to account with the structural information in order to make a partition of the multilevel network and produce a positional system for actors and events. For instance, the clustering information from Figure 8.1 for actors and events in $\mathcal{X}_{G20}$ where there are four (or two) classes of actors and three classes of organizations can be the basis for constructing a smaller multilevel role structure.
- There is also the possibility that the partial order structure in the Concept lattice provides the clustering information for a positional system of both actors and events. In this way, we can

reduce the network and pursue the network role structure further as semigroup systems of role relations.

- Finally, for the analysis of multilevel structures it is possible to focus on sets of actors who are connected through trade flows with different types, and which is flexible enough to incorporate node attributes in the analysis. For example, if we take a trade flow of commodity X and commodity Y, we can include elements of the co-domain or actor attributes in the same manner. That is, if, for instance, relation A links a set of actors to the category named "advanced economy," the compound XAY will represent a trade flow relation made of these two types of commodities between countries through a network member having a cutting-edge economy.

An important aspect comes with the substantial interpretation of a trade system incorporating a political dimension in the analysis in a systematic manner, which is an important aspect when it comes to comparing complex economic systems such as the G20 countries network, or a single multifaceted structure through time, or the raw network and its reduced structure.

10.8   Multilevel Networks: Summary

Multilevel configurations combine one and two-mode network structures, and the ability to analyze these systems in their complexity remains a current challenge within the social network analysis discipline. Visualizing multilevel systems is an important step in the analysis, where the events or elements in the networks co-domain can be depicted either as vertices or as undirected edges for common affiliations in the multilevel graph. In the case an edge representation, the intensity of the relationships across domains reflects the size or prominence of the event. This implies that the multilevel network becomes a valued structure where a valued multigraph represents the edges across domains with a score value.

Naturally, the relations within the domain can be a valued as well in a multiplex configuration such as the multilevel structure of the G20 countries network. In this case, weights in the relations were included both in the graph and in matrix representations, which included a multilevel positional system used further analyses in algebraic terms. Hence, for the construction of the algebra for multilevel networks, it was required to perform a reduction of the two domains of the network where potential structurally equivalent actors and events were linked through role relations. We benefited from the partially ordered semigroup and the Concept lattice to represent the relationships both within and across the domains of the multilevel positional system, and it was required the creation of different visualization strategies for multilevel networks for the study of such complex systems even if they are reduced multilevel structures.

10.9   Learning Multilevel Networks by Doing

10.9.1   *Multilevel Network 'Clients and Attorneys'*

```
# relations within domains (Clients and attorneys)
R> n1 <- transf(list(A = c("C1, C2","C1, C3","C2, C1","C3, C1","C3, C2","C3, C4"),
+    B = c("A1, A3","A2, A3","A3, A2")), type = "toarray")

# relations across domains. Add elements in domain and in codomain
R> n2 <- transf(list(C = c("A1, C2","A1, C4","A3, C4"), D = c("C2, A1","C3, A3")),
+    type = "toarray2", add = list("A2", "C1"), adc = list(c("C1","C3"), "A2"))
```

```
# construct multilevel structure specifying domain and codomain labels
R> n12 <- mlvl(n1, n2, lbs = list(c("C1","C2","C3","C4"), c("A1","A2","A3"))))
```

```
# scope multilevel network with customized scope and edge shapes
R> scpmlm <- list(cex = 6, lwd = 2, pos = 0, ecol = 1, bwd = .5, pch = 21:22, vcol = 1,
+    collRecip = TRUE, fcol = "#FFFFFF", fsize = 12, rot = 90)

# plot Fig. 10.1 with multilevel network of Clients and attorneys as multigraph
R> multigraph(n12, layout = "force", seed = 1, scope = scpmlm, lty = c(3:4,1:2))
```

10.9.1.1   Bipartite Layout for Multilevel Structures

```
# one-mode structure among Attorneys
R> naa <- transf(c("A1, A3", "A2, A3", "A3, A2"), type = "toarray", sort = TRUE)

# affiliation network with Attorneys in the domain
R> nac <- transf(c("A1, C2", "A1, C4", "A3, C4"), type = "toarray2", add = "A2",
+    adc = c("C1","C3"), sort = TRUE)

# one-mode structure among Clients
R> ncc <- transf(c("C2, C1", "C3, C1", "C1, C2", "C3, C2", "C1, C3", "C3, C4"),
+    type = "toarray", sort = TRUE)

# another affiliation network with Clients in the domain
R> nca <- transf(c("C2, A1", "C3, A3"), type = "toarray2", sort = TRUE, adc = "A2",
+    add = c("C1","C4"))
```

```
# customized scope for multilevel network as bipartite graphs
R> scpmlb <- list(ecol = 1, cex = 8, vcol = 1, lwd = 6, pos = 0, fcol = "#FFFFFF",
+      fsize = 16)

# plot Fig. 10.1 with multilevel network as bipartite graphs
R> mlgraph(mlvl(naa, nac), layout = "bip", scope = scpmlb, lty = c(1,4),
+      pch = 22:21)
R> mlgraph(mlvl(ncc, nca), layout = "bip", scope = scpmlb, lty = c(2,3),
+      pch = 21:22, mirrorX = TRUE)
```

10.9.2   Multilevel Structure of G20 Network with Bridges

To perform an analysis of $\mathcal{X}_{G20b}^M$, we use 'G20net', which is the Pathfinder network of the G20 trade network from Chapter 9, and also object 'G20' that records the affiliation network data from Chapter 8. Also from Chapter 8, we have actor attributes in 'acc' and in 'acb', which is for the event "bridges" in 'G20'.

```
# choose countries in bridges
R> bridges <- which(acb!="none")

# extract of the G20 m&h multiplex network (exclude)
R> G20Bnet <- G20net[bridges, bridges, ]

# extract of the G20 bipartite network
R> G20b <- G20[bridges, c(1:5)]

      P5 G4 G7 BRICS MITKA
AUS    0  0  0     0     1
BRA    0  1  0     1     0
CHN    1  0  0     1     0
DEU    0  1  1     0     0
FRA    1  0  1     0     0
GBR    1  0  1     0     0
IDN    0  0  0     0     1
IND    0  1  0     1     0
JPN    0  1  1     0     0
KOR    0  0  0     0     1
MEX    0  0  0     0     1
RUS    1  0  0     1     0
TUR    0  0  0     0     1
USA    1  0  1     0     0
```

```
# multilevel with co-affiliation of actors
R> G20Bcn2 <- mlvl(G20Bnet, G20b, type = "cn2")

# multilevel with binomial projection
R> G20Bbpn <- mlvl(G20Bnet, G20b, type = "bpn")
```

```
# define multilevel scopes
R> scpml <- list(cex = 5, lwd = 2, pos = 0, lty = 1, pch = c(21,15), vcol0 = 8,
+    ecol = c("#00BFFF","#CD5B45","#FF7F24"), ffamily = "mono", fsize = 10)

# vertex colors for co-affiliations
R> vtcc <- list(vcol = c("#BCEE68","#BDB76B","#66CDAA","#838B8B")
# vertex colors for binomial projection
R> vtcb <- list(vcol = c("#BCEE68","#BDB76B","#66CDAA","#838B8B","#FF7F00")

# number of radii according to the amount of actors and events in 'G20bpn'
R> nr <- c(rep(1,length(G20Bbpn$lbs$dm)), rep(2,length(G20Bbpn$lbs$cdm)))

[1] 1 1 1 1 1 1 1 1 1 1 1 1 1 1 1 2 2 2 2 2
```

```
# Fig. 10.3: valued multigraph for co-membership with concentric layout
R> mlgraph(G20Bcn2, layout = "conc", valued = TRUE, scope = c(scpml,vtcc),
+    clu = list(acc[bridges], rep(1,nrow(G20b))) )

# Fig. 10.3: concentric graph with 2 radii for organizations as vertices
R> mlgraph(G20Bbpn, layout = "conc", nr = nr, scope = c(scpml,vtcb),
+    clu = list(acc[bridges], rep(1,nrow(G20b))) )
```

10.9.3 Multilevel Structure of G20 Trade and Affiliation Networks

```
# multilevel system with a binomial projection of actors and events
R> G20bpn <- mlvl(G20net, G20, type = "bpn")

# multilevel system with actor co-affiliations
R> G20cn2 <- mlvl(G20net, G20, type = "cn2")
```

```
# plot Figure 10.2 and Figure 10.3
R> scpG20tt <- list(cex = 4, vcol = c("#FFD700","#CD853F","#CD5B45","#66CDAA"),
+    ecol = c("#CD5B45","#00BFFF","#FF7F00"), lwd = 2, pos = 0, lty = 1, fsize = 7,
+    undRecip = TRUE, collRecip = TRUE)

# coordinates of multilevel system in 'G20bpn' with a force-directed layout
R> crdG20 <- frcd(G20bpn$mlnet, seed = 123, maxiter = 100)[1:19,]
```

```
# multilevel multigraph with actor co-affiliations and clustering information
R> multigraph(G20cn2, coord = crdG20, scope = scpG20tt, clu = acc)
```

```
# multilevel graph with a binomial projection and clustering information
# for actors and events
R> mlgraph(G20bpn, coord = crdG20, scope = scpG20tt, bwd = .25, hds = .5,
+    clu = list(ac, rep(1,nrow(G20))))
```

10.9.4   Positional System for the Algebraic Analysis

```
# clustering information from countries affiliations
R> club <- vector(); length(club) <- nrow(G20b)
# counter
R> k <- 1L
# take just the non-overlapping affiliations in 'G20b'
R> for(i in seq(3,5)) {
+    club[which(G20b[,i]==1)] <- k
+    k <- k + 1L
+    }
```

```
[1] 3 2 2 1 1 1 3 2 1 3 3 2 3 1
```

```
# positional system for actors with clustering information
R> PSG20bmh <- reduc(G20Bnet, valued = TRUE, lbs = c("G7.C","BRICS.C","MITKA.C"),
+    clu = club)

, , M

          G7.C BRICS.C MITKA.C
G7.C     19249   22865    3538
BRICS.C      0       0       0
MITKA.C    752    7572     412

, , H

          G7.C BRICS.C MITKA.C
G7.C      3050     296     293
BRICS.C  29577     584    1187
MITKA.C  13477     860       0
```

```
# positional system for events ('row' option for two-mode networks)
R> PSG20b <- reduc(G20b, valued = TRUE, lbs = c("G7.C","BRICS.C","MITKA.C"),
+    clu = club, row = TRUE)

         P5 G4 G7 BRICS MITKA
G7.C      3  2  5     0     0
BRICS.C   2  2  0     4     0
MITKA.C   0  0  0     0     5
```

```
# construct a multilevel positional system with symmetric co-domain
R> G20PSmhbpn <- mlvl(PSG20bmh, PSG20b, lbs = c("M","H","O"), symCdm = TRUE)
```

```
# scopes for multilevel
R> scpPSml <- list(cex = 6, pch = 21:22, ecol = c("#00BFFF","#CD5B45","#FF7F24"),
+    clu = c(1:3,rep(4,5)), pos = 0, fsize = 11, rot = 29, swp = TRUE)
R> scpPSml2 <- list(vcol = c("#BCEE68","#66CDAA","#A0A17B","#FF7F00"), vcol0 = 8)
```

```
# plot the binomial proyection of the multilevel positional system
R> mlgraph(G20PSmhbpn, valued = TRUE, layout = "force", seed = 1,
+    scope = c(scpPSml, scpPSml2))
```

10.9.5   *Relational Structure of Multilevel Configurations*

10.9.5.1   **Incorporate Affiliations as Diagonal Matrices**

```
# first read 'PSG20b' and convert the data frame into array
R> G20PSaa <- read.srt(PSG20b, attr = TRUE, rownames = TRUE, dichot = TRUE)
```

```
# then remove G4 from array
R> G20PSaa <- G20PSaa[,,c(1,3:5)]

# and rename P5 and MITKA
R> dimnames(G20PSaa)[[3]][1] <- "P"
R> dimnames(G20PSaa)[[3]][4] <- "A"
```

```
# bind trade relations with actors affiliations into a single array
R> G20BPSat <- zbind(PSG20bmh, G20PSaa)
```

```
, , P                                       , , BRICS

          G7.C BRICS.C MITKA.C                        G7.C BRICS.C MITKA.C
G7.C       1      0       0        G7.C       0       0       0
BRICS.C    0      1       0        BRICS.C    0       1       0
MITKA.C    0      0       0        MITKA.C    0       0       0


, , G7                                      , , A

          G7.C BRICS.C MITKA.C                        G7.C BRICS.C MITKA.C
G7.C       1      0       0        G7.C       0       0       0
BRICS.C    0      0       0        BRICS.C    0       0       0
MITKA.C    0      0       0        MITKA.C    0       0       1
```

10.9.5.2 Partially Ordered Semigroup

```
# semigroup with simplified strings
R> G20BPSatS <- semigroup(G20BPSat, "symbolic", smpl = TRUE)

# strings partial order and equations with dichotomous relations
R> G20BPSatP <- partial.order(strings(G20BPSat, valued = FALSE, smpl = TRUE))

# string relations
R> strings(G20BPSat, valued = FALSE, smpl = TRUE, equat = TRUE, k = 3)
```

```
# labeling for Cayley graph and inclusion lattice comes from the semigroup
R> lbsc <- lbsp <- G20BPSatS$st

# significant equations to show in the Cayley graph
R> lbsc[which(lbsc=="m")] <- "m = mh"; lbsc[which(lbsc=="PA")] <- "Bm"
R> lbsc[which(lbsc=="hm")] <- "hh = hm"; lbsc[which(lbsc=="Pm")] <- "Pm = Gm = Gh"

# significant equations to show in the inclusion lattice
R> lbsp[which(lbsp=="m")] <- "m,mh"; lbsp[which(lbsp=="PA")] <- "Bm"
R> lbsp[which(lbsp=="hm")] <- "hh,hm"; lbsp[which(lbsp=="Pm")] <- "Pm,Gm,Gh"
```

```
# plot Cayley graph
R> ccgraph(G20BPSatS, cex = 3, pos = 0, col = 8, lbs = lbsc, fsize = 7, rot = -55)

# plot inclusion lattice
R> diagram(G20BPSatP, lwd = 2, lbs = lbsp, fsize = 9)
```

10.9.5.3 Concept Lattice with MITKA, G4, and P5

```
# Galois connections in the positional system for events
R> PSG20bgc <- galois(PSG20b, labeling = "full")

# partial order in 'PSG20bgc'
R> PSG20bpo <- partial.order(PSG20bgc, type = "galois")

# plot lattice diagram
R> diagram(PSG20bpo, lwd = 2)
```

10.9.6   Two-class Multilevel Positional System

```
# extract G7 and BRICS countries for positional system
R> G7BRPSat <- zbind(PSG20bmh[1:2,1:2,], G20PSaa[1:2,1:2,c(1,2,4)])
```

```
, , m                                    , , P

         G7.C BRICS.C                              G7.C BRICS.C
G7.C    19249   22865                    G7.C         1       0
BRICS.C     0       0                    BRICS.C      0       1

, , h                                    , , G7

         G7.C BRICS.C                              G7.C BRICS.C
G7.C     3050     296                    G7.C         1       0
BRICS.C 29577     584                    BRICS.C      0       0

                                         , , BRICS

                                                   G7.C BRICS.C
                                         G7.C         0       0
                                         BRICS.C      0       1
```

10.9.6.1   Two-class and Attributes Partially Ordered Semigroup

```
# semigroup of role relations with simplified labels
R> Sml <- semigroup(G7BRPSat, type = "symbolic", smpl = TRUE)

# string equations with dichotomous relations
R> stml <- strings(G7BRPSat, smpl = TRUE, equat = TRUE, k = 3, valued = FALSE)

# partial order structure
R> Pml <- partial.order(stml)
```

```
# new labels of the two structures in the partially ordered semigroup
R> nlbs <- mlbs <- Sml$st

# significant equations are shown in Cayley graph and inclusion lattice
R> nlbs[which(nlbs=="GB")] <- "Bm";  mlbs[which(mlbs=="GB")] <- "Bm"
R> nlbs[which(nlbs=="m")] <- c("m = mh");  mlbs[which(mlbs=="m")] <- c("m,mh")
R> nlbs[which(nlbs=="h")] <- c("h = hm");  mlbs[which(mlbs=="h")] <- c("h,hm")
```

```
# plot semigroup as Cayley graph
R> ccgraph(Sml, cex = 3, pos = 0, col = 8, fsize = 8, bwd = .7, lbs = nlbs, rot = 25)

# plot inclusion lattice with customized lables
R> diagram(Pml, lwd = 2, fsize = 16, lbs = mlbs)
```

10.9.6.2   Concept Lattice without MITKA and P

```
# construct manually the formal context from the multilevel network
R> PSG20bg <- transf(c("G7.C, G7", "BRICS.C, BRICS", "G7.C, P", "BRICS.C, P"),
+    type = "toarray2")

          G7 BRICS P
G7.C       1     0 1
BRICS.C    0     1 1
```

```
# plot the Concept lattice diagrams with full and reduced formats
R> diagram(partial.order(galois(PSG20bg, labeling = "full"), type = "galois"))
R> diagram(partial.order(galois(PSG20bg, labeling = "reduced"), type = "galois"))
```

11

Comparing Relational Structures

When analyzing different types of network relations, either coming from the same set of actors measured at different points of time or else from structures of diverse population networks, it is possible to make a comparison of their relational structures. The comparison of such multiplex systems implies that the generating relations have the same relational content, or at least measured ties having a similar substantive meaning.

In this chapter, we are going to perform a comparison of three multiplex networks, which are Incubators where actors have ties with the same relational content. The arrangements to compare are the Incubator networks A and B we have studied so far in Chapters 5 and 6, plus another complex structure with the same nature as the other two, which is named ***Incubator network C*** or $\mathscr{X}_C$. The basis for comparing the structures of these multiplex networks lies in the algebraic constraints that are governing either their respective Role structures or else aggregated Role structures after a progressive factorization. That is, the restrictions will help us to look at common patterns occurring in the algebraic representation of the relational structures.

Typically, relational structures for multiplex networks are reduced configurations or Role structures with relationships among classes of relations coming from Positional systems. The different Positional systems are constructed using a structural correspondence type that is able to combine the different kinds of relations occurring between social actors.

As we have seen in Chapter 10, multiplex networks can integrate different types of relations between network members with their attributes. By combining different types of relations in the network reduction process, we seek not only to preserve the multiplicity of the ties but also to produce a positional system where correspondent actors carry out similar *types* of role relations. This principle applies in the positional analysis performed in Chapter 5 Role Structure in Multiplex Networks, where we used the notion of Compositional equivalence to multiplex networks to find related positions made of correspondent actors.

Establishing or "finding" the shared and divergent patterns among different role relations usually constitutes a demanding task, and is so even if the networks have the same relational content, such as Incubator networks $\mathscr{X}_A$, $\mathscr{X}_B$, and $\mathscr{X}_C$. Although the difficulty lies mostly in the interpretation of the resultant structures, another reason is due to the high level of abstraction that relational structures entail. For the interpretation in substantial terms of relational structures

Algebraic Analysis of Social Networks: Models, Methods and Applications using R,
First Edition. J. A. R. Ostoic. Companion website: www.wiley.com/go/ostoic/algebraicanalysis.

and the various kinds of logic of interlock, we count the different kinds of algebraic constraints that are governing these relational structures.

Hence, the resultant Image matrices express one type of algebraic constraint (cf. Table 3.2 of Chapter 3 Multiplex Network Configurations) that represent the network Role structure or else the aggregated Role structure after a progressive factorization. However, there are two more constrictions associated with these Role structures, namely the set of equations and the set of inclusions. In the structure of the Image matrices, equations and inclusions are among the string relations occurring at an aggregated level, and in the comparison process, the set of common and differing patterns involves all three kinds of algebraic constraints. For the substantial interpretation of the different multiplex networks, as said, we benefit from these constrictions.

11.1   Comparing Structures with Algebraic Constraints

Recall that algebraic constraints are structural restrictions on the relational or Role structures of the networks to compare. The significance of the algebraic constraints is that the "shared" or "common" patterns in each of these constrictions allows making a substantial interpretation of the role interlock in the aggregated relational systems of the compared networks.

In the case of Role structures $\mathcal{Q}_A$, $\mathcal{Q}_B$, and $\mathcal{Q}_C$, which correspond to the three Incubator networks to compare, there are these aggregated semigroup structures that reflect the three algebraic constraints. That is, apart from the aggregated relational structure itself, which involves sets of equations, all cases correspond to partially ordered configurations with hierarchies among representative strings. We are aware, though, that the different aggregation processes are the product of a decomposition procedure by means of a subdirect representation, and two of three aggregated Role structures yielded different overlapping Factor representations. It is only Role structure $\mathcal{Q}_B$ that has a unique non-trivial quotient semigroup representing the aggregated relational structure.

Since the different relational structures have "the same generating set", the results of comparison between these arrangements are straightforward in substantial terms both for the set of equations and for the hierarchy of the strings. However, as we know from the synthesis rules section in Chapter 6, any decomposition product of a subdirect representation may result in overlapping reduced configurations. We will see later on that the comparison of semigroup tables, which relates string relations or string role relations in the Incubator networks, does not always have a similar interpretation.

Table 11.1 provides the three different algebraic constraints in relational structures with conditions of "what constitutes a common structure among different systems". Each type of constraint depends and is a product of an associated operation among the elements involved in the type

Table 11.1   Algebraic constraints with criteria of "shared structure" among relational systems. *Equation and inclusion are operators among strings of Role structures.* CSS *stand for Common Structure Semigroup and* JNTHOM *for Joint Homomorphic reduction of Role structures.*

Constraint	Shared Structure	Operation
Equality	Common set of equations	$=$
Hierarchy	Common set of inclusions	$\leq$
(Role) Structure	JNTHOM or CSS between (role) tables	$\circ$

of restriction. All together, they characterize the relational interlock of the system when is represented by a partially ordered semigroup. For instance, the partial ordering of string relations produces a hierarchy, equality in the system is produced by the set of equations, and the composition of representative strings produces a relational or a Role structure, depending on whether the generators of the relational system are ties between actors or ties between categories of actors.

As we can see in Table 11.1, a common structure among relations is given by shared sets of equations and inclusions of strings in the aggregated relational structures. However, there are two criteria for the shared structure between Role structures, which are the Joint Homomorphic reduction or JNTHOM and the Common Structure Semigroup or CSS. Image matrices representing the Role structures are given in terms of semigroup tables, and we will look at the two options for network comparison where we try to assess some advantages and disadvantages of both of them.

However, before making the comparison of relational structures with the different algebraic constraint options, we briefly look at Incubator network B and Incubator network C. We are particularly interested in the positional analyses to produce their respective positional systems, and the construction and decomposition of their Role structures. For networks $\mathscr{X}_B$ and $\mathscr{X}_C$, the whole process is given in Appendix B and in Appendix E, where we can see that it follows the same logic used in the analysis of Incubator network A made in Chapters 4, 5 and 6, and it is from these reduced structures that the comparison takes place.

11.2   Incubator Networks B and C

As said before, it is less difficult to compare two relational structures when they have the same relational content, like the different Incubator networks, than networks of substantially different sets of generator relations. Figures 11.1 and 11.2 depict the multigraphs of Incubator network B, $\mathscr{X}_B$, and Incubator network C, $\mathscr{X}_C$, respectively, and these configurations correspond to the other two multiplex network structures with the same relational content as $\mathscr{X}_A$. That is to say that they are measured with relations C, F, and K that stand for formal collaboration, informal friendship, and perceived competition relations among the entrepreneurial firms, respectively. Networks $\mathscr{X}_B$ and $\mathscr{X}_C$ have the same actor attributes as well: relations A and B, which stand for the adoption of LinkedIn and Facebook, the two Web innovations in the year 2010.

As we clearly see in the multigraphs of Figures 11.1 and 11.2, Incubator network B is made of a single dense component and two isolated actors, whereas Incubator network C has three components and a similar number of isolated actors. It is also worth mentioning that relation K in network $\mathscr{X}_C$ is present in just a single component whereas the dyadic configuration is made of relation F only. The fact that some types of ties are not occurring in one or more components may or may not have significant implications in the relational and Role structure of multiple networks.

11.2.1   Positional Analysis of $\mathscr{X}_B$ and $\mathscr{X}_C$

The positional analysis for the other Incubator networks $\mathscr{X}_B$ and $\mathscr{X}_C$ has a similar process than the one made for $\mathscr{X}_A$ in Chapter 5 Role Structure in Multiplex Networks. That is, the class assignment of the actors is based on Compositional equivalence, which is applied with

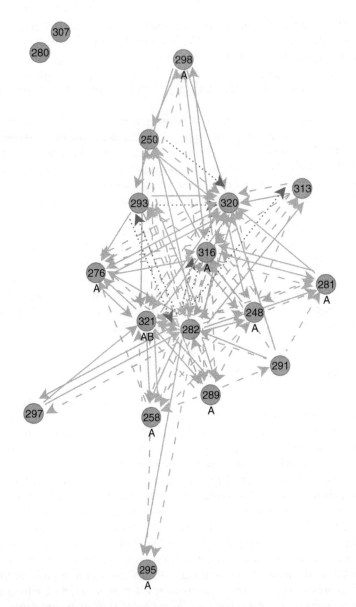

Figure 11.1 Multigraph of Incubator network B, $\mathscr{G}_B^+$. *Solid arcs are collaboration, dashed arcs are friendship, dotted arcs are perceived competition, and actor attributes are represented by* A *and* B, *which stand for the adoption of LinkedIn and Facebook in 2010.*

relational contrast to the directed social relations. The partition of the multiplex network is based on the partial order structure of the Cumulated Person Hierarchy product of the equivalence applied, and a decisive factor is choosing which are actors who belong to the same class in the network.

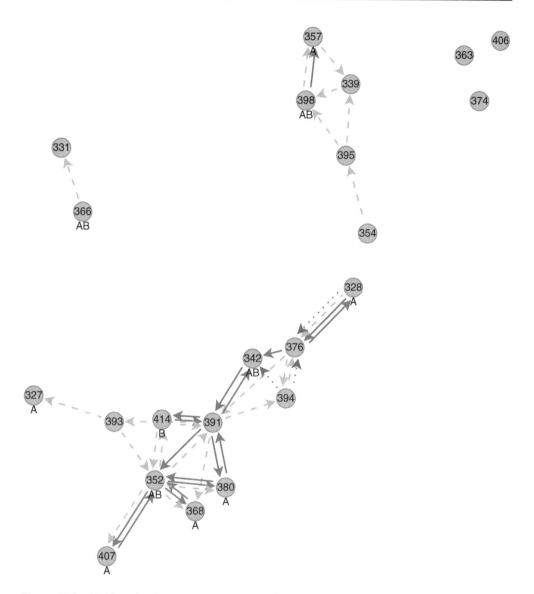

Figure 11.2   Multigraph of Incubator network C, $\mathcal{G}_C^+$. *As in Figure 11.1, solid arcs are collaboration relations, dashed arcs are friendship ties, dotted arcs for perceived competition, and letters represent innovations type adoption.*

Recall from Table 5.4 in Chapter 5 that in the case of Incubator network A the criterion for the network partition and the class assignment of the actors is based on a partitioned Cumulated Person Hierarchy table for identifying blocks of ties between and within the different "groups" of actors. Ideally, all within class blocks should be filled and the between class blocks should be empty in such table structure. However, inconsistencies occur almost always, and we try to find the best fit possible based on the information provided in $\mathscr{H}$.

In the next points we are going to comment briefly the positional analysis process for the other two Incubator networks, $\mathscr{X}_B$ and $\mathscr{X}_C$, which have the same relational content as $\mathscr{X}_A$. The reason to construct the Role structures of the two positional systems is that the comparison of the networks relational structures is performed on role relations between classes of actors, which in this case are structurally equivalent in terms of the Compositional equivalence criterion.

11.2.1.1 Incubator Network C

We start with Incubator network C, and Appendix B contains the different steps in the positional analysis for this system (see Ostoic, 2020, for the construction of the Role structure of this multiplex network). On the other hand, Appendix B provides the steps in the positional analysis of the other Incubator network, $\mathscr{X}_B$, which will be used in the comparison of relational structures.

Hence, if we look at Appendix B, we can see in the class assignment for Incubator network C that the positional system has three classes of actor rather than four as with $\mathscr{X}_A$ from Chapter 5 Role Structure in Multiplex Networks. However, like the positional system of $\mathscr{X}_A$, the actor attributes do not have a structuring effect in the Role structure of network $\mathscr{X}_C$, and this is because the classes representing both types of actor attributes yield the identity matrix.

The inclusion lattice of the Cumulated Person Hierarchy $\mathscr{H}_C$ product of Compositional equivalence to the actors in $\mathscr{X}_C$ (given in Ostoic, 2020) reveals significant information. Primary is that there is a set of incomparable elements in the lattice structure of $\mathscr{H}_C$, which is not made only of isolated actors but also with network members that are connected in the system at all levels such as actor 328 (cf. Figure 11.2). Hence, the set of incomparable elements in the lattice structure makes one class of actors, whereas the rest of the actors make two categories in the positional system according to their inclusion relations.

The Role structure for Incubator network C, which is represented by $\mathscr{Q}_C$, is given in Appendix B as well, and Figure B.3 provides the partially ordered semigroup as an inclusion lattice diagram. In this case, however, the Role structure is not given as a multiplication table, but rather as a Word and as a *right multiplication table*, which is another denomination of the Edge table. The Word and Edge tables are configurations that reflect the complete structure of the semigroup of relations, whereas the attached inclusion lattice serves to represent the partial order structure of the semigroup.

There is only a single generator relation without any structuring effect in the positional system for $\mathscr{X}_C$ that corresponds to the actor attributes, and this is because they equal to the identity relation; i.e. A = B = I. There is also a couple of significant equations in the Role structure $\mathscr{Q}_C$ that are given at the bottom of Figure B.3, and such equations correspond to some compound role relations of length 2. There exist also equations with larger compounds, but in this case (and with most other cases in multiplex networks) larger compounds do not provide very useful information for a substantial interpretation of the network Role structure.

11.2.1.2 Incubator Network B

With respect to Incubator network B, the positional analysis is given in Appendix B, and the decomposition of the Role structure for this Incubator network B has been given already given in Chapter 6 Decomposition of Role Structures. Although this positional system has four classes of actors as with $\mathscr{X}_A$, there is one kind of actor attribute that has structuring effect in the Role structure $\mathscr{X}_B$, and this is because string relation B is represented by a single entry in the diagonal matrix with the other cells empty. This means that not all classes of actors in this multiplex structure adopted innovation B at that point of time, and hence network $\mathscr{X}_B$ differs from the other two Incubator networks in the sense that it is the only configuration where actor attributes information has a structural effect in the relational structure.

Figure B.1 in Appendix B provides the multiplication table of the aggregated maximal non-trivial homomorphic image of $\mathscr{Q}_B$, and we evidence that that the identity element in this Role structure in this case corresponds to compound KB. A picture of this non-trivial aggregated Role structure that corresponds to network $\mathscr{X}_B$ was given in Figure 6.8 (Chapter 6 Decomposition of Role Structures) with both the inclusion lattice with representative string role relations, and the Cayley graph of the reduced role table. Even though $\mathscr{X}_B$ is the densest network of the three Incubators in terms of ties, the Role structure $\mathscr{Q}_B$ at an aggregated level resulted, having an unambiguous and symmetric character.

Based on the information given in Chapters 5, 6, Appendix A, and Appendix B, we can state that two of the three positional systems for the Incubator networks are made of four classes of actors while the reduced structure for Incubator network C is made of only three positions. As said before, there is only one positional system, which corresponds to Incubator network B, that has an actor attribute (B) with a structuring effect in the Role structure, while in the case of relation A there is not an structuring effect in any of the three positional systems found.

As a result, the comparison of relational structures is going to be made in the three Incubator networks $\mathscr{X}_A$, $\mathscr{X}_B$, and $\mathscr{X}_C$ having a similar relational content.

11.3   Equality

Equality is one significant algebraic constraint in the relational or role structure of a multiplex network that is represented by the *set of equations* among the elements of the semigroup object. Recall that an equation between two or more string relations occurs when they connects precisely the same individuals in the network. For different networks with a similar relational content, the "common" set of equalities among the string relations in their relational structures constitutes a "shared" configuration.

11.3.1   Set of Equations in Incubator Role Structures

Since there has been a decomposition process of the different role structures of the three Incubator networks $\mathscr{Q}_A$, $\mathscr{Q}_B$, and $\mathscr{Q}_C$ by factorization, the set of equations includes first the different equations to the representative strings in the semigroup of relations, and then it becomes gradually larger as the decomposition of the relational structure progress. This is because with the factorization of the Role structures the distinct partially ordered structures of the Factors become smaller and therefore closer to the universal element in the Lattice of Congruence relations where all string relations become equal.

This is the same as there is a single set of equations with all the string of role relations in the universal structure where all the relations are equated. However, since a 1-element semigroup lacks structure, we try avoiding this situation in the analysis and we stop with the minimal non-trivial homomorphic images where there are at least two sets of equations, even if one of the sets is with a single element from the semigroup of relations.

The analysis with the set of equations is first and foremost restricted to the measured types of ties including actor attributes in Role structures $\mathcal{Q}_A$, $\mathcal{Q}_B$, and $\mathcal{Q}_C$. However, we bear in mind that we added the converse of the measured relations to achieve relational contrast in the network relational structure. This means that the converse of relations constitute generators of the semigroup, and they are part of a different compound, of the semigroup structure, and part of the set of equations as well.

11.3.1.1   Set of Equations in $\mathcal{Q}_A$

We start by reviewing the set of equations in the Role structure of Incubator network A, $\mathcal{Q}_A$, and we recall from Chapter 6 Decomposition of Role Structures that the factorization process for this system brought four maximal homomorphic images. These homomorphic reductions correspond to the decomposition of three Factors, which are given in Figure 6.7, and where Factor 1 has two representation forms.

The set of equations among the different representative strings in the Factors for $\mathcal{Q}_A$ is where the interpretation of the role interlock for this system takes place, and the equations among generators are:

$$F_1\mathcal{Q}_A: \quad \{C = F = K\}, \{A = L\}, \{D\}, \{G\}$$

$$F_2\mathcal{Q}_A: \quad \{C\}, \{F\}, \{K = A\}, \{D = G = L\}$$

$$F_3\mathcal{Q}_A: \quad \{C = F = A = D = G\}, \{K = L\}$$

We evidence with the three equations that the interpretation of the role interlock with this algebraic constraint is not univocal, and there is an overlapping of different logics in Role structure $\mathcal{Q}_A$. For instance, only Factor 2 separates the three measured social relations, and the string representing both actor attributes is equated to perceived competition. In Factor 1, however, all measured social relations are equal, and there is any equation between perceived competition and actor attributes in F_1 and F_3.

Hence, we can only say in substantial terms that "part" of the logic in the interlock of Role structure $\mathcal{Q}_A$ is

"Classes of actors are equally related both with formal and informal role relations."

The additional equations that are the product of the progressive factorization process are given below, where parentheses serve to indicate the additional equations in the steps product of a progressive factorization.

$$F_1^1\mathcal{Q}_A: \quad (C = (D = G)), A$$

$$F_1^2\mathcal{Q}_A: \quad (D = G), (C = A)$$

$$F_2\mathcal{Q}_A: \quad ((C = F) = K), D$$

$$F_3\mathcal{Q}_A: \quad —$$

Since a further factorization of the last Factor produces a trivial representation of the Role structure, there is no need to have additional equations in F_3. However, there are additional equations in the other two Factors, and while the tie converses are equated with Factor 1, C gets separated from the actor attribute relation in one representation, whereas these two types of tie are equated in the other representation of this factor. Hence, there is not a clear separation of ties in Factor 1 and the interpretation is not straightforward. Factor 2, on the other hand, equates C and F, and differentiates these ties from K or perceived competition, which is equated with relation D. An extra round in the factorization process of Factor 2, however, equates the three measured social ties.

As a result, the progressive factorization implies different ways to model the role table for $\mathscr{Q}_A$, and at this level of aggregation there are several ways to interpret the logic of role interlock in the Role structure that corresponds to Incubator network A that we need to account for.

11.3.1.2   Set of Equations in $\mathscr{Q}_B$

The decomposition process of Incubator network B is a special one, and the outcomes from the factorization process differ from the previous case. First, there are in total 38 maximal homomorphic images with an order of 11 that correspond to the decomposition of six factors of the Role structure with a progressive factorization that is given in Appendix E. The important aspect in this case is that all of these maximal homomorphic images are isomorphic with each other, which means that there is a univocal interpretation of the role interlock $\mathscr{Q}_B$.

The partially ordered semigroup structure of the non-trivial aggregated Role structure that corresponds to $\mathscr{Q}_B$ was given graphically in the two plots of Figure 6.8 in Chapter 6. Because all factors are isomorphic with each other, there is no reason to look at the particular equations among primitives in each factor.

Instead, we give the additional equations among representative string generators in the factors that are product of the progressive factorization of this Role structure that can be summarized as:

$$F_{1-6}\mathscr{Q}_B: \qquad (C = F = G),\ (K = L); \qquad (A \neq B)$$

As a result, the two types of actor attributes are not equal in the Role structure for this Incubator network. Besides, the formal and informal role relations are not contained in each other, which means in substantial terms that at an aggregated level in Role structure $\mathscr{Q}_B$

"At that time, some classes of actors adopted just one innovation type but not both."

and this is presumably mostly A.

With respect to the other types of social role relations in $\mathscr{Q}_B$

"Classes of actors in general cooperate and cultivate friendship relations at the same time."

11.3.1.3   Set of Equations in $\mathscr{Q}_C$

In the case of the third Role structure that corresponds to Incubator network C, the factorization of the semigroup brought two factors that are given in Figure B.2 in Appendix B. With Role structure $\mathcal{Q}_C$, the two factor representations have the same order and the partial order structures are isomorphic when they are unlabeled.

Each Factor in $\mathcal{Q}_C$ has a single equation among its generators, and the classes of role relations in these representations are

$$F_1\mathcal{Q}_C: \qquad \{C = G\}, \{F\}, \{K\}, \{A\}, \{L\}$$

$$F_2\mathcal{Q}_C: \qquad \{C = F\}, \{K\}, \{A\}, \{G\}, \{L\}$$

The progressive factorization of these two factors of $\mathcal{Q}_C$ is very similar. There is a two-step reduction (first with a single atom) that produces two maximal homomorphic images for each factor where one image is the 1-element semigroup where all strings are equated, and the other structure is a 3-element aggregated role table. The only aspect that differs corresponds to the content of a pair of equations.

The two non-trivial maximal homomorphic tables with the equations is given at the bottom of the Figure B.2 in Appendix B, and we can see that the relational content of the generators involved in the equations is similar as well. That is, formal collaboration equals with either perceived competition or the transpose of this type of tie, and informal friendship or its transpose is equated to the actor attributes.

After the progressive factorization procedure, the additional equations among generator strings for the Role structure of $\mathcal{S}_C$ are:

$$F_1^1\mathcal{Q}_C: \qquad (C = L), (A = F), K$$

$$F_1^2\mathcal{Q}_C: \qquad ((C = F), (K = L), A)$$

$$F_2^1\mathcal{Q}_C: \qquad (C = K), (A = G), L$$

$$F_2^2\mathcal{Q}_C: \qquad ((C = G), (K = L), A)$$

where for the second factor we had the equation $\{C = F\}$.

The implications of the role interlock in substantial terms mean that at an aggregated level of the Role structure for the positional system of Incubator network C

> *"Classes of actors that cooperate with each other cultivate friendship relations or are perceived as competitors."*

However, there is ambiguity in the interpretation of actor attributes in this case, and with respect to perceived competition as well.

11.4   Hierarchy of Relations

Another type of algebraic constraint that complement the set of equations is the *hierarchy of relations* with the sets of inclusions in the strings of relations. The hierarchy in the relations—or role relations in the case of the Incubator networks—is an outcome from the partial ordering of string role relations occurring in the systems, and it is derived from the decomposition of their Role structures. This means that the hierarchy of relations is a result from the culmination of an aggregation process, and includes the representative strings of the maximal homomorphic images after the aggregation.

In the case of the three Incubator networks, the hierarchies in the relations serve for the comparison of the network Role structures $\mathscr{Q}_{A-B-C}$ that correspond to the resultant positional systems $\mathscr{S}_{A-B-C}$. Recall that the factorization of the Role structures of the three Incubator networks were made through a subdirect representation, and in two of the three cases, there is some ambiguity that arises in the inclusion order of the strings.

For example, after the aggregation of the Role structure $\mathscr{Q}_A$ (cf. Figure 6.6 in Chapter 6), it is no clear whether relation K (i.e. perceived competition) is included in relation A (i.e. actor attribute) or the other way around. Certainly, this can be resolved by referring to the partial order table of this Role structure, which is given both in this figure and also in Table 6.3. Here, apparently, A is in the upper bound of relation K, but none of these ties are covering other relations in the partial order. Hence, the two possibilities, i.e. $K \leq A$ and $A \leq K$, are worth candidates for the assessment of shared patterns.

In the other two Role structures of the Incubator networks, $\mathscr{Q}_B$ and $\mathscr{Q}_C$, there is no contradiction in the order inclusion of the measured role relations as with $\mathscr{Q}_A$. There is just one non-trivial role interlock for network B and therefore the interpretation of the role interlock has a more conclusive character. In the case of Incubator network C, $\mathscr{Q}_C$ (cf. Appendix B), there exists some ambiguity among generators in the two factors, but no contradiction.

Having ambiguity and contradiction has been the cost of a factorization procedure by means of a subdirect representation, and this implies that factors have different logics overlapping with each other. However, the main benefit of the factorization is counting with smaller constituents of the semigroup structure where it is possible to make a substantial interpretation of the network's relational interlock.

11.4.1   Set of Inclusions in Incubator Networks

The decomposition of the Role structures for $\mathscr{Q}_A$, performed in Chapter 6, and for $\mathscr{Q}_C$ and $\mathscr{Q}_B$ in Appendix B, produce a different hierarchy in the structure of role relations for each Incubator network. If we account for the measured primitive relations, the distinct hierarchies are expressed by the following set of inclusions:

$$\mathscr{Q}_A: \quad (K \leq (A \text{ or } A \leq K) \leq C \leq F$$
$$\mathscr{Q}_B: \quad (K \leq B \text{ or } B \leq K) \leq A \leq (C = F)$$
$$\mathscr{Q}_C: \quad K \leq A \leq F \leq C$$

where in the case of Role structures $\mathscr{Q}_A$ and $\mathscr{Q}_C$ A equals B.

The set of hierarchies given above implies that the possible inclusions among pairs of strings that are common to the Role structures for the Incubator networks are:

$$\mathscr{Q}_A, \mathscr{Q}_B, \mathscr{Q}_C: \quad K \leq C; \quad K \leq F; \quad A \leq C \text{ and } A \leq F$$

On the other hand, the particular inclusions in the Role structures not shared among Incubator networks are the following:

$$\mathscr{Q}_A: \quad C \leq F$$
$$\mathscr{Q}_B: \quad B \leq A$$
$$\mathscr{Q}_C: \quad F \leq C$$

Such ordering of the network-measured relations is then an algebraic constraint that allows us to make a straightforward interpretation of the role interlock of each Incubator network where the comparison takes place in substantial terms.

11.5   Shared Structure by Role Tables

A third algebraic constraint that is useful for the comparison of multiplex networks corresponds to the semigroup structures in the form of multiplication tables. Since we are dealing with reduced Role structures, we refer to these configurations rather as *role tables*, which in many cases are *aggregated* role tables. Role tables that are a product of a factorization process like $\mathcal{Q}_{A-B-C}$ imply that the aggregated structures are partially ordered. This means that apart from the set of equations, the algebraic constraint of the hierarchy of role relations summarizes the outcome.

Before making the comparison of the Role structures through the composition of the representative strings that corresponds to the three Incubator networks, however, we start by looking at critical theoretical concepts that are useful for the comparison of semigroup and partial order structures. After that, we are going to see that there are competing candidates in establishing what is "shared" between partially ordered semigroups for the comparison process.

For instance, consider first two semigroups of relations; say $S(R)$ and $T(R)$, where the last structure is the homomorphic image of the first one. We denote this fact by using the partial order relation as $T \leq S$; however, we need to apply the concept of *generator-preserving homomorphism* from one semigroup onto another, which supposes that there is a one-to-one correspondence between generators. If $\mathcal{R}$ is the common generator set of S and T, the generator-preserving homomorphism acts as an identity mapping on generators where $h(R) = R$; for all $R \in \mathcal{R}$, which means that two generators of S cannot be mapped into the same element of T (Boorman and White, 1976, p. 1419; cf. also Bonacich (1980)).

Certainly, for semigroups that represent role structures there may be some generators, which stand for more than one primitive relation from the original relational structure. This is since these primitive string relations are equated during the decomposition and aggregation process.

11.5.1   Lattice of Homomorphisms of the Semigroup

Generator-preserving homomorphism is one of the premises behind the comparison of different algebraic systems, and there are other related concepts that are significant as well for the comparison by relational structures. Recall that the free semigroup generated by the set of relations $\mathcal{R}$ consists of all possible finite strings of letters from $\mathcal{R}_i \in \Sigma$ with the operation of the concatenation of strings. Since the set of semigroups generated by $\mathcal{G}^+$ may be endowed with a lattice ordering, the set of all partitions of the semigroup of relations forms a complete lattice (Birkhoff, 1967). In this sense, provided that two semigroups have the same generator set, they may be interpreted as a partition of the corresponding free semigroup of relations, which is part of the Lattice of Homomorphisms of the Semigroup $L(S)$.

The Lattice of Homomorphisms of the Semigroup is dual to the Lattice of Congruence relations $L_\pi(S)$ for partially ordered structures, and both the Lattice of Homomorphisms of the Semigroup and the Lattice of Congruence relations are essential to address the comparison of two Role structures. The crucial aspect for that is the establishment of what is "shared" or "common" between the various patterns present in their respective role tables. Considering that two Role structures are located in the same lattice structure, the semigroup structures to be compared usually lie between $F(S)$ and the 1-element semigroup. Indeed, if both structures resemble the semigroup with a single element, it means that they share all their patterns, and each of their strings becomes equal. On the other hand, if they resemble the free semigroup, then we obtain the opposite result where the string relations maintain a unique configuration.

As a result of this, there are two candidates for the characterization of "shared patterns" between semigroups S and T: One is the supremum or the least upper bound of the join of the two structures in the lattice structure of the semigroup, which means that the reduction is made with $\sup\{S, T\}$. On the other hand, the other candidate is the infimum or greatest lower bound of the meet of these structures, or the reduction of the two semigroup structures is instead made with $\inf\{S, T\}$. These are competing possibilities for the comparison of relational or Role structures, and that is why we take a closer look at both options next and apply them to the comparison of the role tables that correspond to the three Incubator networks.

The Lattice of Homomorphisms of the Semigroup is crucial to understand the comparison of relational structures, since it will help us in the decision of what constitutes a "shared" structure among different relational systems, even though such an arrangement should typically be of immense size. Figure 11.3 has an illustration that shows the Lattice of Homomorphisms

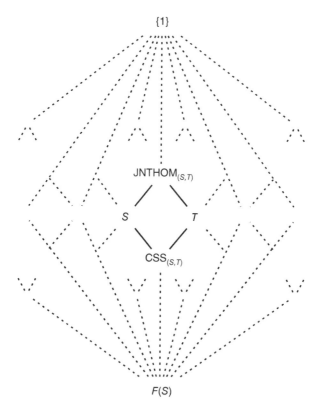

Figure 11.3   Illustration of the Lattice of Homomorphisms of the Semigroup, $L(S)$, with the JNTHOM and CSS of structures S and T.

of the Semigroup $L(S)$ with two notions of shared structure between two semigroup of relations.

The JNTHOM and CSS of two semigroup structures are concepts that are explained later. For now, it is essential to mention that the maximum element in the Lattice of Homomorphisms of the Semigroup is the free semigroup $F(S)$ that encompasses all combinations of the semigroup generators. In contrast, the minimum element in the lattice structure corresponds to the 1-element semigroup with the universal relation where all strings are equal. As a result, the Common Structure Semigroup for semigroups S and T is closer to $F(S)$ while the JNTHOM of these two structures is more distant to the free semigroup and closer to the 1-element semigroup.

11.5.2   Joint Homomorphic Reduction, JNTHOM

In their original approach to comparing two Role structures Boorman and White (1976) proposed the *joint reduction* as the "intersection" of the role tables and the one that "is the largest table consistent with both original role tables" (p. 1406). If we consider the Lattice of Homomorphisms of the Semigroup having the universal semigroup as the supremum and free semigroup of relations as its infimum, the *Joint Homomorphic reduction* or JNTHOM of two semigroup structures S and T is the lattice *union* or the join of these structures; i.e. $S \vee T$ (Birkhoff, 1967; Boorman and White, 1976) (cf. also Bonacich and McConaghy, 1980).

This means that the joint reduction $S \vee T$ is a homomorphic simplification of both semigroup structures, and relations that are congruent in either S or T must be congruent in their joint reduction. In this sense, the joint reduction of two structures "will be smaller [data algebra]" (Boorman and Arabie, 1980, p. 168) than the constituent structures, and this implies that the resultant Joint Homomorphic reduction configuration is coarser or a least refined structure than the two constitutive semigroups.

When considering role tables as the resultant structures for the Incubator networks role systems, for example, the Joint Homomorphic reduction implies that all the equations holding true in *either* of the original Role structures must hold true in their joint reduction. This implies that some elements not equal in one of the semigroup structures, or even in either of two homomorphic images, can be equated in their joint product. This is why one of the reasons why there have been critical voices regarding the joint reduction technique (e.g. Bonacich and McConaghy, 1980), and they proposed an alternative way to establish what is common between two semigroup structures by increasing the conditions imposed to the equations for a shared structure. Hence, before considering the analysis of the Incubator networks Role structures with JNTHOM, we will first take a brief look at the alternative method to joint reduction.

11.5.3   Common Structure Semigroup, CSS

With the JNTHOM, the emphasis is on the reduced image matrices and on the congruence classes of the semigroups that are subject to comparison, where the shared structure in this case results from a simultaneous reduction of the two images. However, in the task of comparing semigroup structures, Bonacich and McConaghy (1980) proposed an alternative to Joint Homomorphic reduction with a notion they called the *Common Structure Semigroup* or CSS (cf. also Bonacich, 1980).

In the Common Structure Semigroup procedure, an equation holds in the shared structure if and only if it holds true in *both* original structures. Consequently, we obtain the smallest configuration that is homomorphism of the two semigroups in such terms, which means that there are any strings of relations of semigroup tables that are forced to be equal with the CSS. This form for homomorphism then produces the most refined structure containing the largest number of distinct elements in the two semigroups. As a result, with the CSS procedure the focus is more on the equations and on the congruence classes in the semigroups rather than in the image tables alone as with JNTHOM.

In terms of the Lattice of Homomorphisms of the Semigroup, the Common Structure Semi-group considers the lattice *intersection* of the semigroups, rather than taking the union of two structures as with the JNTHOM. This means that for semigroups S and T, the common structure or "meet semigroup" is the greatest lower bound of the meet of the semigroups or else $S \wedge T$.

A method to produce the Common Structure Semigroup of two systems is by considering the disjoint union of the matrices representing the relations of these systems (cf. Pattison, 1981; McConaghy, 1981*b*). Hence, for two Role structures S_1 and S_2, the meet semigroup or CSS is represented as:

$$CSS_{S_1 - S_2} = \begin{pmatrix} A(RS_1) & 0 \\ 0 & A(RS_2) \end{pmatrix}$$

In this way, the CSS does not force any equation in its structure, since the assumption with this formula is that the involved systems do not interact for the Common Structure Semigroup (cf. Wu, 1984). Hence, for a total of t relational systems $S_1, S_2, \ldots S_t$, the Common Structure Semigroup will be:

$$CSS_{S_t} = \begin{pmatrix} A(RS_1) & \cdots & 0 \\ \vdots & \ddots & \vdots \\ 0 & \cdots & A(RS_t) \end{pmatrix}.$$

That is, a diagonal meta-array with an adjacency matrix for each type of relation in a collection of systems.

11.5.4   What Constitutes a "Shared" Structure?

Regarding the controversy generated (Bonacich, 1980; Boorman and Arabie, 1980, also McConaghy (1981*b,a*); Pattison (1981)) over which of the two procedures, the Joint Homo-morphic reduction or the Common Structure Semigroup, is the most appropriate way to represent a simplified version of two Role structures, we are going to mention two significant aspects.

1. One concern has to do with Bonacich (1980, p. 161) when he cites Birkhoff (1967) and points out that even if all homomorphisms between two semigroups produce a lattice structure, the generation-preserving homomorphisms need not to be a complete lattice, but form only a sublattice. This may be the case, since while each semigroup with generator R_i corresponds to a partition of the lattice of the free semigroup of relations, not every partition of the lattice

of homomorphisms corresponds to a semigroup, which means that a generator-preserving simplification of each of the two given semigroups may not exist. Regarding this point, Boorman and Arabie (1980) assert that the lack of definition in the joint reduction, which results from its failure to be generator-preserving homomorphism, means that the algebras being compared are in fact very different, and that is an informative substantive outcome (p. 173).

2. The other concern deals with the ideal situation when considering different semigroup of relations, in which is preferable to count with a model that contains a string of ties that are shared or are in common between the two semigroup structures. Pattison (1981) argues that part of the dispute lies in the function of the types of features of which one considers a semigroup to consist; while the Joint Homomorphic reduction regards the homomorphic images as the descriptive characteristics of the semigroup, the Common Structure Semigroup takes focus on the set of equations. In consequence, the selection is a matter of interpretation and a substantive point of view is needed to consider which of the strategies is more useful.

11.6   Semigroup Tables with Joint Homomorphic Reduction

Following the presentation of JNTHOM and CSS, and the discussion about what is a shared configuration between two relational structures according to these perspectives, we now start the comparison of the Role structures of Incubator networks $\mathscr{X}_A$, $\mathscr{X}_B$, and $\mathscr{X}_C$. Both with JNTHOM or with CSS, the resultant homomorphic images by factorization of the Role structures to compare are isotone to role semigroup tables, and this is because their factor representations are partially ordered structures.

As an illustration of the JNTHOM, we start by looking at the shared patterns inside the particular Role structures of the Incubator networks, and then we perform the comparison across the Role structures with CSS. The analysis with JNTHOM, however, will only take place with the set of maximal homomorphic images for $\mathscr{Q}_A$ and $\mathscr{Q}_C$, and this is because all aggregated images in $\mathscr{Q}_B$ are isomorphic with each other and there is a single decomposed structure.

11.6.1   JNTHOM of Aggregated Role tables $\mathscr{Q}_A$

If we consider the Role structure of Incubator network A, $\mathscr{Q}_A$, the factorization of this system produced in the first place three maximal homomorphic images, which are given in Table 6.3 in Chapter 6. However, as we can see in these decomposed role tables, two factors have image matrices that can be reduced further with a progressive factorization, and in the same Chapter Figure 6.7 gives the three maximal non-trivial homomorphic images with order 2, which is the same order as the third Factor.

The four maximal non-trivial homomorphic Image matrices of $\mathscr{Q}_A$ are given in Table 11.2 with the equations among primitive relations. We can see that in all cases C equals F, which means that for this system there is a differentiation between these strings with relation K and the two actor attributes. On the other hand, the second homomorphic Image matrix of the first factor can be considered as a trivial representation where all primitives are equated. Hence, we omit from the analysis the second aggregated matrix from Factor 1, F_1^2, and this is because all measured relations are equated in this configuration. The other string corresponds to a transpose of a primitive role relation and hence the structure is not clearly defined at this level of aggregation.

Table 11.2 Maximal non-trivial homomorphic Image matrices of $\mathscr{Q}_A$. *Equations with transpose relations are omitted where Factor F_1^2 constitutes a trivial representation where all measured relations collapse.*

$\circ$	C	A
C	C	C
A	C	A

$\circ$	D	C
D	D	D
C	D	C

$\circ$	C	K
C	C	C
K	C	K

$\circ$	K	C
K	K	K
C	K	C

$C = F = K$ $\quad\quad$ $C = F = K = A$ $\quad\quad$ $C = F$ and $K = A$ $\quad\quad$ $C = F = A$

F_1^1 $\quad\quad\quad\quad$ F_1^2 $\quad\quad\quad\quad\quad$ F_2 $\quad\quad\quad\quad\quad$ F_3

One significant result of this aggregation process is that we are able to obtain a single arrangement allows a straightforward comparison of different logics in the Role structure to be made. This reduced structure scorrespond to a core-periphery pattern, as we recall from Chapter 3 Multiplex Network Configurations. In the context of relational structures, the Strength of Weak Ties theory implies that strong ties are at the core while weak ties are in the periphery, and this arrangement can guide in the interpretation of the different role interlocks in the system.

Although it is clear that relations C and F are equated in all aggregated structures, in two of the three image matrices these role relations constitute a strong type of tie, while in another configuration they are a weak type of role relation. Relation K also constitutes a strong type of role relation in two of the three images but most of the time perceived competition is in opposition to the formal and informal roles of friendship and collaboration among classes of actors. A different situation occurs with the actor attributes. Since this type of relation always acts as the identity element in the semigroup structure, A constitutes a weak type of role relation in the logic of interlock for $\mathscr{Q}_A$.

The class assignment of the distinct measured relations in the three factor representations of $\mathscr{Q}_A$ are

$$F_1^1 \mathscr{Q}_A: \quad [1] = C, F, K \quad\quad [2] = A$$
$$F_2 \mathscr{Q}_A: \quad [1] = C, F \quad\quad [2] = K, A$$
$$F_3 \mathscr{Q}_A: \quad [1] = C, F, A \quad\quad [2] = K$$

The joint table has brought significant information for the analysis of the shared features in the role structure that corresponds to this Incubator network, and the outcome in this case coincides with the logic expressed in the hierarchy of relations. In this sense, the algebraic constraints allow us to gain an insight into the relational structure of the network and this is made by considering the different aspects of the role interlock inherent in the relational system.

11.6.2 JNTHOM of Aggregated Role tables $\mathscr{Q}_B$ and $\mathscr{Q}_C$

Because there is a single representation of the Role structure that corresponds to Incubator network B, there is no need to look at the shared characteristics in the factor representation with JNTHOM. This is because, after many additional steps in the factorization of this partially ordered structure, it is evident that there is a lack of additional differentiations in the classes of relations. The maximal homomorphic image of this aggregated role structure results in there being the 1-element semigroup that categorizes all string relations together into a single class, and this trivial representation with no structure constitutes at the end the JNTHOM of all relational systems in the Lattice of Homomorphisms of the Semigroup $L(S)$.

For Incubator network C, on the other hand, the factorization of the Role structure $\mathcal{Q}_C$ initially produced two factors with an equal order of 7, which are given as partially ordered semigroups at the top and the middle of Figure B.2 in Appendix B. From these factors, the progressive factorization of the factors in $\mathcal{Q}_C$ produces four maximal homomorphic images, all having an order of 3, and which are given at the bottom of the picture in the Appendix together with the additional equations of the primitive measured relations. A further decomposition of these partially ordered images matrices would produce the 1-element semigroup through the factorization procedure, and therefore these structures could be regarded as the smallest non-trivial representations of $\mathcal{Q}_C$.

However, even if all representations of $\mathcal{Q}_C$ are small, the four homomorphic images have congruence classes of elements, which means that the semigroup structure can be further reduced through congruence relations. Below is given the class assignment of elements in the four homomorphic images product of congruence relations, and to the right are the different equations produced with the progressive factorization procedure.

$$F_1^1\mathcal{Q}_C: \quad [1] = C, K \qquad [2] = A \qquad\qquad (C = F)$$

$$F_1^2\mathcal{Q}_C: \quad [1] = C, K \qquad [2] = F \qquad\qquad (F = A)$$

$$F_2^1\mathcal{Q}_C: \quad [1] = C, L \qquad [2] = A \qquad\qquad ((C = F) = K)$$

$$F_2^2\mathcal{Q}_C: \quad [1] = C, K \qquad [2] = A \qquad\qquad (C = F)$$

11.6.3   Joint Table for Incubator Networks

Recall that the maximal homomorphic image of $\mathcal{Q}_B$, which is product of the factorization procedure, constitutes both the joint homomorphic reduction and the common structure semigroup of this role table. Hence, the JNTHOM structure shared by both $\mathcal{Q}_A$, $\mathcal{Q}_B$, and $\mathcal{Q}_C$ is made of two classes of role relations that follow a core-periphery pattern. Recall that this table relates an identity element with an absorbing element, but the representative ties that correspond to the classes in the joint table structure are different.

As a result, at an aggregated level in the three Incubator networks formal collaboration and informal friendship constitute strong types of tie, whereas perceived competition occurs mostly in $\mathcal{Q}_C$. On the other hand, actor attributes do not have a central structuring effect in any of these networks when it comes to the joint table representation.

Table 11.3 provides a prototype of a *joint table* with classes [1] and [2] that stand, respectively, for the zero and the identity element of a semigroup structure, and this is a single arrangement with a core-periphery pattern that typically results from an aggregation process. We obtain a joint table with only two classes of ties also by applying congruence relations to the homomorphic

Table 11.3 Joint table with two classes of elements. *This configuration represents among other things a core-periphery pattern or the relationships between strong and weak ties.*

∘	[1]	[2]
[1]	[1]	[1]
[2]	[1]	[2]

images of the Role structures of all three Incubator networks where one has a neutral character and the other element is an absorbing constituent.

It is clear from the equations and the configuration of the joint reduction table that at an aggregated level of the network relational system, string relations C and K that correspond to cooperation and competition ties are determinant in the establishment of the Role structure, and the actor attributes have almost no influence in the role interlock.

If we consider the informal friendship role relations represented by F, we have to emphasize that this type of tie is equated with the actor attributes in one image representation, and with the rest of social relations in the other image factors. In substantial terms, this means that the character of the friendship informal ties is not so determinant as the cooperation and the competition ties in the constitution of the Role structure for this network, but surely more significant than innovation adoption for this Incubator network data.

11.7 Comparison Across Networks with Common Structure Semigroup

Besides the JNTHOM approach, another strategy to find a shared structure across networks with a similar relational content is the Common Structure Semigroup or CSS. With CSS, the common structure of two relational systems is expected to be larger and more complex than with the JNTHOM. As stated in its definition, the CSS takes the lattice intersection of two semigroups in the Lattice of Homomorphisms of the Semigroup $L(S)$, and one way to produce the CSS of two semigroups is by constructing a meta-array with the semigroup generators of both systems.

While the joint reduction of the factorized Role structures from the Incubator networks we saw in the last section is a product of the resultant decomposition tables, the Common Structure Semigroup is a direct outcome from the generator relations in the structures to compare. In this case, we take the aggregated roles from the different systems and produce the disjoint union of the positional systems as defined in Equation 11.1, and the common patterns in the systems to compare are then recorded in the semigroup structure of such disjoint union configuration. Wu (1984) has successfully applied this method in his analysis of the relations among the elite in two different communities, which data was gathered by Laumann and Pappi (1976), and it complements the algebraic analysis made before by Breiger and Pattison (1978) on this particular network.

It is possible to reduce the difficulty in the computation of the Common Structure Semigroup for the Role structures of Incubator networks $\mathcal{Q}_{A-B-C}$, either by reducing the number of comparisons or by considering only the positional systems of measured relations in the meta-array structure of the CSS; that is, by dropping transpose relations and actor attributes.

The four alternatives for comparing different relational structures of the Incubator networks are:

1. CSS $_{\mathcal{Q}_A-\mathcal{Q}_B}$
2. CSS $_{\mathcal{Q}_A-\mathcal{Q}_C}$
3. CSS $_{\mathcal{Q}_B-\mathcal{Q}_C}$
4. CSS $_{\mathcal{Q}_A-\mathcal{Q}_B-\mathcal{Q}_C}$

However, even though the simplest option for comparing multiplex networks is considering just two relational structures with two types of relations, the CSS with these alternatives do not necessarily produce the smallest semigroup structure.

Figure 11.4 provides the meta-matrices of four types of role relations in the three Incubator networks with a shadow background. These matrices correspond to the three measured social relations (C, F, K) and the actor attribute B that has a structuring effect. On the other hand, role

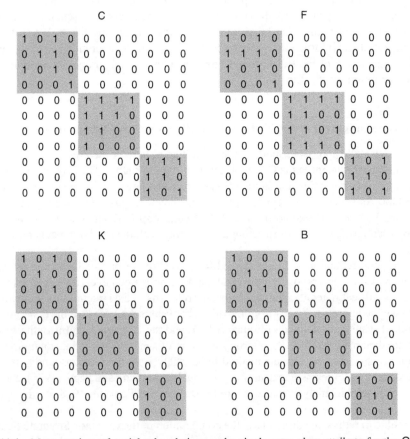

Figure 11.4 Meta-matrices of social role relations and a single actor class attribute for the CSS Role structure. *Relational structures of $\mathcal{Q}_A$, $\mathcal{Q}_B$, and $\mathcal{Q}_C$ in the meta-matrix are shaded. Generator A representing the identity relation is omitted from the picture.*

relation A is represented by the identity matrix in the three Role structures, which means that it is the identity element is in the meta-matrix as well.

Naturally, one can also incorporate the transposes of the social relations in the CSS structure, which will possibly produce a larger and more complex semigroup. However, this is something we try to avoid in the first place, and the substantial interpretation of the tie transposes in the Role structure can be deduced from the measured social relations.

The meta-matrices resemble the diagonal matrices used to represent actor attributes, and this type of configuration assures that the constituent parts of the system do not interact. In order to represent the CSS structure of the Incubator networks, the meta-matrices now constitute the generators of the semigroup—in a similar way we did it in Chapter 5 Role Structure in Multiplex Networks—and this algebraic structure serves to make the comparison. The fact that we are able to construct partially ordered semigroups with the meta-matrices also means that we can establish a hierarchy of the role relations, and the resultant CSS structure will have the three algebraic constraints in partially ordered semigroups.

It is evident from the meta-matrices in Figure 11.4 that systems $\mathscr{Q}_A$ and $\mathscr{Q}_C$ are identical in all role relations except for K, and in fact perceived competition results are the most distinctive type of relation in all Incubator networks. However, we use the meta-matrices to construct the CSS structure of the three positional systems of Incubator networks, and then perform the substantial interpretation of the resultant algebraic system.

11.7.1 CSS for Incubator networks A, B, and C

We continue with the CSS for different Role structures, and the comparison is made to the three Incubator networks $\mathscr{Q}_A$, $\mathscr{Q}_B$, and $\mathscr{Q}_C$ having the same relational content. Comparing relational structures such as $CSS_{\mathscr{Q}_A - \mathscr{Q}_B - \mathscr{Q}_C}$ through the Common Structure Semigroup is possible with the generators (and transposes if needed) of the semigroup tables. This is because these multiplex networks have the same relational content, and the first step is to construct the meta-matrix where the positional systems of the Role structures to compare are located.

Figure 11.5 presents the Cayley colour graph of the entire semigroup CSS structure of $\mathscr{Q}_A$, $\mathscr{Q}_B$, and $\mathscr{Q}_C$ with all measured relations, and we use again the meta-matrices given in Figure 11.4 to construct the Common Structure Semigroup that corresponds to the three relational systems. The arcs in the Cayley graph have the same shape, and this is to distinguish it from the multi-graphs $\mathscr{G}^+_{A,B,C}$.

As one would expect, the Common Structure Semigroup that corresponds to the three role systems of the Incubator networks becomes very large even without considering tie transposes. In this case, the order of the CSS semigroup, which is the number of vertices in the graph, is 65, with a generator acting as the identity element and the largest word made of three letters. Although the CSS structure is large, we can focus the analysis on just some parts of the semi-group structure, like the system for particular elements in the semigroup, or where the social relations in the CSS Role structure are located.

For instance, because relation A constitutes the identity element of the semigroup, it implies that this actor attribute relates to all the other generators including A itself. The vertex representing the identity string relation then is connected by a directed edge to all the rest of generators, with their respective color representing these strings. One consequence of this is that the identity element assures that the Cayley graph representing the relational system connects as a single

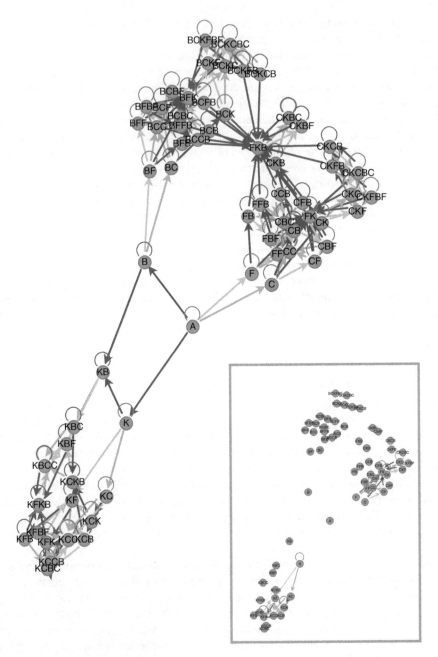

Figure 11.5   Cayley graphs of the CSS $_{\mathcal{Q}_A - \mathcal{Q}_B - \mathcal{Q}_C}$ Role structure for Incubator networks. *Arcs in green are for* C, *khaki3 for* F, *red for* K, *and blue variant for* B. *Bottom-right: Cayley graph with arcs only for the measured social relations in the* CSS *Role structure.*

component. On the other hand, there is no absorbing element in the CSS semigroup structure, and this means that there is no node in the Cayley graph having just incoming ties.

At the bottom-right of Figure 11.5, there is the Cayley graph with the same layout as above, but this time only with the edges involving the social relations in the semigroup structure. This means that the edges representing actor attributes are removed from the picture, and the Cayley graph becomes "disconnected" into two components with the measured social relations.

Figure 11.6 provides the Cayley colour graph for social relations, which is a small portion of the entire Common Structure Semigroup as the adjacent graph shows it. Because we are just focusing on the measured social relations in Incubator networks, which are collaboration C, friendship F, and perceived competition K among the actors in the network, actor attributes are ignored in this configuration.

The identity element A is acting as a bridge in the CSS $_{\mathscr{D}_A-\mathscr{D}_B-\mathscr{D}_C}$ semigroup structure, and we can see in the Cayley graph of Figure 11.5 that the system is divided into two components. One component involves relation K and the product of this string by right multiplication, and the other component is for string role relations C and F and the compounds of these with right

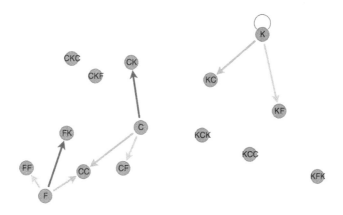

(a) Cayley graph with links of measured social relations.

∘	C	F	K
C	CC	CF	CK
F	CC	FF	FK
K	KC	KF	K

(b) CSS semigroup fragment table for three generators in $\mathscr{L}_{A-B-C}$.

Figure 11.6   Extract of CSS $_{\mathscr{D}_A-\mathscr{D}_B-\mathscr{D}_C}$ representations of generators for Role structure of Incubator networks. *The Cayley graph above is isomorphic to the one at the bottom-right of Figure 11.5 when counting all representative strings of the semigroup, where "isolated strings" in the graph are products involving a compound relation.*

multiplication. However, to see the role interlock of the generator relations of the Incubator networks, we pay attention only to the primitive ties in $CSS_{\mathcal{Q}_A - \mathcal{Q}_B - \mathcal{Q}_C}$. There is an extract of the semigroup table for all measured relations with the products of these semigroup generators in Figure 11.6b with the respective compounds where we can see the reflexive character of K.

11.7.2   CSS Order Role Structure for $\mathcal{Q}_{A-B-C}$

The set of inclusions is another algebraic constraint in Common Structure Semigroups derived from partially ordered relational structures. As said before, inclusions among string relations appear in directed multiplex networks and they produce a hierarchy of the ties.

In the case of the three Incubator networks, the common set of inclusion among generators produces the following ordering

$$((B \leq A) \text{ and } (K \leq A \text{ or } A \leq K)) \leq (C = F).$$

This means that in the $CSS_{\mathcal{Q}_A - \mathcal{Q}_B - \mathcal{Q}_C}$ order structure there is not found a univocal lineal order in their common structure and some ambiguity prevails for the three systems. However, we corroborate with the Common Structure Semigroup the fact that perceived competition occurred within a system of formal and informal ties in all cases. In addition, actor attribute A encompasses actor attribute B in this structure, and this relationship has significant consequences in the substantial interpretation for the comparison of the three Incubator networks in this case.

11.8   Comparing Structures in Substantial Terms

The three algebraic constraints, which are equality or the set of equations, hierarchy in string relations or the set of inclusions, and the role tables will help us in making a comparison of the three Incubator networks by means of a substantial interpretation of the involved systems. For instance, the set of equations allows making a common statement for the associated string relations, while the statements coming from the hierarchy of ties serve to contextualize the strings included in other types of relations. On the other hand, the organizing principles in the Role structures expressed in the role tables may reveal some theoretical arrangement common in the systems to compare with a substantial interpretation.

11.8.1   Hierarchy of Social Relations and Actor Attributes

A substantial interpretation seeks to concretize an abstract statement that in this case is given in mathematical terms. One major concern with the substantial interpretation of the inclusions among the primitive role relations is to look at the different narratives suggested by this kind of algebraic constraint in a given multiplex network.

That is to say that the set of inclusions, for example, together with other logics in the interlock of the role relations, translate into an expected behavior of the actors or different classes of actors generalizing the individual firms in the case of the different Incubator networks. It is important to mention, however, that most of these implications for the behavior of the individual actors in the different relational systems are conclusive statements grounded on the system of role relations from the measured types of tie. This means that these allegations do not belong to a probabilistic regime, but they are deterministic in nature, although uncertainty is an intrinsic part of reality in social systems.

For the substantial interpretation of the different systems, we are going to refer sometimes to strings C and F as *formal* and *informal* relations, respectively, and their corresponding letters K, A, and B to represent perceived competition and the two actor attributes.

11.8.1.1 Social Relations

The logic of role interlock in Incubator network A, $\mathcal{Q}_A$ considers collaboration and friendship ties and the inclusion of the primitive relations $C \leq F$ states that:

"*Actors seek collaboration within their acquaintances*" ($\mathcal{X}_A$).

This statement implies that the informal system of ties prevails over the formal relations in these Incubator networks and there can be good reasons for this type of situation. Collaboration among start-ups companies gathered in business Incubator centers typically begins with informal contacts, and that situation is expected to happen especially in new-created settings as this one at that point of time.

Although in the Role structure of Incubator network A there are the informal ties that include a system of formal collaboration relations, in the case of $\mathcal{Q}_C$ in Incubator network C, on the other hand, the inclusion $F \leq C$ suggests that:

"*Acquaintance relations among actors occur within a network of collaboration*" ($\mathcal{X}_C$).

For Incubator network C, social links inside the network are then more institutionalized and more stable over time. One of the reasons could be due to a high level of specialization of the firms here or just because this Incubator center is more established than the previous one.

Since there is no inclusion among strings C and F in the partial order structure, the two statements given above apply to Incubator network B as well. This means that there is an ambiguous situation concerning formal and informal ties in the Role structure that corresponds to $\mathcal{Q}_B$, and this suggests a level of institutionalization of $\mathcal{X}_B$.

Furthermore, if we relate the formal and informal relations with the cognitive ties of perceived competition, then the inclusions $K \leq C$ and $K \leq F$ apply for $\mathcal{Q}_A$, $\mathcal{Q}_B$, and $\mathcal{Q}_C$. In substantial terms, this means that:

"*Actors that perceive other actors as competitors are part of the same formal and informal network*" ($\mathcal{X}_A$, $\mathcal{X}_B$, $\mathcal{X}_C$).

This corroborates the fact that perceived competition is scarce or even absent in the Incubator networks. Such situation can be beneficial when it comes to a collective collaboration effort where each of the parts complements to each other, but on the other hand competition can also encourages neck-and-neck firms to innovate (Aghion, Bloom, Blundell, Griffith and Howitt, 2005), and it can be a driven force in the creation of novel products. One element to take into account is that perceived competition is a cognitive tie, which means that even though competition is present, it is either not an object for reflection by the actors, or it is not manifested because of cultural or idiosyncratic reasons.

11.8.1.2   Social Relations and Actor Attributes

It is also possible to make the interpretation of the string inclusions involving the actor attributes with social relations. In this sense, in all Incubator networks there is the inclusion of actor attributes with the formal and informal ties as A ≤ C and A ≤ F where A = B. In substantial terms this implies that:

"Actors with similar attributes have formal and informal relations" ($\mathscr{X}_A$, $\mathscr{X}_B$, $\mathscr{X}_C$).

Generally, formal and informal contacts between the actors are at this point of time through other means than social services from the Internet like LinkedIn or Facebook, which are actor attributes A and B. The firms are after all located in the same physical setting, and such Web utilities seem to be more oriented toward the custumers, which are not necessary other firms located in the same Incubator center.

Moreover, inside the network of formal and informal ties there is also an inclusion relation between perceived competition and the adoption of innovations. In this sense, the inclusion K ≤ A means for Incubator networks B and C that:

"Actors that adopted Web innovations are (were) not perceived as competitors" ($\mathscr{X}_B$, $\mathscr{X}_C$).

In consequence, in $\mathscr{Q}_B$ and $\mathscr{Q}_C$ even though some actors operating in these Incubators have adopted a similar innovative utility to carry out their promotion activities, this does not necessarily mean that they regard each other as competitors. However, the other way around is true for $\mathscr{Q}_C$, and we can verify in the data of this Role structure that there is just one exception to this assertion. For example, actor 342 adopted both innovations and it is regarded as competitor by actor 394, cf. Figure 11.2. On the other hand, in the case of the Role structure that corresponds to Incubator network A such inclusion is, however, not univocal.

Finally, there is an ordering relationship between the two types of innovations or actor attributes. In the case of the role relation of Incubator network B, there is the inclusion relation B ≤ A, meaning that at this point of time (i.e. year 2010):

"Firms that had an active Facebook profile are (were) also in LinkedIn" ($\mathscr{X}_B$).

When considering the inclusion relation between the two types of innovations then, apart from this Incubator network, the other Role structures consider just a single category for both LinkedIn and Facebook, which means that the algebraic analysis did not make any differentiation between these innovation types. This is because what has been more significant in the algebraic analysis has been the relationship between the attributes of the actors and the other types of measured relations in the modeling.

The hierarchy in the relations permits the making of important inferences structures about the organizational logic in the network that is based on the actors' relations. This resembles to the institutional system of practices considered by Breiger and Mohr (2004), and which has a natural connection with the interlock of relations existing in the system. However, there are additional algebraic constraints, which are part of the relational interlock of the network, and that complement the information provided by the hierarchy of relations.

11.8.2 Set of Equations or Equality in $\mathcal{Q}_A$, $\mathcal{Q}_B$, and $\mathcal{Q}_C$

Contrary to the hierarchy of relations, there is some ambiguity with the resultant set of equations after the progressive factorization in most of the image representations that corresponds to the Incubator networks. The ambiguity in the outcomes of the factorization procedure of these networks has been the price to pay for the subdirect decomposition of the Role structures where part of the resultant configurations overlap with each other. The benefit of the factorization procedure, however, has been to count with a considerable reduction of the relational structures of the Incubator multiplex networks. Hence, the two algebraic constraints that we have seen so far separately on the decomposed Role structures allow the making of substantial interpretations of the social systems by taking into consideration all kinds of relations involved.

We should continue with the comparison of the resultant set of equations and hierarchy in the role relations that correspond to $\mathcal{Q}_A$, $\mathcal{Q}_B$, and $\mathcal{Q}_C$. However, rather than looking just at the coincidences and divergences in the set of equations, for example, we are going to continue with the implied image matrices, and this is because the maximal homomorphic images are the consequences of such equations. In this way, by comparing the algebraic structures representing the role tables we are able to assess the rest of the algebraic constraints that are common to these structures.

In our task of comparing the outcomes that correspond to the different Incubator networks of the study, an assessment was possible because all the relational systems are the product of the same generator relations. Consequently, the modeling of these structures has been performed by applying the same principles and methods both to the actors and also their diverse types of relations. In this sense, we have considered the comparison between different relational structures the coincidences in patterns in a shared structure that can be expressed either in the joint table or in the common semigroup.

As a result, the comparison across structures is based on the algebraic constraints that are inherent to these relational systems, and which comprise the hierarchy in the ties, the different sets of equations among the relations, and the representative structures of the role tables. The coincidences and the divergences among the different structures can gradually be exposed for each type of algebraic constraint, and the following substantial explanations can be made for the different types of tie.

11.9 Structuring Effect of Role Relations in Incubators

Some remarks of the structuring effects of the role relations by type including actor attributes are:

- *Formal and informal ties*: In the case of Incubator network A the informal ties have been the dominating form for relations among the network members, and also between the different categories of actors. The exception to this reality was Incubator network C where the essential relational structure had a more formal character than a configuration made of informal links as the other systems. In terms of relational composition, such characterization is also valid for the three Incubator networks, and in the Role structure of $\mathscr{X}_A$, for example, the combination of friendship and collaboration produced just friendship or CF = FC = F.

If we consider the sets of equations and the congruence relations in the distinct role relations, then the particularities of these types of tie disappeared in the maximal homomorphic images. Formal and informal ties were part of the same category of ties, namely that one, which effectively shaped the structure of the system.

- *Perceived competition*: Regarding the cognitive relation of perceived competition among firms, this was the least occurring type of tie in Incubator networks. If perceived competition had taken place, then it was in the context of either collaboration or friendship relations. Further algebraic constraints involving this type of tie are, however, limited to its relationship with the adoption of innovations.
- *Innovation adoption*: The treatment of the attributed-based information such as the adoption of innovations was made in a relational manner, and we were able to establish an integrated system where the analysis of actor attributes and the different social ties have taken place. However, the difference between the two types of attributes disappeared in the process of modeling the actors in two of three Incubator networks, and even more dramatically is when the differentiation of actor attributes among the classes of actors vanishes as with Role structures $\mathscr{Q}_A$ and $\mathscr{Q}_C$. This suggests that innovation adoption did not have a structuring effect in the configuration of these two Incubator networks in case we consider their Role structures.

The exceptions to the lack of influence of innovation adoption have been Incubator network B, where there has been a differentiation among the classes of firms in its Role structure $\mathscr{Q}_B$. However, the way the attribute-based relation was embedded in the Role structure has been limited, and it ended up in the joint table being part of the neutral element after the transformations. With this Incubator network, the two types of innovations were represented by different structures in the role relation, and this clearly differentiated this system from the rest of the Incubator networks. In this case, both the adoption of LinkedIn and Facebook had an independent interlock from the rest of the ties but apart from the fact that perceived competition occurred among firms with a LinkedIn profile, the algebraic constraints involving innovation adoption in $\mathscr{Q}_B$ were not very decisive in shaping the relational structure than the rest of the types of tie. In substantial terms, the relation between the two actor attributes implies that those firms in each one of the three Incubator networks who adopted innovation B or Facebook in year 2010 also tended to adopt innovation A or LinkedIn by that period of time. By deduction, most likely they adopted A "before" B.

At the bottom line of the analysis, we differentiated in the diverse Role structures between one kind of role relation that is structuring the configuration of the system, and another class that had a much more limited part in the structuring process. The algebraic constraints showed that the Role structure of the different networks was driven by a combination of formal and informal relations in all cases where the adoption of innovation had a more restricted impact. However, in Incubator network B the adoption of innovations resulted in being more significant than the rest of the role systems; with respect to the role of perceived competition, the structuring power of this cognitive type of relation was partial in each of the Incubator networks.

It is worth noting that the decomposition of $\mathcal{Q}_B$ was made with congruence classes by substitution, and the decomposition of $\mathcal{Q}_C$ was by means of factorization (see Appendix B and Appendix E for other details in the factorization of these Role structures).

11.10   Comparing Relational Structures: Summary

Relational structures represented configurations made of several types of relations in multiplex networks, and the ability to compare these systems constituted an important aspect in the analysis of different networks. Semigroups stand for the structure of role relations, and the three algebraic constraints guided the comparison of different relational structures. One condition for network comparison is that networks needed to have their same generating set of relations or at least with a similar relational content, and we looked at the coincidences and divergences of the algebraic constrains to make a substantial interpretation of relations in the involved networks.

In the case of role tables, the arrangement of the Lattice of Homomorphisms of the Semigroup allowed the establishment of diverse kinds of shared structures between two semigroups. For instance, JNTHOM and CSS that represent either the intersection or the union of semigroups served to describe common structure. The joint reduction implies that all the equations holding true in either of the original Role structures must hold true in their joint reduction, whereas the meet common structure considers the lattice intersection of the two semigroups.

In substantial terms, we differentiated in the Incubator networks one category of role relations having a structuring effect on the configuration of the Role structures, and another class of role relation with a much more limited part in the structuring of the system. The algebraic constraints showed that the Role structures of the Incubator networks were driven by a combination of formal and informal relations, but the adoption of Web innovations had a more restricted impact except for a single Incubator network. With respect to the character of perceived competition, the structuring power of this cognitive type of role relation was partial for each of the compared systems.

11.11   Learning Comparing Relational Structures by Doing

The visualization and the construction of the Role structure with decompositions corresponding to Incubator network A, $\mathscr{X}_A$ have been made in Chapters 5 and 6, and we look at this process here for the rest of Incubator networks.

11.11.1   Visualization of Incubator Networks B and C

```
# load data Incubator networks B and C
R> data(incB)
R> data(incC)
```

```
# define a common scope for 𝒳_B and 𝒳_C
R> scp <- list(directed = TRUE, cex = 3, ecol = c("#00C000", "#FFFF00", "#FF0000"),
+     lwd = 2, vcol = "#3399FF", vcol0 = "#808080", pos = 0, bg = 8)
# define particular scopes for networks B and C
R> scpb <- list(bwd = .75, hds = .75, swp = TRUE)
R> scpc <- list(bwd = .50, hds = .50)

# plot multigraphs with multiple scopes
R> multigraph(incB, layout = "force", seed = 123, scope = c(scp, scpb))
R> multigraph(incC, layout = "force", seed = 123, scope = c(scp, scpc))
```

11.11.2   Positional Analysis and Role Structure for $\mathscr{X}_B$ and $\mathscr{X}_C$

11.11.2.1   Incubator B

```
# Relation-Box R_B(W_3) with tie transposes for social relations
R> rb_B <- rbox(incB$net, transp = TRUE, tlbs = c("D","G","L",NA,NA))
# positional system 𝒮_B with defined cluster information
R> ps_B <- reduc(rb_B$w, clu = c(2,2,1,4,4,1,2,2,3,2,3,2,4,4,1,2,2,4))
# multiplication table of Role structure 𝒬_B
R> S_B <- semigroup(ps_B, type = "symbolic")
# partial order table of 𝒬_B
R> P_B <- partial.order(strings(ps_B))
```

11.11.2.2   Incubator C

```
# Relation-Box R_C(W_3) with tie transposes
R> rb_C <- rbox(incC$net, transp = TRUE, tlbs = c("D","G","L",NA,NA))
# positional system 𝒮_C with defined cluster information
R> ps_C <- reduc(rb_C$w, clu = c(1,2,2,2,1,3,2,2,2,2,1,2,1,1,3,1,1,2,2,2,1,3))
# multiplication table of 𝒬_C
R> S_C <- semigroup(ps_C, type = "symbolic")
# partial order table of 𝒬_C
R> P_C <- partial.order(strings(ps_C))
```

11.11.3   Decomposition of $\mathscr{Q}_B$ and $\mathscr{Q}_C$

11.11.3.1   Incubator B

Objects S_B and P_B with the partially ordered semigroup representing $\mathscr{Q}_B$ are from the previous section.

```
# factorization of 𝒬_B with meet complements of Atoms
R> ii_B <- fact(S_B, P_B, atmc = TRUE)
# partition (π) relations of the factorization with the poset
R> pr_B <- pi.rels(ii_B, po.incl = TRUE)
```

```
# decomposition of 𝒬_B with the meet-complements of Atoms in π-relations
# with additional equations to the string relations
R> sB <- decomp(S_B, pr_B, type = "mca", reduc = TRUE, force = TRUE)
# look at the Factor orders in aggregated structure
R> sB$ord


[1]  12 30 15 23 15 23
```

Then we perform a decomposition of the six Factors.

```
# list objects for induced inclusions, π-relations, and aggregated structures
R> ii_Bfc <- list(); pr_Bfc <- list(); sB2 <- list()
# perform factorization on Images in 'sB'
R> for(k in seq_along(sB$ord)) {
+    ii_Bfc[[k]] <- fact(sB$IM[[k]], sB$P[[k]], atmc = TRUE, patm = TRUE)
+    pr_Bfc[[k]] <- pi.rels(ii_Bfc[[k]], po.incl = TRUE)
+    }
# and perform decomposition of aggregated structures from factorization
R> for(k in seq_len(length(pr_pr_Bfc))) {
+    temp <- as.semigroup(sB$IM[[k]])
+    sB2[[k]] <- decomp(temp, pr_Bfc[[k]], type = "mca", reduc = TRUE)
+    print(sB2[[k]]$ord)
+    }


[1] 1
[1] 21 18 18
[1] 1
[1] 23
[1] 1
[1] 23
```

We can see that Images 2, 4, and 6 can be further decomposed. After long progressive factorizations of these Factors, which is given in Appendix E, the 39 aggregated Images from the three Factors result, being isomorphic to each other with an order of 11.

11.11.3.2 Incubator C

Partially ordered semigroup with objects S_C and P_C is from the previous section.

```
# factorization of the partially ordered semigroup with the Atoms in $L_\pi(\mathcal{Q}_C)$
R> fct_C <- fact(S_C, P_C)


...


$atm
$atm$`17, 8`
[1] "17, 8" "21, 8"

$atm$`13, 15`
[1] "13, 15" "20, 15"


...
attr(,"class")
[1] "Ind.incl"
```

```
# Image matrices and partial order tables from the decomposition of $L_\pi(\mathcal{Q}_C)$ with
# meet-complements of Atoms as $\pi$-relations
R> decomp(S_C, pi.rels(fct_C), type = "mca", reduc = TRUE)
```

```
$IM                                       $PO
$IM[[1]]                                   $PO[[1]]
     C  F  K  A  L CC CK                        C F K A L CC CK
C   CC  C CK  C CC CC CK                    C   1 0 0 0 0  1  0
F   CC  F CK  F CC CC CK                    F   1 1 0 0 0  1  0
K    L  K  K  K  L  L  K                    K   1 1 1 1 1  1  1
A    C  F  K  A  L CC CK                    A   1 1 0 1 0  1  0
L    L  L  K  L  L  L  K                    L   1 0 0 0 1  1  0
CC  CC CC CK CC CC CC CK                    CC  0 0 0 0 0  1  0
CK  CC CK CK CK CC CC CK                    CK  1 1 0 0 0  1  1

$IM[[2]]                                   $PO[[2]]
     C  K  A  G  L CC LC                        C K A G L CC LC
C   CC  K  C CC  K CC CC                    C   1 0 0 0 0  1  0
K   CC  K  K CC  K CC CC                    K   1 1 0 0 0  1  0
A    C  K  A  G  L CC LC                    A   1 0 1 1 0  1  0
G    C  K  G  G  L CC LC                    G   1 0 0 1 0  1  0
L   LC  L  L LC  L LC LC                    L   1 1 1 1 1  1  1
CC  CC  K CC CC  K CC CC                    CC  0 0 0 0 0  1  0
LC  LC  L LC LC  L LC LC                    LC  1 0 0 1 0  1  1

$ord
[1] 7 7


attr(,"class")
[1] "Decomp"  "Pi.rels"  "mca"
```

11.11.4   Equalities in Incubator Networks

11.11.4.1   Set of Equations in Incubator A

From Chapter 6 we have the decomposition of $\mathcal{Q}_A$ with the meet-complements in π-relations in object SsA where the equality is based on three Factors.

```
# set of equations of measured relations in aggregated structure of 𝒬ₐ
R> SsA$eq

$eq
$eq[[1]]
$eq[[1]][[1]]
[1] C   F   K   CK CL KL

$eq[[1]][[2]]
[1] A L


...

$eq[[2]]
$eq[[2]][[1]]
[1] C    CK   DC   GC   KDC DCK

$eq[[2]][[2]]
[1] F    CD   CL   DF   CKD DCL
$eq[[2]][[3]]
[1] K A


...

$eq[[3]]
$eq[[3]][[1]]
[1] C   F   A   D   G   CD DC DF GC

$eq[[3]][[2]]
 [1] K   L    CK  CL  KD  KL  DK   CKD KDC DCK DCL
```

11.11.4.2   Set of Equations in Incubator B

From Appendix E, we obtain SsB where there is a single measured relation with equations.

```
R> SsB$eq

[1] C   F   CBF
```

11.11.4.3   Set of Equations in Incubator C

From Appendix B, we obtain SsC.

```
R> SsC$eq

$eq
$eq[[1]]
$eq[[1]][[1]]
[1] C  G  CF

$eq[[1]][[2]]
[1] F  FF

$eq[[1]][[3]]
[1] K   KF  LK  LKF

...

$eq[[2]]
$eq[[2]][[1]]
[1] C  F  GC

$eq[[2]][[2]]
[1] K  CK CL KL
```

Appendix A

Datasets

Datasets of the following networks are used in this book:

- Kariera kinship (Radcliffe-Brown, 1913; White, 1963).
- Incubators A, B, C (Ostoic, 2013).
- Florentine families (Padgett and Ansell, 1993), retrieved from http://moreno.ss.uci.edu/data# padgett.
- Clients and attourneys (Wasserman and Iacobucci, 1991).
- Group of twenty (Wikipedia, 2017, 2019).

Kariera kinship

The **Kariera kinship** network is an elementary system with two types of relations, one for marriage and the other for descent relations among four clans of the Kariera society in Western Australia, *Banaka*, *Burung*, *Karimera*, and *Palyeri*.

There are two types of relations F and G in the Kariera kinship network that stand for rules of marriage and descent, respectively, and which define this kinship system that is represented by a Group structure.

Incubators A, B, C

Incubator networks A, B, C, $\mathscr{X}_A$, $\mathscr{X}_B$, and $\mathscr{X}_C$ are made of three types of directed relations C, F, K, and two kinds of actor attributes A and B. These represent collaboration, friendship, perceived competition, and the adoption of two Web innovations (Linkedin and Facebook) among Danish entrepreneurial firms in year 2010.[1]

---

[1] Linkedin (http://www.linkedin.com) is a professional oriented Web service launched in year 2003, whereas Facebook (https://www.facebook.com) started in year 2004 as a general networking site.

Algebraic Analysis of Social Networks: Models, Methods and Applications using R,
First Edition. J. A. R. Ostoic. Companion website: www.wiley.com/go/ostoic/algebraicanalysis.
© 2021 John Wiley & Sons Ltd. Published 2021 by John Wiley & Sons Ltd.

The relational structure of the Incubator networks is represented by a partially ordered semi-group with contrasting relations D,G and L of the directed ties.

Note that the **Small configuration** $\mathscr{X}_Z$, which was used to illustrate the composition of two relations, is a multiplex directed network that is isomorphic to a component of $\mathscr{X}_C$.

Florentine families

The **Florentine families** network $\mathscr{X}_F$ is made of two kinds of undirected relations, which are marriage M and business B relations among families. There is also a pair of actor attributes in this network that is the financial wealth W and the number of priorates P the families held.

The relational structure is represented by a partially ordered semigroup of relations.

Clients and attorneys

The **Clients and attorneys** network is a multilevel structure with two domains. There are two types of directed relations within each domain "respects" and "recommends" among attorneys and clients, respectively, and two types of directed links across domains that are "mails a letter" and "hires".

Group of twenty

The **Group of twenty** or **G20 Countries** network comprises different types of structures made of relations among a number of countries. Network $\mathscr{X}_{G20}$ represents the entire structure, while $\mathscr{X}_{G20b}$ is a subset of $\mathscr{X}_{G20}$.

The acronyms of the members of $\mathscr{X}_{G20}$ in Chapters 8, 9, and 10 are:

      P5: Permanent Five security, government, military
      G4: Group of Four
      G7: Group of Seven
   BRICS: Brazil, Russia, India, China, South Africa
   MITKA: Mexico, Indonesia, Thailand, South Korea, Australia
     DAC: OECD Development Assistance Committee
    OECD: Organization for Economic Development
    Cwth: Commonwealth countries
     N11: Next-11[2]

---

[2] Besides countries already in $\mathscr{X}_{G20}$, also Bangladesh, Egypt, Iran, Nigeria, Pakistan, Philippines and Vietnam.

- ARG: Argentina
- AUS: Australia
- BRA: Brazil
- CAN: Canada
- CHN: China
- DEU: Germany
- FRA: France
- GBR: Great Britain
- IDN: Indonesia
- IND: India

- ITA: Italia
- JPN: Japan
- KOR: South Korea
- MEX: Mexico
- RUS: Russia
- SAU: Saudi Arabia
- TUR: Turkey
- USA: United States
- ZAF: South Africa

The distinct structures and data related to network $\mathscr{X}_{G20}$ are the following:

G20 Affiliations network

Two-mode structures $\mathscr{X}_{G20}^{B}$ and $\mathscr{X}_{G20b}^{B}$ for affiliations of the G20 countries to supranational organizations. $\mathscr{X}_{G20b}^{B}$ is the configuration for countries in $\mathscr{X}_{G20}$ that are affiliated with bridge organizations.

G20 Indicators

Economic and socio-demographic indicators of countries in $\mathscr{X}_{G20}^{B}$ and $\mathscr{X}_{G20b}^{B}$, and which are treated as many-valued contexts and analyzed within Formal Concept Analysis.

G20 Trade network

The G20 Trade network $\mathscr{X}_{G20}^{V}$ and subset $\mathscr{X}_{G20b}^{V}$ are valued and multiplex configurations. The relations are two kinds of goods exchanged between G20 countries and which are the trade of fresh milk m and honey h in year 2017.

G20 Multilevel network

Multilevel structures $\mathscr{X}_{G20}^{M}$ and $\mathscr{X}_{G20b}^{M}$ combine G20 Affiliations and Trade networks of these two systems. These configurations, apart from being multilevel, are also multiplex and valued.

Appendix B

Role structures of Incubator networks

Role Structure of $\mathcal{X}_A$

The role structure of Incubator network A is subject of Chapters 5 and 6.

Role Structure of $\mathcal{X}_B$

Aggregated versions of Role structure for Incubator network B are possible with all the additional equations made on the strings of relations in $\mathcal{Q}_B$ with a number of logics of interlock.
   We start with the depiction of the multigraph of Fig. 11.1 in Chapter 11.

```
# load the data
R> data(incB)

# making the scope
R> scp <- list(directed = TRUE, ecol = c("#00C000","orange","#FF0000"), lwd = 2,
   + pos = 0, cex = 3, vcol = "#3399FF", vcol0 = "#808080", bwd = .75, swp = TRUE)

# plotting Fig. 11.1 with a force-directed layout
R> multigraph(incB, layout = "force", seed = 123, scope = scp)
```

Positional system of Incubator network B

As with Incubator network A, the positional analysis of $\mathcal{X}_B$ is based on the notion of Compositional equivalence of the actors in the network. The basis for Composition equivalence lies on the Relation-Box of the network $\mathbf{R}(\mathcal{X}_B)$, which in this case has both actor attributes and transposes of the social relations. On the basis of $\mathbf{R}(\mathcal{X}_B)$, we continue with the construction of the Cumulated Person Hierachy of Incubator network B $\mathcal{H}_B$.

Algebraic Analysis of Social Networks: Models, Methods and Applications using R,
First Edition. J. A. R. Ostoic. Companion website: www.wiley.com/go/ostoic/algebraicanalysis.
© 2021 John Wiley & Sons Ltd. Published 2021 by John Wiley & Sons Ltd.

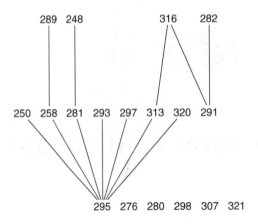

Figure B.1   Cumulated Person Hierarchy $\mathcal{H}_B$.

The Relation-Box is stored in rb_B, and both the set of relations and the resulted Image matrices are given in object incB. The adjacency matrices representing the network relations are stored in component net of incB, and NA indicates that the respective string generator does not need to have a tie transpose. This is because, in this case, diagonal matrices stand for the two actor attributes.

```
# construct the Relation-Box for 𝒳_B with tie transposes of social relations
R> rb_B <- rbox(incB$net, transp = TRUE, tlbs = c("D", "G", "L", NA, NA))
```

Function diagram allows us to plot this partial order structure as a Hasse diagram as depicted in Figure B.1.

```
# inclusion lattice of the Cumlated Person Hierarcy in Fig. B.1
R> require("Rgraphviz")
R> diagram(cph(rb_B))
```

The clustering information is based on the information from $\mathcal{H}_B$.

```
R> clsB <- c(2, 2, 1, 4, 4, 1, 2, 2, 3, 2, 3, 2, 4, 4, 1, 2, 2, 4)
```

```
# clustering information for 𝒟_B
R> as.table(rbind(dimnames(incB$net)[[1]], clsB))
```

```
A 248 250 258 276 280 281 282 289 291 293 295 297 298 307 313 316 320 321
B 2   2   1   4   4   1   2   2   3   2   3   2   4   4   1   2   2   4
```

We can see in the Hasse diagram of the Cumulated Person Hierarchy $\mathcal{H}_B$ that there are both comparable and incomparable elements in the partial order structure. As with the modeling of Incubator network C, the incomparable actors make their own class and we concentrate the analysis on the directly related actors.

Within the comparable elements, actors 291 and 295 make a class since they do not cover any other element in $\mathcal{P}_B$. The remaining actors in the partial order structure are in two categories. First are the actors that are not covered by any other actors, and the other class is made of actors who are both covering and covered in the lattice diagram.

As a result, the positional system $\mathcal{S}_B$ by means of Compositional equivalence is made of four classes of actors. We perform the reduction of the network with all social relations, their transposes, and the two kinds of actors attributes.

```
# reduction of generators with clustering information
R> ps_B <- reduc(rb_B$w, clu = clsB)


, , C                                              , , A

     [,1] [,2] [,3] [,4]                                [,1] [,2] [,3] [,4]
[1,]    0    1    0    1                           [1,]    1    0    0    0
[2,]    1    1    1    1                           [2,]    0    1    0    0
[3,]    0    1    0    0                           [3,]    0    0    1    0
[4,]    1    1    0    1                           [4,]    0    0    0    1


, , F                                              , , B

     [,1] [,2] [,3] [,4]                                [,1] [,2] [,3] [,4]
[1,]    0    1    1    1                           [1,]    0    0    0    0
[2,]    1    1    1    1                           [2,]    0    0    0    0
[3,]    1    1    0    1                           [3,]    0    0    0    0
[4,]    0    1    0    1                           [4,]    0    0    0    1


, , K                                              , , D
                                                  ...
     [,1] [,2] [,3] [,4]
[1,]    0    0    0    0
[2,]    1    1    0    0
[3,]    0    0    0    0
[4,]    0    0    0    0
```

A permutation of the positional system is possible with a clustering information in the permutation function perm().

```
# .S_B as in Fig. 11.4
R> perm(ps_B, clu = c(3,1,4,2))
```

Role tables in $\mathcal{Q}_B$

This configuration that expresses the logic of interlock of $\mathcal{Q}_B$ is given in Chapter 6, and is subject to comparison in Chapter 11.

Role Structure of $\mathscr{X}_C$

Positional system of Incubator network C

The construction of the role structure as a partially ordered semigroup relies on the Cumulated Person Hierarchy with the chosen length of strings, and now we review the whole process of establishing the network positional system and role structure of an empirical network represented by Incubator network C, netC. In this case, the ties of the network are directed, and for digraphs is suggested to generate first "relational contrast" in the system, which is achieved by including the "tie transposes" in the construction of the Relation-Box . Thus, if relation C in the network represents "collaborates with," then the tie transpose will stand for "pointed as collaborator by." A similar construction occurs with the other kinds of tie occurring in the system.

```
R> data(incC)
R> rb_C <- rbox(incC$net, transp = TRUE)
```

With Compositional equivalence, structurally correspondent actors will have a similar set —or rather lack— of inclusions in the partial order structure, and visualizing the poset often provides useful insight for the partition of the network, especially when the structure is relatively large.

The inclusion lattice of the Cumulated Person Hierarchy of Incubator C $\mathscr{H}_C$ can be plotted with the two flavors, one that includes all elements in the partial order structure, and another without the incomparable elements in the poset.

```
R> diagram(cph(rb_C), incmp = FALSE)
```

As with many cases, there is no univocal solution in the network partition, and the researcher needs to make a judgment of the correspondence between actors that is based on theory or some other criteria. The simplest way of partitioning this particular network would be making two classes of actors, one with those who have an inclusion relation in the Cumulated Person Hierarchy, and another class with the incomparable actors. However, most of the times this is a trivial solution since it ends up with the identity and universal matrices, which either have none or an annihilating structuring effect in the role structure, and we need to differentiate equivalent actors who are linked in the inclusion lattice structure as well.

Based on the output from $\mathscr{H}_C$, we categorize the actors into 3 classes, one for the incomparable actors in the poset, and record the information as a vector in clsC.

```
# clustering information for 2_C
R> clsC <- c(2,3,3,3,2,1,3,3,3,3,2,3,2,2,1,2,2,3,3,3,2,1)

R> as.table(rbind(dimnames(cph(rb_C))[[1]], clsC))
```

	A	B	C	D	E	F	G	H	I	J	K	L	M	N	O	P	Q	R	S	T	U	V
	327	328	331	339	342	352	354	357	363	366	368	374	376	380	391	393	394	395	398	406	407	414
clsC	2	3	3	3	2	1	3	3	3	3	2	3	2	2	1	2	2	3	3	3	2	1

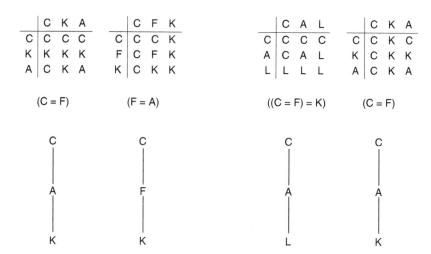

Figure B.2   Progressive Factorization of $\mathcal{Q}_C$ with meet complements and maximal non-trivial Images with equations until length $k = 2$.

Function **perm** allows making a permutation of the matrix representing the Cumulated Person Hierarchy with the clustering information allocated to the clu argument and we verify that actors 352, 391, and actor 414 (who covers the first one) clearly differentiate from the rest of the network members. Such differentiation is because these three actors contain the rest of the elements in the poset structure without being contained by them, and the output below involving the linked actors in the lattice structure serves as the basis for the network partition.

```
R> perm(cph(rb_C), clu = clsC)

    352 391 414 327 342 368 376 380 393 394 407 328 331 339 354 357 363 366 374 395 398 406
352   1   0   1   0   0   0   0   0   0   0   0   0   0   0   0   0   0   0   0   0   0   0
391   0   1   0   0   0   0   0   0   0   0   0   0   0   0   0   0   0   0   0   0   0   0
414   0   0   0   0   0   0   0   0   0   0   0   0   0   0   0   0   0   0   0   0   0   0
327   1   1   1   1   0   1   1   0   1   1   0   0   0   0   0   0   0   0   0   0   0   0
342   1   1   1   0   1   0   0   1   0   0   0   0   0   0   0   0   0   0   0   0   0   0
368   1   1   1   0   0   1   1   0   1   1   0   0   0   0   0   0   0   0   0   0   0   0
376   1   1   1   0   0   1   0   0   0   0   0   0   0   0   0   0   0   0   0   0   0   0
380   1   1   1   0   1   0   0   1   0   0   0   0   0   0   0   0   0   0   0   0   0   0
393   1   1   1   0   0   1   1   0   1   1   0   0   0   0   0   0   0   0   0   0   0   0
394   1   1   1   0   0   1   1   0   1   1   0   0   0   0   0   0   0   0   0   0   0   0
407   1   1   1   0   1   1   1   1   1   1   1   0   0   0   0   0   0   0   0   0   0   0
...
```

The reduction of the network is then made accroding to the clustering information given in clsC, and produces the positional system:

C	F	K	A	D	G	L	B
1 1 1	1 0 1	1 0 0	1 0 0	1 1 1	1 1 1	1 1 0	1 0 0
1 1 0	1 1 0	1 0 0	0 1 0	1 1 0	0 1 0	0 0 0	0 1 0
1 0 1	1 0 1	0 0 0	0 0 1	1 0 1	1 0 1	0 0 0	0 0 1

S El.	WordTable Word	N.	G.	EdgeTable 1	2	3	4	5	6
1	C	–	–	7	8	9	1	7	10
2	F	–	–	7	11	9	2	7	10
3	K	–	–	12	13	3	3	12	14
4	A	–	–	1	2	3	4	5	6
5	G	–	–	15	1	9	5	16	17
6	L	–	–	18	18	19	6	18	6
7	CC	1	1	7	7	9	7	7	10
8	CF	1	2	7	8	9	8	7	10
9	CK	1	3	7	20	9	9	7	10
10	CL	1	6	7	7	9	10	7	10
11	FF	2	2	7	11	9	11	7	10
12	KC	3	1	12	12	3	12	12	14
13	KF	3	2	12	13	3	13	12	14
14	KL	3	6	12	12	3	14	12	14
15	GC	5	1	7	7	9	15	7	10
16	GG	5	5	15	15	9	16	16	17
17	GL	5	5	21	21	22	17	21	17
18	LC	6	1	18	18	19	18	18	6
19	LK	6	3	18	23	19	19	18	6
20	CKF	9	2	7	20	9	20	7	10
21	GLC	17	1	21	21	22	21	21	17
22	GLK	17	3	21	24	22	22	21	17
23	LKF	19	2	18	23	19	23	18	6
24	GLKF	22	2	21	24	22	24	21	17

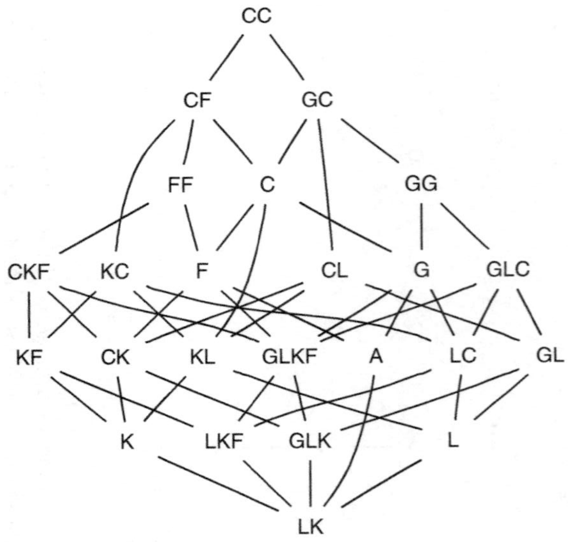

Figure B.3 Role Structure for Incubator network C, $\mathcal{D}_C$ A = B, CC = FC, CK = FK

Which produces a partially ordered semigroup of the Role structure of Incubator network $\mathcal{Q}_C$ where the two actor attributes are equated, $A = B$, and $C = D$. Figure B.3 represents this Role structure as reported in Ostoic (2013).

The positional system and role structure of $\mathcal{Q}_C$ is obtained from the Relation-Box with contrasting relations.

```
# Relation-Box
R> rb_C <- rbox(incC$net, transp = TRUE, tlbs = c("D","G","L",NA,NA))
# positional system
R> ps_C <- reduc(rb_C$w, clu = c(1,2,2,2,1,3,2,2,2,2,1,2,1,1,3,1,1,2,2,2,1,3))
# role structure
R> S_C <- semigroup(ps_C, type = "symbolic")
R> P_C <- partial.order(strings(ps_C))
```

The first stage in the decomposition is given in Chapter 6, and Figure B.2 presents the progressive Factorization of the two Factors of $\mathcal{Q}_C$ with the equations among the primitive relations below the homomorphic Images. The factorization by induced inclusions in the partial order of the semigroup structure is given next

```
# induced inclusions
R> ii_C <- fact(S_C, P_C, atmc = TRUE, patm = TRUE)
# π-relations from induced inclusions
R> pr_C <- pi.rels(ii_C, po.incl = TRUE)
```

The decomposition of S_C in SsC brings two atoms with order 7.

```
# decomposition of 𝒬_C with meet-complements in π-relations
R> SsC <- decomp(S = S_C, pr = pr_C, type = "mca", reduc = TRUE)
```

Appendix C

Valued data in G20 Trade network

Group of Twenty Indicators

Six economic and socio-demographic indicators (with units of measure given afterward) of the G20 countries according to Wikipedia (2019) are

	Trade	Nom._GDP	GDP_PC	HDI	Population	Area
ARG	142370	477743	10667	0.825	44570000	2780400
AUS	496700	1417003	56698	0.939	25182000	7692024
BRA	484600	3453000	16396	0.828	210400000	8515767
CAN	947200	1739110	46733	0.926	37078000	9984670
CHN	4201000	14216503	9633	0.752	1396982000	9572900
DEU	2866600	3963880	48670	0.936	82786000	357114
FRA	1212300	2761633	42931	0.901	65098000	640679
GBR	1189400	2829163	42261	0.922	66466000	242495
IDN	346100	1100911	3789	0.694	265316000	1904569
IND	850600	2971996	2016	0.640	1334221000	3287263
ITA	948600	2025866	34349	0.880	60756000	301336
JPN	1522400	5176205	40106	0.909	126431000	377930
KOR	1170900	1656674	32046	0.903	51665000	100210
MEX	813500	1241450	9614	0.774	124738000	1964375
RUS	844200	1610381	10950	0.816	143965088	17098242
SAU	521600	762259	23187	0.853	33203000	2149690
TUR	417000	706237	8716	0.791	71867000	783562
USA	3944000	21344667	62517	0.924	328116000	9526468
ZAF	200100	371298	6560	0.699	57420000	1221037

Algebraic Analysis of Social Networks: Models, Methods and Applications using R,
First Edition. J. A. R. Ostoic. Companion website: www.wiley.com/go/ostoic/algebraicanalysis.
© 2021 John Wiley & Sons Ltd. Published 2021 by John Wiley & Sons Ltd.

Commodities in G20 Trade valued network

The United Nations COMTRADE code Commodity Classifications (United Nations, 2017) for commodities are:

- *milk* that corresponds to code commodity classification 19: "Preparations of cereals, flour, starch or milk; pastrycooks' products"
- *honey* that corresponds to code commodity classification 04: "Dairy produce; birds' eggs; natural honey; edible products of animal origin, not elsewhere specified or included"

The data of the Trade valued network is from year 2017, which is the last available data online for all G20 countries at the UN COMTRADE database by the end of year 2019.

Units of measure of G20 country data

- Trade: ×1000 USD (2014)
- Nom. GDP: ×1000 USD (2019)
- Nom. GDP per capita: USD (2019)
- HDI: (2017)
- Population: (2018)
- Area: km^2

G20 Trade valued network and salient structures

The G20 Trade valued network of fresh milk and honey with a matrix format are given in Tables C.1 and C.2, while Tables C.3 and C.4 provide the salient structures of these valued structures after the application of the Triangle inequality principle to $\mathscr{X}^V_{G20mh}$.

Table C.1 Trade flow of Commodity Fresh Milk – Year 2017. *Total trade in USD ×10000.*

	ARG	AUS	BRA	CAN	CHN	DEU	FRA	GBR	IDN	IND	ITA	JPN	KOR	MEX	RUS	SAU	TUR	USA	ZAF
ARG	0	0	0	0	29	0	0	0	0	0	0	0	72	0	0	0	0	0	0
AUS	0	0	0	0	58286	0	0	0	1119	42	0	490	4119	0	0	771	0	417	0
BRA	0	0	0	0	0	0	0	0	0	0	0	0	0	0	0	0	0	589	0
CAN	0	0	0	0	648	0	6	0	0	0	0	0	0	0	0	0	0	2733	0
CHN	0	0	0	0	0	0	58	0	0	0	0	0	0	0	0	0	0	54	0
DEU	0	0	0	54	144174	0	73581	9281	21	0	252237	160	1157	0	0	1423	1641	88	0
FRA	1	32	107	177	67141	75533	0	13451	4429	297	177520	13	9208	51	0	9596	13	2537	0
GBR	0	90	0	323	17341	6415	11701	0	0	0	82	63	349	0	0	272	0	1016	0
IDN	0	0	0	0	0	0	0	0	0	0	0	0	0	0	0	0	0	0	0
IND	0	0	0	0	0	0	0	0	0	0	0	0	0	0	0	0	0	0	0
ITA	0	27	0	6	12637	2457	3553	573	0	14	0	0	202	0	0	0	0	298	0
JPN	0	0	0	0	0	0	0	0	0	0	0	0	386	0	0	0	0	0	0
KOR	0	0	0	0	17433	0	0	0	0	0	0	0	0	0	0	0	0	0	0
MEX	0	0	0	0	0	0	0	0	0	0	0	0	0	0	0	0	0	7517	0
RUS	0	0	0	0	0	0	0	0	0	0	0	4	0	0	0	0	0	1	0
SAU	0	0	0	0	0	0	0	0	0	0	0	0	0	0	0	0	337	0	0
TUR	0	0	0	0	0	1	0	0	0	0	0	0	0	0	0	624	0	0	0
USA	0	42	41	24925	1832	14	0	0	0	16	0	80	166	21737	0	137	0	0	0
ZAF	0	0	0	0	0	0	0	0	0	0	0	0	0	0	0	0	0	0	0

Table C.2 Trade flow of Commodity Honey – Year 2017. *Total trade in USD* ×10000.

	ARG	AUS	BRA	CAN	CHN	DEU	FRA	GBR	IDN	IND	ITA	JPN	KOR	MEX	RUS	SAU	TUR	USA	ZAF
ARG	0	5443	0	873	0	24633	527	1325	2694	0	3739	14421	0	0	0	3190	0	98133	0
AUS	0	0	0	3308	8604	47	5	1130	581	154	290	669	276	0	10	774	0	761	0
BRA	2	1298	0	5470	426	7812	1256	4497	0	0	180	19	0	0	0	0	0	57542	0
CAN	0	0	0	0	1974	9	173	0	0	0	0	12752	588	0	0	0	0	35997	0
CHN	0	11869	0	955	0	12987	15845	47435	82	560	8741	56440	0	0	0	1159	0	10	0
DEU	2	1	5	379	2964	0	0	5766	415	58	4929	606	47	158	0	13779	31	2948	0
FRA	0	52	0	95	1479	3311	0	1830	35	133	2180	1002	4	12	120	342	0	2191	0
GBR	0	0	75	0	691	532	657	0	7	100	1124	97	53	83	8	630	0	343	0
IDN	0	0	0	0	0	0	0	0	0	0	0	0	0	0	0	0	0	0	0
IND	0	119	0	781	0	146	2	15	0	0	0	136	0	0	0	2725	0	109056	0
ITA	0	23	13	19	199	15880	12449	2521	0	8	0	1056	16	0	13	1375	6	1184	0
JPN	0	0	0	0	28	0	0	0	0	0	0	0	3	0	0	0	0	0	0
KOR	0	2	0	0	68	0	0	0	0	0	0	30	0	0	0	0	0	0	0
MEX	0	2038	0	76	343	76658	858	15551	0	0	160	1963	0	0	0	13027	0	21400	0
RUS	0	0	0	196	5837	79	0	0	0	0	0	3	5	0	0	42	0	564	0
SAU	0	30	0	288	263	0	0	0	1321	0	0	0	0	0	0	0	0	148	0
TUR	0	10	0	374	402	5509	548	31	0	0	0	14	0	0	0	546	0	15645	0
USA	90	48	11	6064	1119	14	0	76	26	1917	23	2620	2929	138	0	251	0	0	0
ZAF	0	0	0	0	0	0	0	0	0	0	0	0	0	0	0	0	0	0	0

Table C.3 Valued network $\mathscr{D}_{G20mh}^{-V}$ with Commodity Fresh Milk – Year 2017. Total trade in USD ×10000 and dropping values less than 250. Values in gray correspond to deleted edges after the application of triangle inequality.

	ARG	AUS	BRA	CAN	CHN	DEU	FRA	GBR	IDN	IND	ITA	JPN	KOR	MEX	RUS	SAU	TUR	USA	ZAF
ARG	0	0	0	0	0	0	0	0	0	0	0	0	0	0	0	0	0	0	0
AUS	0	0	0	0	5829	0	0	0	0	0	0	0	412	0	0	0	0	0	0
BRA	0	0	0	0	0	0	0	0	0	0	0	0	0	0	0	0	0	0	0
CAN	0	0	0	0	0	0	0	0	0	0	0	0	0	0	0	0	0	273	0
CHN	0	0	0	0	0	0	0	0	0	0	0	0	0	0	0	0	0	0	0
DEU	0	0	0	0	14417	0	7358	928	0	0	25224	0	0	0	0	0	0	0	0
FRA	0	0	0	0	6714	7553	0	1345	443	0	17752	0	921	0	0	960	0	254	0
GBR	0	0	0	0	1734	641	1170	0	0	0	0	0	0	0	0	0	0	0	0
IDN	0	0	0	0	0	0	0	0	0	0	0	0	0	0	0	0	0	0	0
IND	0	0	0	0	0	0	0	0	0	0	0	0	0	0	0	0	0	0	0
ITA	0	0	0	0	1264	0	355	0	0	0	0	0	0	0	0	0	0	0	0
JPN	0	0	0	0	0	0	0	0	0	0	0	0	0	0	0	0	0	0	0
KOR	0	0	0	0	1743	0	0	0	0	0	0	0	0	0	0	0	0	0	0
MEX	0	0	0	0	0	0	0	0	0	0	0	0	0	0	0	0	0	752	0
RUS	0	0	0	0	0	0	0	0	0	0	0	0	0	0	0	0	0	0	0
SAU	0	0	0	0	0	0	0	0	0	0	0	0	0	0	0	0	0	0	0
TUR	0	0	0	0	0	0	0	0	0	0	0	0	0	0	0	0	0	0	0
USA	0	0	0	2493	0	0	0	0	0	0	0	0	0	2174	0	0	0	0	0
ZAF	0	0	0	0	0	0	0	0	0	0	0	0	0	0	0	0	0	0	0

Table C.4 Valued network $\mathcal{X}^V_{G20mh}$ with Commodity Honey – Year 2017. Total trade in USD ×10000 and dropping values less than 250. Values in gray correspond to deleted edges after the application of triangle inequality.

	ARG	AUS	BRA	CAN	CHN	DEU	FRA	GBR	IDN	IND	ITA	JPN	KOR	MEX	RUS	SAU	TUR	USA	ZAF
ARG	0	544	0	0	0	2463	0	0	269	0	374	1442	0	0	0	319	0	9813	0
AUS	0	0	0	331	860	0	0	0	0	0	0	0	0	0	0	0	0	0	0
BRA	0	0	0	547	0	781	0	450	0	0	0	0	0	0	0	0	0	5754	0
CAN	0	0	0	0	0	0	0	0	0	0	0	1275	0	0	0	0	0	3600	0
CHN	0	1187	0	0	0	1299	1585	4743	0	0	874	5644	0	0	0	0	0	0	0
DEU	0	0	0	0	296	0	0	577	0	0	493	0	0	0	0	1378	0	295	0
FRA	0	0	0	0	0	331	0	0	0	0	0	0	0	0	0	0	0	0	0
GBR	0	0	0	0	0	0	0	0	0	0	0	0	0	0	0	0	0	0	0
IDN	0	0	0	0	0	0	0	0	0	0	0	0	0	0	0	0	0	0	0
IND	0	0	0	0	0	0	0	0	0	0	0	0	0	0	0	273	0	10906	0
ITA	0	0	0	0	0	1588	1245	252	0	0	0	0	0	0	0	0	0	0	0
JPN	0	0	0	0	0	0	0	0	0	0	0	0	0	0	0	0	0	0	0
KOR	0	0	0	0	0	0	0	0	0	0	0	0	0	0	0	0	0	0	0
MEX	0	0	0	0	0	7666	0	1555	0	0	0	0	0	0	0	1303	0	2140	0
RUS	0	0	0	0	584	0	0	0	0	0	0	0	0	0	0	0	0	0	0
SAU	0	0	0	0	0	0	0	0	0	0	0	0	0	0	0	0	0	0	0
TUR	0	0	0	0	0	551	0	0	0	0	0	262	293	0	0	0	0	1565	0
USA	0	0	0	606	0	0	0	0	0	0	0	0	0	0	0	0	0	0	0
ZAF	0	0	0	0	0	0	0	0	0	0	0	0	0	0	0	0	0	0	0

Appendix D

Layout visualization algorithms

The following code is written in R (R Core Team, 2015) as implemented in package ***multigraph*** (Ostoic, 2019). Package ***stats***, which is a core library of R, is for random data generation with a seed based on the current clock of the computer. A previous function `layout_spring_adj` from Dunning, I. (2020) written in julia language has more details on the trigonometry of the two forces.

Force-directed

The Force-directed layout is described in Fruchterman and Reingold (1991).

```
## FORCE-DIRECTED LAYOUT ALGORITHM
## FUNCTION    forced()
## INPUT:      x (square and binary matrix with network data)
## ARGUMENTS:  seed (seed for the initial random distribution)
##             maxiter (number of iterations)


R> forced <- function (x, seed = NULL, maxiter = 100) {

# matrix size is the number of rows
n = nrow(x)

# initial coordinates X and Y with a seed
set.seed(seed)

# record the generated random uniformed distributed data as data frame 'crd'
crd = data.frame(X = round(stats::runif(n)*1L,5),Y = round(stats::runif(n)*1L,5))

# 1st and 2nd columns of 'crd' are "location" coordinates X and Y
locx = crd[,1]
locy = crd[,2]
```

```r
# a factor of the square root of the ratio between product of the difference of
# max and min values of the two locations to the matrix size
k = .75 * sqrt(((max(locx) - min(locx)) * (max(locy) - min(locy))) / n)

# vectors with "forces" on locations X and Y are initial zero
forcex = rep(0, n)
forcey = rep(0, n)

# iteration begins from 1 to the indicated 'maxiter'
for(niter in seq_len(maxiter)) { #

 # start with rows 1 to n
 for(i in seq_len(n)) { #

  # scalar forces for X and Y are initial zero
  forcevx = 0
  forcevy = 0

  # for columns reverse n to 1
  for(j in rev(seq_len(n))) {

   # define distance of elements i and j in x if these are not equal
   if(i != j) {

   dx = locx[j] - locx[i]
   dy = locy[j] - locy[i]
   # distance formula
   d = sqrt(dx^2 + dy^2)

   # apply a quadratic force when i and j are linked
   if(x[i, j] != 0) {
       force = (d / k) -k^2 / d^2
   # or a repulsive force otherwise
       } else if(x[i, j] == 0) {
       force = -k^2 / d^2
   }

   # update scalars with distances
   forcevx = forcevx + force * dx
   forcevy = forcevy + force * dy
   }

  }

  # update force vectors with scalars
  forcex[i] = forcevx
  forcey[i] = forcevy
```

```
}

# update n locations in X and Y and cool down with a 'scala'
for(i in seq_len(n)) {

    # distance between forces
    fd = sqrt(forcex[i]^2 + forcey[i]^2)
    scala = min(fd, 2L / niter) / fd

    locx[i] = locx[i] + (forcex[i] * scala)
    locy[i] = locy[i] + (forcey[i] * scala)

}

}

# initial coordinates are updated
crd[,1] = locx
crd[,2] = locy

return(crd)
}
```

The output in object crd represents the coordinates for plotting graphs with a force-directed layout algorithm.

Stress-majorization

The energy of the layout or stress function is described in Kruskal and Seery (1980) (and in Gansner et al. (2005)) where the classical Multidimensional Scaling (Kruskal, 1964) stress function is optimized via majorization to guarantee converge.

```
## STRESS-MAJORIZATION LAYOUT ALGORITHM
## FUNCTION    stsm()
## INPUT:      net (square and binary matrix with network data)
## ARGUMENTS:  seed (seed for the initial random distribution)
##             maxiter (number of iterations)

R> stsm <- function (net, seed = NULL, maxiter = 100) {

    # network order and matrix weights distances
    n = dim(net)[1]
    mwd = delta^(-2)
    mwd[which(mwd == Inf)] = 0
```

```
# compute matrix weights of ideal pairwise distances
lpmwd = as.matrix(mwd)
for (i in seq_len(n)) {
    for (j in seq_len(n)) {
        ifelse(isTRUE(as.matrix(mwd)[i, j] != 0) == TRUE,
            lpmwd[i, j] = -1 * as.matrix(mwd)[i, j], NA)
    }
}
diag(lpmwd) = apply(as.matrix(mwd), 1, sum)
s = svd(lpmwd)
p = (s$d > max(.Machine$double.eps^(2/3) * s$d[1], 0))
if (all(p)) {
    pilpmwd = s$v %*% (1/s$d * t(s$u))
}
else if (any(p)) {
    pilpmwd = s$v[, p, drop = FALSE] %*% (1L/s$d[p] * t(s$u[,
        p, drop = FALSE]))
}
else {
    pilpmwd = matrix(0, nrow = ncol(lpmwd), ncol = nrow(lpmwd))
}
# initial and updated coordinates from normal random variables
set.seed(seed)
vec = stats::rnorm(n * 2L)
X0 = cbind(vec[seq_len(n)], vec[(n + 1):length(vec)])
Xs = as.matrix(X0)
# compute new stress and weighted Laplacian internal functions
newstss = sts(X0, delta = delta, mwd = mwd)
for (iter in seq_len(maxiter)) {
  if (any(is.nan(lz(X0, delta, mwd))) == FALSE) {
  X = pilpmwd %*% (lz(X0, delta, mwd) %*% X0)
  X[which(X == Inf)] = 0
  oldstss = newstss
  newstss = sts(X, delta = delta, mwd = mwd)
  Xs = X
  # smallest positive floating-point number for absolute/relative tolerances
  abstols = abstolx = reltols = sqrt(.Machine$double.eps)
  ifelse(isTRUE(abs(newstss - oldstss) < (reltols *
      newstss)) == TRUE, break, NA)
  ifelse(isTRUE(abs(newstss - oldstss) < abstols) ==
      TRUE, break, NA)
  ifelse(isTRUE(norm(X - X0, type = "F") < abstolx) ==
      TRUE, break, NA)
  X0 = X
  }
}
# split network components and isolates
cmps = multiplex::comps(netd)
```

```
nds = Xs
ifelse(isTRUE(sum(nds) == 0) == TRUE, rat = 1L, rat = (max(nds[,
    1]) - min(nds[, 1]))/(max(nds[, 2]) - min(nds[, 2])))
if (isTRUE(length(cmps$isol) > 1) == TRUE) {
    nds[, 1] = (nds[, 1] - min(nds[, 1]))/(max(nds[, 1]) -
        min(nds[, 1]))
    ifelse(isTRUE(rat > 0) == TRUE, nds[, 2] = ((nds[, 2] -
        min(nds[, 2]))/(max(nds[, 2]) - min(nds[, 2]))) *
        (1L/rat), nds[, 2] = ((nds[, 2] - min(nds[, 2]))/(max(nds[,
        2]) - min(nds[, 2]))) * (rat))
    nds = as.matrix((nds))
    ndst = nds[which(nds[, 1] != 0),]
    # internal function popl() for populating coordinates for isolates
    tmpi = popl(length(cmps$isol), seed = seed)/(length(cmps$isol) *
        2) * length(cmps$isol)
    if (is.null(cmps$com) == FALSE) {
        locx = ((tmpi[, 1]/3L) - (min(ndst[, 1])) - 0)
        ifelse(isTRUE(rat > 0) == TRUE, locy = ((min(ndst[,
            2])) - (tmpi[, 2]/3L) - 0), locy = ((max(ndst[,
            2])) + (tmpi[, 2]/3L) + 0))
        ndst.chull = grDevices::chull(ndst)
        ndst.chull = ndst[ndst.chull,]
        ifelse(isTRUE(length(which(ndst.chull[, 1] < mean(ndst.chull[,
            1]))) > length(which(ndst.chull[, 1] > mean(ndst.chull[,
            1])))) == TRUE, locx = locx + (1L/n), locx = locx +
            ((1L/n) * -1))
        ifelse(isTRUE(length(which(ndst.chull[, 2] < mean(ndst.chull[,
            2]))) > length(which(ndst.chull[, 2] > mean(ndst.chull[,
            2])))) == TRUE, locy = locy - (1L/n), locy = locy -
            ((1L/n) * -1))
    }
    else {
        locx = (tmpi[, 1])
        locy = (tmpi[, 2])
    }
    nds[which(lbs %in% cmps$isol), ] = (cbind(locx, locy))
}
else if (isTRUE(length(cmps$isol) == 1L) == TRUE) {
    ndst = nds[which(nds[, 1] != 0),]
    locx = max(ndst[, 1]) + (1L/n)
    ifelse(isTRUE(rat < 0) == TRUE, locy = max(ndst[, 2]) +
        (1L/n), locy = min(ndst[, 2]) - (1L/n))
    nds[which(lbs %in% cmps$isol), ] = (cbind(locx, locy))
}
# update coordinates
nds[, 1] = (nds[, 1] - min(nds[, 1]))/(max(nds[, 1]) - min(nds[,
    1])) + (1L/n)
ifelse(isTRUE(rat > 0) == TRUE, nds[, 2] = ((nds[, 2] -
```

```
          min(nds[, 2]))/(max(nds[, 2]) - min(nds[, 2]))) * (1L/rat),
          nds[, 2] = ((nds[, 2] - min(nds[, 2]))/(max(nds[, 2]) -
              min(nds[, 2]))) * (rat) + (1L/n))
      nds[, 2] = nds[, 2] * -1

  return(nds)
  }
```

The output in object **nds** represents the coordinates for plotting graphs with stress-majorization layout algorithm.

Laplacian Function

```
## WEIGHTED LAPLACIAN
## FUNCTION   lz()
## INPUT:     x (data frame with current coordinates)
## ARGUMENTS: delta (ideal pairwise distances)
##            w (weights)

R> lz <- function (x, delta, w) {
    ifelse(missing(delta) == TRUE, delta = matrix(1, nrow(x),
        nrow(x)), NA)
    L = matrix(0, nrow = attr(w, "Size"), ncol = attr(w, "Size"))
    for (i in seq_len(nrow(x))) {
        D = 0
        for (j in seq_len(nrow(x))) {
            if (isTRUE(i != j) == TRUE) {
                nrmz = norm(x[i,] - x[j,], type = "2")
                delt = as.matrix(w)[i, j] * as.matrix(delta)[i, j]
                L[i, j] = ((-delt)/nrmz)
                D = D - ((-delt)/nrmz)
            }
            else {
                nrmz = 0
            }
        }
        L[i, i] = D
    }
    L[which(L == Inf)] = 0
    L
}
```

New stress internal function

```
## NEW STRESS INTERNAL
## FUNCTION    sts()
## INPUT:      nds (node coordinates)
## ARGUMENTS:  delta (ideal pairwise distances)
##             mwd (matrix weights distances)

R> sts <- function (nds, delta = NULL, mwd = NULL) {
    ifelse(is.null(delta) == TRUE, delta = matrix(1, nrow(nds),
        nrow(nds)), NA)
    strss = 0
    mnds = as.matrix(nds)
    if (is.null(mwd)) {
        mwd = delta^(-2)
        mwd[which(mwd == Inf)] = 0
    }
    for (j in seq_len(nrow(nds))) {
        for (i in seq_len(j - 1)) {
            if (isTRUE(i > 0 && j > 1) == TRUE) {
                strss = strss + as.matrix(mwd)[i, j] %*% (norm(mnds[i,
                    ] - mnds[j,], type = "2") - as.matrix(delta)[i, j])^2
            }
            else {
                NA
            }
        }
    }
    strss
}
```

Appendix E

Role structure workflow

Decomposition of Role structure $\mathcal{Q}_B$

This appendix provides the workflow for the construction and decomposition of the Role structure $\mathcal{Q}_B$, which is depicted in the cover of this book. However, any decomposition of a Role structure requires having the Role structure first, and that is why we start with the positional analysis of the network by creating classes of compositional equivalent actors since this network is multiplex. A next step is the construction of its Role structure, which results being a large system that requires a further decomposition, in this case by a factorization procedure.

Incubator network B

```
# load Incubator network B dataset
R> require("multiplex")
R> data(incB)
```

Positional analysis and Role structure

```
# Relation-Box
R> rb_B <- rbox(incB$net, transp = TRUE, tlbs = c("D","G","L",NA,NA))
# positional system
R> ps_B <- reduc(rb_B$w, clu = c(2,2,1,4,4,1,2,2,3,2,3,2,4,4,1,2,2,4))
# partially ordered semigroup
R> S_B <- semigroup(ps_B, type = "symbolic")
R> P_B <- partial.order(strings(ps_B))
```

Factorization

```
# induced inclusions from factorization of the partially ordered semigroup;
# include meet-complements of atoms, and potential atoms
R> ii_B <- fact(S = S_B, P = P_B, atmc = TRUE, patm = TRUE)
```

```
# π-relations from the induced inclusions
R> pr_B <- pi.rels(ii_B, po.incl = TRUE)
#Warning message:
```

```
# decomposition of π-relations with meet-complement of the atoms
# (force extra iteration)
R> sB <- decomp(S = S_B, pr = pr_B, type = "mca", reduc = TRUE, force = TRUE)
```

```
# all induced inclusions and π-relations in 'sB'
R> ii_B1 <- list()
R> pr_B1 <- list()
R> for(k in seq_along(sB$ord)) {
     ii_B1[[k]] <- fact(sB$IM[[k]], sB$P[[k]], atmc = TRUE, patm = TRUE)
     pr_B1[[k]] <- pi.rels(ii_B1[[k]], po.incl = TRUE)
}
```

```
# perform a decomposition of the six image matrices
R> sB2 <- list()
R> for(k in seq_len(length(pr_B1))) {
     sB2[[k]] <- decomp(as.semigroup(sB$IM[[k]]), pr_B1[[k]], type = "mca",
+      reduc = TRUE)
}
```

```
# how many quotient semigroups?
R> length(sB2)
#[1] 6

# orders of images matrices
R> sB2[[1]]$ord
#[1] 1
R> sB2[[2]]$ord
#[1] 21 18 18
R> sB2[[3]]$ord
#[1] 1
```

```
R> sB2[[4]]$ord
#[1] 23
R> sB2[[5]]$ord
#[1] 1
R> sB2[[6]]$ord
#[1] 23
```

Then decompose further Images 2, 4, and 6.

Progressive factorization of Factors

Image matrices 2, 4, and 6 now correspond to Factors 1, 2, and 3 where T means TRUE.

Factor 1

```
###################### FACTOR 1  (Image 2)


###################### B21 (2nd image in 'sB2' is Factor 1)
S3B21 <- sB2[[2]]$IM[[1]]
P3B21 <- sB2[[2]]$PO[[1]]
#
sB2$ord
#[1] 21

###################### progressive factorisation of 'S3B21'
#
ii_B21 <- fact(S3B21, P3B21, atmc=T, patm=T)
fct_B21 <- decomp(as.semigroup(S3B21), pi.rels(ii_B21, po.incl=T), type="a", reduc=T)
fct_B21$ord
#[1] 20
#
ii_B211 <- fact(fct_B21$IM, fct_B21$PO, atmc=T, patm=T)
fct_B211 <- decomp(as.semigroup(fct_B21$IM), pi.rels(ii_B211, po.incl=T),
+    type="a", reduc=T)
# two atoms ...
fct_B211$ord
#[1] 15 15

#### B2111
ii_B2111 <- fact(fct_B211$IM[[1]], fct_B211$PO[[1]], atmc=T, patm=T)
fct_B2111 <- decomp(as.semigroup(fct_B211$IM[[1]]), pi.rels(ii_B2111, po.incl=T),
+    type="a", reduc=T)
fct_B2111$ord
#[1] 11
#

#### B2112
ii_B2112 <- fact(fct_B211$IM[[2]], fct_B211$PO[[2]], atmc=T, patm=T)
fct_B2112 <- decomp(as.semigroup(fct_B211$IM[[2]]), pi.rels(ii_B2112, po.incl=T),
+    type="a", reduc=T)
fct_B2112$ord
#[1] 11
#


###################### B22
S3B22 <- sB2[[2]]$IM[[2]]
P3B22 <- sB2[[2]]$PO[[2]]
#
ii_B22 <- fact(S3B22, P3B22, atmc=T, patm=T)
fct_B22 <- decomp(as.semigroup(S3B22), pi.rels(ii_B22, po.incl=T), type="a", reduc=T)
# no equations
ii_B22x <- fact(fct_B22$IM, fct_B22$PO, atmc=T, patm=T)
fct_B22x <- decomp(as.semigroup(fct_B22$IM), pi.rels(ii_B22x, po.incl=T), type="a",
+    reduc=T)
# two atoms
fct_B22x$ord
#[1] 14 15
```

```
#### B22x1
ii_B22x1 <- fact(fct_B22x$IM[[1]], fct_B22x$PO[[1]], atmc=T, patm=T)
fct_B22x1 <- decomp(as.semigroup(fct_B22x$IM[[1]]), pi.rels(ii_B22x1, po.incl=T),
+    type="at", reduc=T)
#$ord
#[1] 11
#


#### B22x2
ii_B22x2 <- fact(fct_B22x$IM[[2]], fct_B22x$PO[[2]], atmc=T, patm=T)
fct_B22x2 <- decomp(as.semigroup(fct_B22x$IM[[2]]), pi.rels(ii_B22x2, po.incl=T),
+    type="at", reduc=T)
#$ord
#[1] 11
#


##################### B23
S3B23 <- sB2[[2]]$IM[[3]]
P3B23 <- sB2[[2]]$PO[[3]]
#
ii_B23 <- fact(S3B23, P3B23, atmc=T, patm=T)
fct_B23 <- decomp(as.semigroup(S3B23), pi.rels(ii_B23, po.incl=T), type="at", reduc=T)
# no equations
ii_B23x <- fact(fct_B23$IM, fct_B23$PO, atmc=T, patm=T)
fct_B23x <- decomp(as.semigroup(fct_B23$IM), pi.rels(ii_B23x, po.incl=T), type="at",
+    reduc=T)
# two atoms
fct_B23x$ord
#[1] 15 14

#### B23x1
ii_B23x1 <- fact(fct_B23x$IM[[1]], fct_B23x$PO[[1]], atmc=T, patm=T)
fct_B23x1 <- decomp(as.semigroup(fct_B23x$IM[[1]]), pi.rels(ii_B23x1, po.incl=T),
+    type="at", reduc=T)
#$ord
#[1] 11
#

#### B23x2
ii_B23x2 <- fact(fct_B23x$IM[[2]], fct_B23x$PO[[2]], atmc=T, patm=T)
fct_B23x2 <- decomp(as.semigroup(fct_B23x$IM[[2]]), pi.rels(ii_B23x2, po.incl=T),
+    type="at", reduc=T)
#$ord
#[1] 11
#
```

Factor 2

```
##################### FACTOR 2  (Image 4)

##################### B4  (4th image in 'sB2' is Factor 2)
S3B4 <- sB2[[4]]$IM
P3B4 <- sB2[[4]]$PO
#
ii_B4 <- fact(S3B4, P3B4, atmc=T, patm=T)
fct_B4 <- decomp(as.semigroup(S3B4), pi.rels(ii_B4, po.incl=T), type="at", reduc=T)
#$ord
#[1] 23 23
# two atoms, no equations

#### B4x1
ii_B4x1 <- fact(fct_B4$IM[[1]], fct_B4$PO[[1]], atmc=T, patm=T)
fct_B4x1 <- decomp(as.semigroup(fct_B4$IM[[1]]), pi.rels(ii_B4x1, po.incl=T),
+    type="at", reduc=T)
#$ord
#[1] 23 19

### B4x11
ii_B4x11 <- fact(fct_B4x1$IM[[1]], fct_B4x1$PO[[1]], atmc=T, patm=T)
fct_B4x11 <- decomp(as.semigroup(fct_B4x1$IM[[1]]), pi.rels(ii_B4x11, po.incl=T),
+    type="at", reduc=T)
#$ord
#[1] 18 18 19

## B4x111
ii_B4x111 <- fact(fct_B4x11$IM[[1]], fct_B4x11$PO[[1]], atmc=T, patm=T)
fct_B4x111 <- decomp(as.semigroup(fct_B4x11$IM[[1]]), pi.rels(ii_B4x111, po.incl=T),
+    type="at", reduc=T)
#[1] 14 14
```

```
# B4x1111
ii_B4x1111 <- fact(fct_B4x111$IM[[1]], fct_B4x111$PO[[1]], atmc=T, patm=T)
fct_B4x1111 <- decomp(as.semigroup(fct_B4x111$IM[[1]]), pi.rels(ii_B4x1111, po.incl=T),
+    type="at", reduc=T)
#[1]  11
#


# B4x1112
ii_B4x1112 <- fact(fct_B4x111$IM[[2]], fct_B4x111$PO[[2]], atmc=T, patm=T)
fct_B4x1112 <- decomp(as.semigroup(fct_B4x111$IM[[2]]), pi.rels(ii_B4x1112, po.incl=T),
+    type="at", reduc=T)
#[1]  11
#
ii_B4x1112x <- fact(fct_B4x1112$IM, fct_B4x1112$PO, atmc=T, patm=T)
fct_B4x1112x <- decomp(as.semigroup(fct_B4x1112$IM), pi.rels(ii_B4x1112x, po.incl=T),
+    type="at", reduc=T)
#$ord
#[1]  C


## B4x112
ii_B4x112 <- fact(fct_B4x11$IM[[2]], fct_B4x11$PO[[2]], atmc=T, patm=T)
fct_B4x112 <- decomp(as.semigroup(fct_B4x11$IM[[2]]), pi.rels(ii_B4x112, po.incl=T),
+    type="at", reduc=T)
#[1]  14 15

# B4x1121
ii_B4x1121 <- fact(fct_B4x112$IM[[1]], fct_B4x112$PO[[1]], atmc=T, patm=T)
fct_B4x1121 <- decomp(as.semigroup(fct_B4x112$IM[[1]]), pi.rels(ii_B4x1121, po.incl=T),
+    type="at", reduc=T)
#[1]  11
#

# B4x1122
ii_B4x1122 <- fact(fct_B4x112$IM[[2]], fct_B4x112$PO[[2]], atmc=T, patm=T)
fct_B4x1122 <- decomp(as.semigroup(fct_B4x112$IM[[2]]), pi.rels(ii_B4x1122, po.incl=T),
+    type="at", reduc=T)
#[1]  11
#


## B4x113
ii_B4x113 <- fact(fct_B4x11$IM[[3]], fct_B4x11$PO[[3]], atmc=T, patm=T)
fct_B4x113 <- decomp(as.semigroup(fct_B4x11$IM[[3]]), pi.rels(ii_B4x113, po.incl=T),
+    type="at", reduc=T)
#[1]  14 15

# B4x1131
ii_B4x1131 <- fact(fct_B4x113$IM[[1]], fct_B4x113$PO[[1]], atmc=T, patm=T)
fct_B4x1131 <- decomp(as.semigroup(fct_B4x113$IM[[1]]), pi.rels(ii_B4x1131, po.incl=T),
+    type="at", reduc=T)
#[1]  11
#


# B4x1132
ii_B4x1132 <- fact(fct_B4x113$IM[[2]], fct_B4x113$PO[[2]], atmc=T, patm=T)
fct_B4x1132 <- decomp(as.semigroup(fct_B4x113$IM[[2]]), pi.rels(ii_B4x1132, po.incl=T),
+    type="at", reduc=T)
#[1]  11
#


### B4x12
ii_B4x12 <- fact(fct_B4x1$IM[[2]], fct_B4x1$PO[[2]], atmc=T, patm=T)
fct_B4x12 <- decomp(as.semigroup(fct_B4x1$IM[[2]]), pi.rels(ii_B4x12, po.incl=T),
+    type="at", reduc=T)
#$ord
#[1]  19
#
ii_B4x12x <- fact(fct_B4x12$IM, fct_B4x12$PO, atmc=T, patm=T)
fct_B4x12x <- decomp(as.semigroup(fct_B4x12$IM), pi.rels(ii_B4x12x, po.incl=T),
+    type="at", reduc=T)
#$ord
#[1]  14 15

## B4x121
ii_B4x12x1 <- fact(fct_B4x12x$IM[[1]], fct_B4x12x$PO[[1]], atmc=T, patm=T)
fct_B4x12x1 <- decomp(as.semigroup(fct_B4x12x$IM[[1]]), pi.rels(ii_B4x12x1, po.incl=T),
+    type="at", reduc=T)
#[1]  11
#

## B4x122
ii_B4x12x2 <- fact(fct_B4x12x$IM[[2]], fct_B4x12x$PO[[2]], atmc=T, patm=T)
fct_B4x12x2 <- decomp(as.semigroup(fct_B4x12x$IM[[2]]), pi.rels(ii_B4x12x2, po.incl=T),
```

```
+     type="at", reduc=T)
#[1] 11
#


#### B4x2
ii_B4x2 <- fact(fct_B4$IM[[2]], fct_B4$PO[[2]], atmc=T, patm=T)
fct_B4x2 <- decomp(as.semigroup(fct_B4$IM[[2]]), pi.rels(ii_B4x2, po.incl=T), type="at",
+     reduc=T)
#$ord
#[1] 23 18


### B4x21
ii_B4x21 <- fact(fct_B4x2$IM[[1]], fct_B4x2$PO[[1]], atmc=T, patm=T)
fct_B4x21 <- decomp(as.semigroup(fct_B4x2$IM[[1]]), pi.rels(ii_B4x21, po.incl=T),
+     type="at", reduc=T)
#$ord
#[1] 18 18 19


## B4x211
ii_B4x211 <- fact(fct_B4x21$IM[[1]], fct_B4x21$PO[[1]], atmc=T, patm=T)
fct_B4x211 <- decomp(as.semigroup(fct_B4x21$IM[[1]]), pi.rels(ii_B4x211, po.incl=T),
+     type="at", reduc=T)
#$ord
#[1] 14 14


# B4x2111
ii_B4x2111 <- fact(fct_B4x211$IM[[1]], fct_B4x211$PO[[1]], atmc=T, patm=T)
fct_B4x2111 <- decomp(as.semigroup(fct_B4x211$IM[[1]]), pi.rels(ii_B4x2111, po.incl=T),
+     type="at", reduc=T)
#[1] 11
#


# B4x2112
ii_B4x2112 <- fact(fct_B4x211$IM[[2]], fct_B4x211$PO[[2]], atmc=T, patm=T)
fct_B4x2112 <- decomp(as.semigroup(fct_B4x211$IM[[2]]), pi.rels(ii_B4x2112, po.incl=T),
+     type="at", reduc=T)
#[1] 11
#


## B4x212
ii_B4x212 <- fact(fct_B4x21$IM[[2]], fct_B4x21$PO[[2]], atmc=T, patm=T)
fct_B4x212 <- decomp(as.semigroup(fct_B4x21$IM[[2]]), pi.rels(ii_B4x212, po.incl=T),
+     type="at", reduc=T)
#$ord
#[1] 14 15


# B4x2121
ii_B4x2121 <- fact(fct_B4x212$IM[[1]], fct_B4x212$PO[[1]], atmc=T, patm=T)
fct_B4x2121 <- decomp(as.semigroup(fct_B4x212$IM[[1]]), pi.rels(ii_B4x2121, po.incl=T),
+     type="at", reduc=T)
#[1] 11
#


# B4x2122
ii_B4x2122 <- fact(fct_B4x212$IM[[2]], fct_B4x212$PO[[2]], atmc=T, patm=T)
fct_B4x2122 <- decomp(as.semigroup(fct_B4x212$IM[[2]]), pi.rels(ii_B4x2122, po.incl=T),
+     type="at", reduc=T)
#[1] 11
#


## B4x213
ii_B4x213 <- fact(fct_B4x21$IM[[3]], fct_B4x21$PO[[3]], atmc=T, patm=T)
fct_B4x213 <- decomp(as.semigroup(fct_B4x21$IM[[3]]), pi.rels(ii_B4x213, po.incl=T),
+     type="at", reduc=T)
#$ord
#[1] 14 15


# B4x2131
ii_B4x2131 <- fact(fct_B4x213$IM[[1]], fct_B4x213$PO[[1]], atmc=T, patm=T)
fct_B4x2131 <- decomp(as.semigroup(fct_B4x213$IM[[1]]), pi.rels(ii_B4x2131, po.incl=T),
+     type="at", reduc=T)
#[1] 11
#


# B4x2132
ii_B4x2132 <- fact(fct_B4x213$IM[[2]], fct_B4x213$PO[[2]], atmc=T, patm=T)
fct_B4x2132 <- decomp(as.semigroup(fct_B4x213$IM[[2]]), pi.rels(ii_B4x2132, po.incl=T),
+     type="at", reduc=T)
```

```
#[1] 11
#


### B4x22
ii_B4x22 <- fact(fct_B4x2$IM[[2]], fct_B4x2$PO[[2]], atmc=T, patm=T)
fct_B4x22 <- decomp(as.semigroup(fct_B4x2$IM[[2]]), pi.rels(ii_B4x22, po.incl=T),
+    type="at", reduc=T)
#$ord
#[1] 18
#
ii_B4x22x <- fact(fct_B4x22$IM, fct_B4x22$PO, atmc=T, patm=T)
fct_B4x22x <- decomp(as.semigroup(fct_B4x22$IM), pi.rels(ii_B4x22x, po.incl=T),
+    type="at", reduc=T)
#$ord
#[1] 14 15

## B4x221
ii_B4x22x1 <- fact(fct_B4x22x$IM[[1]], fct_B4x22x$PO[[1]], atmc=T, patm=T)
fct_B4x22x1 <- decomp(as.semigroup(fct_B4x22x$IM[[1]]), pi.rels(ii_B4x22x1, po.incl=T),
+    type="at", reduc=T)
#[1] 11
#


## B4x222
ii_B4x22x2 <- fact(fct_B4x22x$IM[[2]], fct_B4x22x$PO[[2]], atmc=T, patm=T)
fct_B4x22x2 <- decomp(as.semigroup(fct_B4x22x$IM[[2]]), pi.rels(ii_B4x22x2, po.incl=T),
+    type="at", reduc=T)
#[1] 11
#
```

Factor 3

```
################### FACTOR 3  (Image 6)


################### B6  (6th image in 'sB2' is Factor 3)
S3B6 <- sB2[[6]]$IM
P3B6 <- sB2[[6]]$PO
#
ii_B6 <- fact(S3B6, P3B6, atmc=T, patm=T)
fct_B6 <- decomp(as.semigroup(S3B6), pi.rels(ii_B6, po.incl=T), type="at", reduc=T)
#$ord
#[1] 23 23
# two atoms, no equations


#### B6x1
ii_B6x1 <- fact(fct_B6$IM[[1]], fct_B6$PO[[1]], atmc=T, patm=T)
fct_B6x1 <- decomp(as.semigroup(fct_B6$IM[[1]]), pi.rels(ii_B6x1, po.incl=T), type="at",
+    reduc=T)
#$ord
#[1] 18 23

### B6x11
ii_B6x11 <- fact(fct_B6x1$IM[[1]], fct_B6x1$PO[[1]], atmc=T, patm=T)
fct_B6x11 <- decomp(as.semigroup(fct_B6x1$IM[[1]]), pi.rels(ii_B6x11, po.incl=T),
+    type="at", reduc=T)
#$ord
#[1] 18
ii_B6x11x <- fact(fct_B6x11$IM, fct_B6x11$PO, atmc=T, patm=T)
fct_B6x11x <- decomp(as.semigroup(fct_B6x11$IM), pi.rels(ii_B6x11x, po.incl=T),
+    type="at", reduc=T)
#$ord
#[1] 15 14

## B6x111
ii_B6x11x1 <- fact(fct_B6x11x$IM[[1]], fct_B6x11x$PO[[1]], atmc=T, patm=T)
fct_B6x11x1 <- decomp(as.semigroup(fct_B6x11x$IM[[1]]), pi.rels(ii_B6x11x1, po.incl=T),
+    type="at", reduc=T)
#[1] 11
#


## B6x112
ii_B6x11x2 <- fact(fct_B6x11x$IM[[2]], fct_B6x11x$PO[[2]], atmc=T, patm=T)
fct_B6x11x2 <- decomp(as.semigroup(fct_B6x11x$IM[[2]]), pi.rels(ii_B6x11x2, po.incl=T),
+    type="at", reduc=T)
#[1] 11
#
```

```
### B6x12
ii_B6x12 <- fact(fct_B6x1SIM[[2]], fct_B6x1SPO[[2]], atmc=T, patm=T)
fct_B6x12 <- decomp(as.semigroup(fct_B6x1SIM[[2]]), pi.rels(ii_B6x12, po.incl=T),
+    type="a", reduc=T)
#Sord
#[1] 19 18 18

## B6x121
ii_B6x121 <- fact(fct_B6x12SIM[[1]], fct_B6x12SPO[[1]], atmc=T, patm=T)
fct_B6x121 <- decomp(as.semigroup(fct_B6x12SIM[[1]]), pi.rels(ii_B6x121, po.incl=T),
+    type="a", reduc=T)
#Sord
#[1] 15 14

# B6x1211
ii_B6x1211 <- fact(fct_B6x121SIM[[1]], fct_B6x121SPO[[1]], atmc=T, patm=T)
fct_B6x1211 <- decomp(as.semigroup(fct_B6x121SIM[[1]]), pi.rels(ii_B6x1211, po.incl=T),
+    type="a", reduc=T)
#[1] 11
#


# B6x1212
ii_B6x1212 <- fact(fct_B6x121SIM[[2]], fct_B6x121SPO[[2]], atmc=T, patm=T)
fct_B6x1212 <- decomp(as.semigroup(fct_B6x121SIM[[2]]), pi.rels(ii_B6x1212, po.incl=T),
+    type="a", reduc=T)
#[1] 11
#


## B6x122
ii_B6x122 <- fact(fct_B6x12SIM[[2]], fct_B6x12SPO[[2]], atmc=T, patm=T)
fct_B6x122 <- decomp(as.semigroup(fct_B6x12SIM[[2]]), pi.rels(ii_B6x122, po.incl=T),
+    type="a", reduc=T)
#Sord
#[1] 15 14

# B6x1221
ii_B6x1221 <- fact(fct_B6x122SIM[[1]], fct_B6x122SPO[[1]], atmc=T, patm=T)
fct_B6x1221 <- decomp(as.semigroup(fct_B6x122SIM[[1]]), pi.rels(ii_B6x1221, po.incl=T),
+    type="a", reduc=T)
#[1] 11
#


# B6x1222
ii_B6x1222 <- fact(fct_B6x122SIM[[2]], fct_B6x122SPO[[2]], atmc=T, patm=T)
fct_B6x1222 <- decomp(as.semigroup(fct_B6x122SIM[[2]]), pi.rels(ii_B6x1222, po.incl=T),
+    type="a", reduc=T)
#[1] 11
#


## B6x123
ii_B6x123 <- fact(fct_B6x12SIM[[3]], fct_B6x12SPO[[3]], atmc=T, patm=T)
fct_B6x123 <- decomp(as.semigroup(fct_B6x12SIM[[3]]), pi.rels(ii_B6x123, po.incl=T),
+    type="a", reduc=T)
#Sord
#[1] 14 14

# B6x1231
ii_B6x1231 <- fact(fct_B6x123SIM[[1]], fct_B6x123SPO[[1]], atmc=T, patm=T)
fct_B6x1231 <- decomp(as.semigroup(fct_B6x123SIM[[1]]), pi.rels(ii_B6x1231, po.incl=T),
+    type="a", reduc=T)
#[1] 11
#


# B6x1232
ii_B6x1232 <- fact(fct_B6x123SIM[[2]], fct_B6x123SPO[[2]], atmc=T, patm=T)
fct_B6x1232 <- decomp(as.semigroup(fct_B6x123SIM[[2]]), pi.rels(ii_B6x1232, po.incl=T),
+    type="a", reduc=T)
#[1] 11
#


#### B6x2
ii_B6x2 <- fact(fct_B6SIM[[2]], fct_B6SPO[[2]], atmc=T, patm=T)
fct_B6x2 <- decomp(as.semigroup(fct_B6SIM[[2]]), pi.rels(ii_B6x2, po.incl=T), type="a",
+    reduc=T)
#Sord
#[1] 19 23
```

```
### B6x21
ii_B6x21 <- fact(fct_B6x2$IM[[1]], fct_B6x2$PO[[1]], atmc=T, patm=T)
fct_B6x21 <- decomp(as.semigroup(fct_B6x2$IM[[1]]), pi.rels(ii_B6x21, po.incl=T),
+    type="at", reduc=T)
#$ord
#[1] 19
#
ii_B6x21x <- fact(fct_B6x21$IM, fct_B6x21$PO, atmc=T, patm=T)
fct_B6x21x <- decomp(as.semigroup(fct_B6x21$IM), pi.rels(ii_B6x21x, po.incl=T),
+    type="at", reduc=T)
#$ord
#[1] 15 14

## B6x211
ii_B6x21x1 <- fact(fct_B6x21x$IM[[1]], fct_B6x21x$PO[[1]], atmc=T, patm=T)
fct_B6x21x1 <- decomp(as.semigroup(fct_B6x21x$IM[[1]]), pi.rels(ii_B6x21x1, po.incl=T),
+    type="at", reduc=T)
#[1] 11
#

## B6x212
ii_B6x21x2 <- fact(fct_B6x21x$IM[[2]], fct_B6x21x$PO[[2]], atmc=T, patm=T)
fct_B6x21x2 <- decomp(as.semigroup(fct_B6x21x$IM[[2]]), pi.rels(ii_B6x21x2, po.incl=T),
+    type="at", reduc=T)
#[1] 11
#


### B6x22
ii_B6x22 <- fact(fct_B6x2$IM[[2]], fct_B6x2$PO[[2]], atmc=T, patm=T)
fct_B6x22 <- decomp(as.semigroup(fct_B6x2$IM[[2]]), pi.rels(ii_B6x22, po.incl=T),
+    type="at", reduc=T)
#$ord
#[1] 19 18 18

## B6x221
ii_B6x221 <- fact(fct_B6x22$IM[[1]], fct_B6x22$PO[[1]], atmc=T, patm=T)
fct_B6x221 <- decomp(as.semigroup(fct_B6x22$IM[[1]]), pi.rels(ii_B6x221, po.incl=T),
+    type="at", reduc=T)
#$ord
#[1] 15 14

# B6x2211
ii_B6x2211 <- fact(fct_B6x221$IM[[1]], fct_B6x221$PO[[1]], atmc=T, patm=T)
fct_B6x2211 <- decomp(as.semigroup(fct_B6x221$IM[[1]]), pi.rels(ii_B6x2211, po.incl=T),
+    type="at", reduc=T)
#[1] 11
#


# B6x2212
ii_B6x2212 <- fact(fct_B6x221$IM[[2]], fct_B6x221$PO[[2]], atmc=T, patm=T)
fct_B6x2212 <- decomp(as.semigroup(fct_B6x221$IM[[2]]), pi.rels(ii_B6x2212, po.incl=T),
+    type="at", reduc=T)
#[1] 11
#


## B6x222
ii_B6x222 <- fact(fct_B6x22$IM[[2]], fct_B6x22$PO[[2]], atmc=T, patm=T)
fct_B6x222 <- decomp(as.semigroup(fct_B6x22$IM[[2]]), pi.rels(ii_B6x222, po.incl=T),
+    type="at", reduc=T)
#$ord
#[1] 15 14

# B6x2221
ii_B6x2221 <- fact(fct_B6x222$IM[[1]], fct_B6x222$PO[[1]], atmc=T, patm=T)
fct_B6x2221 <- decomp(as.semigroup(fct_B6x222$IM[[1]]), pi.rels(ii_B6x2221, po.incl=T),
+    type="at", reduc=T)
#[1] 11
#


# B6x2222
ii_B6x2222 <- fact(fct_B6x222$IM[[2]], fct_B6x222$PO[[2]], atmc=T, patm=T)
fct_B6x2222 <- decomp(as.semigroup(fct_B6x222$IM[[2]]), pi.rels(ii_B6x2222, po.incl=T),
+    type="at", reduc=T)
#[1] 11
#


## B6x223
ii_B6x223 <- fact(fct_B6x22$IM[[3]], fct_B6x22$PO[[3]], atmc=T, patm=T)
fct_B6x223 <- decomp(as.semigroup(fct_B6x22$IM[[3]]), pi.rels(ii_B6x223, po.incl=T),
+    type="at", reduc=T)
```

```
#Sord
#[1] 14 14

# B6x2231
ii_B6x2231 <- fact(fct_B6x223SIM[[1]], fct_B6x223SPO[[1]], atmc=T, patm=T)
fct_B6x2231 <- decomp(as.semigroup(fct_B6x223SIM[[1]]), pi.rels(ii_B6x2231, po.incl=T),
+     type="a", reduc=T)
#[1] 11
#

# B6x2232
ii_B6x2232 <- fact(fct_B6x223SIM[[2]], fct_B6x223SPO[[2]], atmc=T, patm=T)
fct_B6x2232 <- decomp(as.semigroup(fct_B6x223SIM[[2]]), pi.rels(ii_B6x2232, po.incl=T),
+     type="a", reduc=T)
#[1] 11
#
```

Aggregated structure of $\mathcal{Q}_B$

After such long process of progressive factorization, quotient semigroups SsB = fct_B2111, fct_B2112, fct_B22x1, fct_B22x2, fct_B23x1, fct_B23x2, fct_B4x1111, fct_B4x1112, fct_B4x1121, fct_B4x1122, fct_B4x1131, fct_B4x1132, fct_B4x12x1, fct_B4x12x2, fct_B4x2111, fct_B4x2112, fct_B4x2121, fct_B4x2122, fct_B4x2131, fct_B4x2132, fct_B4x22x1, fct_B4x22x2, fct_B6x11x1, fct_B6x11x2, fct_B6x1211, fct_B6x1212, fct_B6x1221, fct_B6x1222, fct_B6x1231, fct_B6x1232, fct_B6x21x1, fct_B6x21x2, fct_B6x2211, fct_B6x2212, fct_B6x2221, fct_B6x2222, fct_B6x2231, and fct_B6x2232 result being isomorphic. A further decomposition of SsB yields into the 1-semigroup with no structure.

```
R> SsB$IM

[[1]]
         C    K    A    B   CK   CB  KC  KB  BC  KCB  BCK
C        C   CK    C   CB   CK   CB   C  KB   C   CB   CK
K       KC    K    K   KB    K  KCB  KC  KB  KB  KCB   KB
A        C    K    A    B   CK   CB  KC  KB  BC  KCB  BCK
B       BC   KB    B    B  BCK    B  KB  KB  BC   KB  BCK
CK       C   CK   CK   KB   CK   CB   C  KB  KB   CB   KB
CB       C   KB   CB   CB   CK   CB  KB  KB   C   KB   CK
KC      KC    K   KC  KCB    K  KCB  KC  KB  KC  KCB    K
KB      KB   KB   KB   KB   KB   KB  KB  KB  KB   KB   KB
BC      BC  BCK   BC    B  BCK    B  BC  KB  BC    B  BCK
KCB     KC   KB  KCB  KCB    K  KCB  KB  KB  KC   KB    K
BCK     BC  BCK  BCK   KB  BCK    B  BC  KB  KB    B   KB
```

The quotient semigroup of the multiplication table representing $\mathcal{Q}_B$ has order of 11.

Bibliography

Aghion, P., Bloom, N., Blundell, R., Griffith, R. and Howitt, P. (2005), 'Competition and innovation: an inverted-U relationship', *The Quarterly Journal of Economics* 120(2), 701–728.

Ahuja, R., Magnanti, T. and Orlin, J. (1993), *Network Flows: Theory, Algorithms, and Applications*, Prentice Hall.

Arabie, P., Boorman, S. and Levitt, P. (1978), 'Constructing blockmodels: How and why', *Journal of Mathematical Psychology* 17, 21–63.

Ardu, S. (1995), '*ASNET*: Algebraic and statistical network analysis—user manual'.

AT&T Labs Research (2019), *Graphviz – Graph Visualization Software*. v2.40.1. **URL:** *https://www.graphviz.org/*

Back, K. (1951), 'Influence through social communication', *Journal of Abnormal and Social Psychology* 46, 9–23.

Bakshy, E., Messing, S. and Adamic, L. (2015), 'Exposure to ideologically diverse news and opinion on Facebook', *Science* 348(6239), 1130–1132.

Baldassarri, D. and Diani, M. (2007), 'The integrative power of civic networks', *American Journal of Sociology* 113(3), 735–780.

Barrat, A., Barthélemy, M. and Vespignani, A. (2008), *Dynamical Processes on Complex Networks*, Cambridge University Press.

Batagelj, V. (1994), 'Semirings for social networks analysis', *Journal of Mathematical Sociology* 19(1), 53–68.

Batagelj, V., Doreian, P. and Ferligoj, A. (1992), 'An optimization approach to regular equivalence', *Social Networks* 14, 121–135.

Batagelj, V., Doreian, P., Ferligoj, A. and Kejzar, N. (2014), *Understanding Large Temporal Networks and Spatial Networks: Exploration, Pattern Searching, Visualization and Network Evolution*, Wiley Series in Computational and Quantitative Social Science, John Wiley & Sons.

Berkowitz, S. D. (1982), *An Introduction to Structural Analysis: The Network Approach to Social Research*, Butterworth.

Birkhoff, G. (1967), *Lattice Theory*, 3rd edn, American Mathematical Society.

Bonacich, P. (1980), 'The 'Common Structure Semigroup,' a replacement for the Boorman and White 'Joint Reduction'', *American Journal of Sociology* 86, 159–166.

Bonacich, P. (1982), 'The common structure graph: Common structural features of a set of graphs', *Mathematical Social Sciences* 2, 275–288.

Bonacich, P. and McConaghy, M. (1980), 'The algebra of blockmodeling', *Sociological Methodology* 11, 489–532.

Boorman, S. and Arabie, P. (1980), 'Algebraic approaches to the comparison of concrete social structures represented as networks: Reply to Bonacich', *American Journal of Sociology* 86, 166–1674.

Boorman, S. and White, H. (1976), 'Social structure from multiple networks. II. Role structures', *American Journal of Sociology* 81(6), 1384–1446.

Borgatti, S., Boyd, J. and Everett, M. (1989), 'Iterated roles: Mathematics and application', *Social Networks* 11, 159–172.

Borgatti, S. and Everett, M. (1989), 'The class of all regular equivalences: Algebraic structure and computation', *Social Networks* 11, 65–88.

Algebraic Analysis of Social Networks: Models, Methods and Applications using R,
First Edition. J. A. R. Ostoic. Companion website: www.wiley.com/go/ostoic/algebraicanalysis.
© 2021 John Wiley & Sons Ltd. Published 2021 by John Wiley & Sons Ltd.

Borgatti, S., Everett, M. and Freeman, L. (2002), '*UCINET* 6 for windows: Software for social network analysis'.
 URL: *http://www.analytictech.com/*

Borgatti, S. and Lopez-Kidwell, V. (2011), Network theory, *in* J. Scott and P. Carrington, eds, '*The SAGE Handbook of Social Network Analysis*', Sage, pp. 40–54.

Borgatti, S. P. (2012), Social network analysis, two-mode concepts in, *in* A. R. Meyers, ed., '*Computational Complexity: Theory, Techniques, and Applications*', Springer-Verlag, pp. 2912–2924.

Borgatti, S. P. and Halgin, D. S. (2011), Analyzing affiliation networks, *in* J. Scott and P. J. Carrington, eds, '*The Sage Handbook of Social Network Analysis*', Sage Publications, pp. 417–433.

Bourgeois, M. and Friedkin, N. (2001), 'The distant core: Social solidarity, social distance, and interpersonal ties', *Social Networks* 23, 245–260.

Boyd, J. P. (1980), 'The universal semigroup of relations', *Social Networks* 2, 91–117.

Boyd, J. P. (1991), *Social Semigroups: A Unified Theory of Scaling and Blockmodelling as Applied to Social Networks*, George Mason University Press.

Boyd, J. P. (2000), 'Redes sociales y semigrupos', *Política y Sociedad* 33, 105–112.

Boyd, J. P. (2002), 'Finding and testing regular equivalence', *Social Networks* 24, 315–331.

Boyd, J. P. and Everett, M. G. (1999), 'Relations, residuals, regular interiors, and relative regular equivalence', *Social Networks* 21, 147–165.

Brass, D. J. and Borgatti, S. P. (2019), Multilevel thoughts on social networks, *in* S. E. Humphrey and J. M. LeBreton, eds, '*The Handbook of Multilevel Theory, Measurement, and Analysis*', American Psychological Association, pp. 187–200.

Breiger, R., Boorman, S. and Arabie, P. (1975), 'An algorithm for clustering relational data with applications to social network analysis and comparison with multidimensional scaling', *Journal of Mathematical Psychology* 12, 328–383.

Breiger, R. L. (1974), 'The duality of persons and groups', *Social Forces* 53(2), 181–190.

Breiger, R. and Mohr, J. (2004), 'Institutional logics from the aggregation of organizational networks: Operational procedures for the analysis of counted data', *Computational and Mathematical Organization Theory* 10, 17–43.

Breiger, R. and Pattison, P. (1978), 'The joint role structure of two communities' elites', *Sociological Methods & Research* 7(2), 213–226.

Breiger, R. and Pattison, P. (1986), 'Cumulated social roles: The duality of persons and their algebras', *Social Networks* 8, 215–256.

Cannon, J. (1971), 'Computing the ideal structure of finite semigroups', *Numerische Mathematik* 18, 254–266.

Cartwright, D. and Harary, F. (1956), 'Structural balance: A generalization of Heider's theory', *Psychological Review* 63, 277–293.

Clifford, A. and Preston, G. (1967), *The Algebraic Theory of Semigroups*, American Mathematical Society.

Davis, J. (1967), 'Clustering and structural balance in graphs', *Human Relations* 20, 181–187.

de Nooy, W., Mrvar, A. and Batagelj, V. (2005), Exploratory Social Network Analysis with Pajek, *Structural Analysis in the Social Sciences*, Cambridge University Press.

Degenne, A. and Forsé, M. (1999), *Introducing Social Networks*, SAGE Publications.

Doreian, P., Batagelj, V. and Ferligoj, A. (2004), *Generalized Blockmodeling*, Vol. 25 of *Structural Analysis in the Social Sciences*, Cambridge University Press.

Doreian, P. and Krackhardt, D. (2001), 'Pre-transitive mechanisms for signed networks', *Journal of Mathematical Sociology* 25, 43–67.

Doreian, P. and Mrvar, A. (1996), 'A partitioning approach to structural balance', *Social Networks* 18, 149–168.

Doreian, P. and Mrvar, A. (2009), 'Partitioning signed social networks', *Social Networks* 31, 1–11.

Dunn, J. and Hardegree, G. (2001), *Algebraic Methods in Philosophical Logic*, Oxford University Press.

Dunning, I. (2020), 'Graph layout algorithms in pure Julia'. [Online; retrieved on 12 February 2020].
 URL: *https://github.com/IainNZ/GraphLayout.jl*

Everett, M. and Borgatti, S. (1991), 'Role colouring a graph', *Mathematical Social Sciences* 21, 183–188. Note in.

Everett, M. and Borgatti, S. (2013), 'The dual-projection approach for two-mode networks', *Social Networks* 35(2), 204–210. Special Issue on Advances in Two-mode Social Networks.

Everett, M., Boyd, J. and Borgatti, S. (1990), 'Ego-centered and local roles: A graph theoretic approach', *Journal of Mathematical Sociology* 15(3–4), 163–172.

Everett, M. G. (1985), 'Role similarity and complexity in social networks', *Social Networks* 7, 353–359.

Everett, M. G. and Borgatti, S. P. (1994), 'Regular equivalence: General theory', *Journal of Mathematical Sociology* 19(1), 29–52.

Festinger, L. (1950), 'Informal social communication', *Psychological Review* 57(5), 271–82.

Fienberg, S. E., Meyer, M. M. and Wasserman, S. S. (1985), 'Statistical analysis of multiple sociometric relations', *Journal of the American Statistical Association* 80(389), 51–67.

Flament, C. (1963), *Applications of Graph Theory to Group Structure*, Prentice-Hall.

Freeman, L. C. (1992), Social networks and the structure experiment, *in* L. Freeman, D. White and A. Romney, eds, 'Research Methods in Social Network Analysis', Transaction Publishers, pp. 11–40.

Freeman, L. and White, D. (1993), 'Using Galois lattices to represent network data', *Sociological Methodology* 23, 127–146.

Friedkin, N. (1984), 'Structural cohesion and equivalence explanations of social homogeneity', *Sociological Methods & Research* 12, 235–261.

Fruchterman, T. and Reingold, E. (1991), 'Graph drawing by force-directed placement', *Software–Practice & Experience* 21(11), 1129–1164.

Gansner, E. R., Koren, Y. and North, S. (2005), *Graph Drawing: 12th International Symposium, GD 2004, New York, NY, USA, September 29-October 2, 2004, Revised Selected Papers*, Springer Berlin Heidelberg, chapter Graph drawing by stress majorization, pp. 239–250.

Ganter, B. and Wille, R. (1996), *Formal Concept Analysis – Mathematical Foundations*, Springer-Verlag.

Granovetter, M. S. (1973), 'The strength of weak ties', *American Journal of Sociology* 78(6), 1360–1380.

Granovetter, M. S. (1982), The strength of weak ties: A network theory revisited, *in* P. Marsden and N. Lin, eds, 'Social Structure and Network Analysis', Sage, pp. 105–130.

Grätzer, G. (2005), *The Congruences of a Finite Lattice: A Proof-by-picture Approach*, Birkhäuser Boston.

Hansen, K. et al. (2018), ***Rgraphviz**: Provides plotting capabilities for R graph objects*. R package version 2.24.0. **URL:** *http://bioconductor.org/packages/Rgraphviz/*

Harary, F. (1994), *Graph Theory*, Mathematics, Addison-Wesley.

Harary, F., Norman, Z. and Cartwright, D. (1965), *Structural Models: An Introduction to the Theory of Directed Graphs*, John Wiley & Sons.

Hartmanis, J. and Stearns, R. (1966), *Algebraic Structure Theory of Sequential Machines*, Prentice-Hall.

Heider, F. (1946), 'Attitudes and cognitive organization', *Journal of Psychology* 21, 107–112.

Heider, F. (2013), *The Psychology of Interpersonal Relations*, Taylor & Francis.

Holland, P. and Leinhardt, S. (1970), 'A method for detecting structure in sociometric data', *American Journal of Sociology* 76, 492–513.

Holland, P. and Leinhardt, S. (1976), 'Local structure in social networks', *Sociological Methodology* 7, 1–45.

Homans, G. (1961), *Social Behavior: Its Elementary Forms*, Harcourt Brace.

Howie, J. M. (1996), *Fundamentals of Semigroup Theory*, Oxford University Press.

International Organization for Standardization (2019), 'Country codes – ISO 3166'. [Online; retrieved on 31 January 2019]. **URL:** *https://www.iso.org/iso-3166-country-codes.html*

Kemeny, J., Snell, J. and Thompson, G. (1974), *Introduction to Finite Mathematics*, 3rd edn, Prentice-Hall.

Kent, D. (1978), *The rise of the Medici: Faction in Florence, 1426-1434*, Oxford University Press.

Kilduff, M. and Tsai, W. (2003), *Social Networks and Organizations*, Sage.

Kim, K. and Roush, F. (1983), *Applied Abstract Algebra*, Ellis Horwood.

König, D. (1936), *Theorie der Endlichen und Unendlichen Graphen*, Akademische Verlagsgesel-Ischaft.

Korn, F. (1973), *Elementary Structures Reconsidered: Lévi-Strauss on Kinship*, Tavistock Publications.

Krushal, J. (1964), 'Multidimensional scaling by optimizing goodness of fit to a nonmetric hypothesis', *Psychometrika* 29, 1–27.

Krushal, J. and Seery, J. (1980), 'Designing network diagrams', In: *Proceedings 1st General Conference on Social Graphics*, pp. 22–50.

Laumann, E. and Pappi, F. (1976), *Networks of Collective Action: A perspective on community influence systems*, Academic.

Lazarsfeld, P., B. B. and Gaudet, H. (1948), *The People's Choice: How the Voter Makes up his Mind in Presidential Campaigns*, Columbia University Press.

Lazega, E. (2001), *The Collegial Phenomenon: The Social Mechanisms of Cooperation among Peers in a Corporate Law Partnership*, Oxford University Press.

Lazega, E. and Pattison, P. (1999), 'Multiplexity, generalized exchange and cooperation in organizations: a case study', *Social Networks* 21(1), 67–90.

Lazega, E. and Snijders, T. (2016), *Multilevel Network Analysis for the Social Sciences*, Vol. 12 of *Methodos*, Springer-Verlag.

Lévi-Strauss, C. (1949), *Les Structures Elementaire de la Parente*, Presses Universitaires de France. With an appendix by A. Weil.

Lévi-Strauss, C. (1967), *Mythologiques, T. II : Du miel aux cendres*, Plon.

Linton, R. (1936), *The Study of Man*, D. Appleton-Century.

Lorrain, F. (1975), *Réseaux Sociaux et Classifications Sociales: Essai sur l'algèbre et la géométrie des structures sociales*, Hermann.

Lorrain, F. and White, H. (1971), 'Structural equivalence of individuals in social networks', *Journal of Mathematical Sociology* 1, 49–80.

Luce, R. and Perry, A. (1949), 'A method of matrix analysis of group structure', *Psychometrika* 14, 95–116.

Maddux, R. D. (2006), *Relation Algebras*, Elsevier.

Mandel, M. J. (1978), 'Roles and Networks: A Local Approach'. B.A. Honours thesis, Harvard University.

Mandel, M. J. (1983), 'Local roles and social networks', *American Sociological Review* 48, 376–386.

McConaghy, M. J. (1981a), 'Negation of the equation – Rejoinder to Pattison', *Sociological Methods & Research* 9(3), 303–312.

McConaghy, M. J. (1981b), 'The common role structure: Improved blockmodeling methods applied to two communities' elites', *Sociological Methods & Research* 9, 267–285.

Merton, R. K. (1957), *Social Theory and Social Structure*, revised and enlarged edition edn, Free Press.

Miller, G. (2012), *Theory and Applications of Finite Groups*, University of Michigan Library.

Nadel, S. (1957), *The Theory of Social Structure*, Free Press.

Newcomb, T. M. (1961), *The Acquaintance Process*, Holt, Rinehart & Winston.

Ostoic, J. A. R. (2013), Algebraic Methods for the Analysis of Multiple Social Networks and Actors Attributes, PhD thesis, University of Southern Denmark.

Ostoic, J. A. R. (2017), 'Creating context for social influence processes in multiplex networks', *Network Science* 5(1), 1–29.

Ostoic, J. A. R. (2018), 'Compositional equivalence with actor attributes: Positional analysis of the Florentine Families network', *Connections* 37, 53–68.

Ostoic, J. A. R. (2019a), **multigraph**: *Plot and Manipulate Multigraphs*. R package devel version 0.92. **URL:** *https://CRAN.R-project.org/package=multigraph*

Ostoic, J. A. R. (2019b), **multiplex**: *Algebraic Tools for the Analysis of Multiple Social Networks*. R package version 2.9.2. **URL:** *https://CRAN.R-project.org/package=multiplex*

Ostoic, J. A. R. (2020), 'Algebraic Analysis of Multiple Social Networks with multiplex', *Journal of Statistical Software* 92(11), 1–41.

Padgett, J. and Ansell, C. (1993), 'Robust action and the rise of the Medici, 1400–1434', *American Journal of Sociology* 98(6), 1259–1319.

Pareto, V. (1916), *Trattato di Sociolgia Generale*, Barbara.

Parsons, T. (1991), *The Social System*, Routledge.

Pattison, P. (1981), 'Equating the 'joint reduction' with blockmodel common role structure: A reply to McConaghy', *Sociological Methods & Research* 9, 286–302.

Pattison, P. and Bartlett, W. (1982), 'A factorization procedure for finite algebras', *Journal of Mathematical Psychology* 25, 51–81.

Pattison, P. E. (1980), An Algebraic Analysis for Multiple Social Networks, PhD thesis, University of Melbourne.

Pattison, P. E. (1993), *Algebraic Models for Social Networks*, Vol. 7 of *Structural Analysis in the Social Sciences*, Cambridge University Press.

Proctor, C. and Loomis, C. (1951), Analysis of sociometric data, *in* M. Jahoda, M. Deutsch and S. Cook, eds, 'Research Methods in Social Relations', Dryden Press, pp. 561–586.

R Core Team (2015), R: *A Language and Environment for Statistical Computing*, R Foundation for Statistical Computing, Vienna, Austria. **URL:** *https://www.R-project.org/*

Radcliffe-Brown, A. R. (1913), 'Three Tribes of Western Australia', *Journal of the Royal Anthropological Institute* 43, 143–194.

Rapoport, A. (1963), Mathematical models of social interaction, *in* R. B. Luce, R.D. and E. Galanter, eds, 'Handbook of Mathematical Psychology', John Wiley & Sons, pp. 493–579.

Reitz, K. and White, D. (1992), Rethinking the role concept: Homomorphisms on social networks, *in* L. Freeman, D. White and A. Romney, eds, 'Research Methods in Social Network Analysis', Transaction Publishers, pp. 429–488.

Sailer, L. D. (1978), 'Structural equivalence: Meaning and definition, computation and application', *Social Networks* 1, 73–90.

Salehi, M., Sharma, R., Marzolla, M., Montesi, D., Siyari, P. and Magnani, M. (2014), 'Diffusion processes on multilayer networks', *CoRR* **abs/1405.4329**.

Sampson, F. S. (1969), A Novitiate in a Period of Change: An Experimental and Case Study of Social Relationships, PhD thesis, Cornell University.

Schvaneveldt, R., Durso, F. and Dearholt, D. (1989), Network structures in proximity data, *in* G. Bower, ed., 'The psychology of learning and motivation: Advances in research & theory', Vol. 24, Academic Press, pp. 249–284.

Shier, D. R. (1991), *Network Reliability and Algebraic Structures*, Oxford Science Publications.

Simmel, G. (1950), *The Sociology of Georg Simmel*, Free Press.

Snijders, T. A. B. (2017), 'Stochastic Actor-Oriented Models for Network Dynamics', *Annual Review of Statistics and Its Application* 4, 343–363.

Tantau, T. (2013), *The TikZ and PGF packages*. **URL:** *http://sourceforge.net/projects/pgf/*

Tönnies, F. and Loomis, C. (1940), *Fundamental Concepts of Sociology: (Gemeinschaft und gesellschaft)*, American Book Company.

United Nations (2017), 'United Nations Commodity Trade Statistics Database (UN Comtrade)'. [Online; retrieved on 28 March 2019]. **URL:** *https://comtrade.un.org/*

Valente, T. W. (2010), *Social Network and Health*, Oxford University Press.

Wasserman, S. (1980), 'Analyzing social networks as stochastic processes', *Journal of the American Statistical Association* 75, 280–294.

Wasserman, S. and Faust, K. (1994), *Social Network Analysis: Methods and Applications*, Vol. 8 of *Structural Analysis in the Social Sciences*, Cambridge University Press.

Wasserman, S. and Iacobucci, D. (1991), 'Statistical modelling of one-mode and two-mode networks: Simultaneous analysis of graphs and bipartite graphs', *British Journal of Mathematical and Statistical Psychology* 44(1), 13–3.

White, D. and Reitz, K. (1983), 'Graph and semigroup homomorphisms on networks of relations', *Social Networks* 5, 193–234.

White, H. (2013), *Chains of Opportunity: System Models of Mobility in Organizations*, Harvard University Press.

White, H., Boorman, S. and Breiger, R. (1976), 'Social structure from multiple networks: I. blockmodels of roles and positions', *American Journal of Sociology* 81, 730–780.

White, H. C. (1963), *An Anatomy of Kinship: Mathematical Models for Structures of Cumulated Roles*, Prentice-Hall.

White, H. C. (1992), *Identity and Control: A Structural Theory of Social Action*, Princeton University Press.

Wikipedia (2017), 'G20—Wikipedia, the free encyclopedia'. [Online; retrieved on 07 July 2017]. **URL:** *https://en.wikipedia.org/wiki/G-20-major-economies*

Wikipedia (2019), 'G20—Wikipedia, the free encyclopedia'. [Online; retrieved on 23 May 2019]. **URL:** *https://en.wikipedia.org/wiki/G20*

Wille, R. (1982), Restructuring lattice theory: an approach based on hierarchies of concepts, *in* I. Rival, ed., 'Ordered Sets', Reidel, pp. 445–470.

Winship, C. (1988), 'Thoughts about roles and relations: An old document revisited', *Social Networks* 10, 209–231.

Winship, C. and Mandel, M. (1983), Roles and positions: A critique and extension of the blockmodelling approach, *in* S. Leinhardt, ed., 'Sociological Methodology', Vol. 14, Jossey-Bass, pp. 314–344.

Wu, L. L. (1984), 'Local blockmodel algebras for analyzing social networks', *Sociological Methodology* 14, 272–313.

INDEX

absorbing element, *see* zero element
actor attributes, 64, 68, 72, 84, 93, 99, 102–104,
    109–114, 119–122, 125–136, 146, 153,
    161–165, 168–174, 188, 192, 205, 209,
    215–216, 218, 285, 287, 292, 301,
    303–310, 316–319, 321–328, 335–336,
    339–341, 345
affiliation networks, 4, 49, 215–234, 249–250, 258–259,
    273–277, 282, 285–286
algebraic constraints, 301, 302, 317, 327–329
atoms, 150–152, 154–162, 168, 234
    meet-complements, 150–152, 154, 156–162, 170,
    172–174, 177, 345
Automorphic equivalence, 88–89
Axiom of Quality, 69, 72, 74, 94, 247

balance, *see* structural balance
balance semiring, *see* semirings
binary operation, 2, 10, 11, 25, 29, 34, 35, 39
binomial projection, 218–220, 223–224, 275–277, 279,
    283, 289, 291
bipartite graph, *see* affiliation networks
bipartite networks, *see* affiliation networks
bundle classes, *see* bundle patterns
bundle patterns, 58–65, 77, 86, 93, 187, 200–201,
    244–245, 261–262, 274, 276,
    284–286
    bundle census, 62, 189, 244, 261
bundles, *see* bundle patterns

Cayley color graph, *see* Cayley graph
Cayley graph, 27–28, 31, 50, 55, 169–171, 247, 248,
    287, 288, 290, 307, 321, 323

closure, 18, 25–26, 36, 38–39, 65–67, 91, 103, 114,
    117, 127, 136, 192–194, 199, 203, 208
cluster, *see* clustering
cluster semiring, *see* semirings
clustering, 14, 90, 197, 218, 222, 291, 340–343
    clustered bipartite graph, 217–218, 223
cohesion, *see* group cohesion
Common Structure Semigroup, 314–316, 319–324
complex networks, *see* complex structures
complex structures, 37, 50, 51, 57, 77, 101, 111, 120,
    188, 246, 284
complex systems, *see* complex structures
Compositional equivalence, 97–101, 103–104, 109,
    114–117, 124, 127, 130, 146, 174, 301, 303,
    306, 341–342
    (*see also* Relation-Box, role structures)
Concept lattices, 226, 229–234, 253–255, 286–289,
    291, 292
    (*see also* lattices)
conceptual scaling, 249–255
congruence by substitution property, 152–153, 170–171
    (*see also* congruence relations)
congruence classes, 19, 146, 159, 171, 174, 314, 315,
    318
    (*see also* congruence relations)
Congruence lattice, 149–159, 168, 174, 177, 307, 312
    (*see also* lattices)
congruence relations, 20, 148–152, 154–156, 158, 159,
    161, 166, 168, 318, 328
    (*see also* congruence classes), (*see also* factorization)
cumulated person hierarchy, *see* person hierarchy

dihedral groups, 30–32

Algebraic Analysis of Social Networks: Models, Methods and Applications using R,
First Edition. J. A. R. Ostoic. Companion website: www.wiley.com/go/ostoic/algebraicanalysis.
© 2021 John Wiley & Sons Ltd. Published 2021 by John Wiley & Sons Ltd.